HEAT MINING

A new source of energy

The first electric power generated from hot dry rock. Approximately 60 kWe with 150°C HDR fluid and R-114 binary cycle. (February 1980, photograph courtesy of R. Pettitt, Los Alamos.)

HEAT MINING

A new source of energy

H. Christopher H. Armstead

BSc, C Eng, FICE, FIMechE, FIEE, FCGI
Consulting Engineer, Dartmouth, UK

and

Jefferson W. Tester

PhD Chem Eng, AIChE

Associate Professor of Chemical Engineering,
Massachusetts Institute of Technology
Cambridge, Massachusetts, USA

LONDON   NEW YORK
E. & F.N. SPON

First published in 1987 by
E. & F.N. Spon Ltd
11 New Fetter Lane, London EC4P 4EE
Published in the USA by
E. & F.N. Spon Ltd
29, West 35th Street, New York NY 10001
© 1987 H.C.H. Armstead and J.W. Tester

Printed in Great Britain at the
University Press, Cambridge

ISBN 0 419 12230 3

British Library Cataloguing in Publication Data

Armstead, H. Christopher H.
  Heat mining: a new source of energy.
  1. Geothermal engineering
  I. Title   II. Tester, Jefferson W.
  621.44   TJ280.7

  ISBN 0–419–12230–3

Library of Congress Cataloging in Publication Data

Armstead, H. Christopher H.
  Heat mining.

  Bibliography: p.
  Includes index.
  1. Energy minerals.   2. Earth temperature.
3. Heat engineering.   I. Tester, Jefferson W.
II. Title.
TN263.5.A76   1987   333.79′2   87–5986
ISBN 0–419–12230–3

Contents

Foreword

There is ample evidence to show that unless timely action is taken, the world seems to be heading for an energy famine in the next century. Frequently, people who consider this matter at all adopt one of three extreme attitudes. There are the Micawbers who are confident that 'something will turn up' and that a policy of *laissez-faire* is therefore justified. Then there are the scaremongers who believe that all is already lost and that we are heading for inevitable and imminent doom. There are also the cynics, who think that the scaremongers are probably right but that the threatened day of reckoning is still fairly remote: their mottos is 'après nous le déluge'.

We feel a more moderate stance is appropriate. We acknowledge that there is indeed a threat of an approaching energy famine; but deplore equally both inaction and panic. Our faith in human ingenuity and its eventual practicality is such that we are confident that timely action *can* be taken, so that disaster may be avoided.

With this motivation, our objectives in writing this book are fourfold: (1) to introduce the concept and basic principles of heat mining, (2) to discuss the magnitude and distribution of the resource in comparison to alternatives, (3) to review the technology for gaining access to the resource and to creating reservoirs within it that are designed to extract energy, and (4) to consider the economic criteria that must be met for the commercialization of heat mining. In this way, we hope that readers will be able to judge for themselves the potentialities of heat mining and will join us as advocates of its rapid development.

In this book, the nature and intensity of the energy famine threat is examined as we consider the available options for meeting that threat, and choose one option which we believe to be promising in magnitude and in attainability – the extraction of stored heat from the earth's crust: in short, *heat mining*. We do not claim this to be the *only* option, indeed we favor the pursuit of all other options that could help push the threat further away and thus ease the urgency. Our choice of heat mining as an option worthy of immediate and vigorous development is based partly upon the immensity and ubiquity of the resource and partly upon the fact that its exploitation requires only the modest improvement of existing technologies rather than the development of new technologies as yet unattained. In fact, many aspects of heat mining have already been shown to be technically feasible. Engineering improvements are constantly being made and a greater understanding is continually being acquired of the processes involved, despite the fact that development work is being done by comparatively few research groups. The authors believe that with a little more push to attain a few technical goals, heat mining could soon become commercially competitive with liquid and gaseous fossil fuels for many applications – and with less pollution.

In Part I of the book basic concepts are introduced: the earth's crustal heat resource is approximately quantified, physical constraints and limitations of extracting energy from it are examined, and the fundamental principles of the heat mining process are described. In addition, heat mining from 'hot dry rock' is contrasted with the development and exploitation of hydrothermal energy which is focused on the recovery of naturally occurring hot fluids contained in underground reservoirs.

In Part II the state of the art of heat mining is covered with descriptions and evaluations presented of the extensive field work that has already been undertaken in a few countries – notably the USA and the United Kingdom. An engineering analysis is presented of the events that have to occur underground in the course of preparing a reservoir for subsequent heat extraction. The need is stressed for lowering the costs of drilling and extending the range of deep crustal penetration, with a summary of what has already been achieved and what is hoped yet to be achieved in this respect. The possibilities of directly recovering heat from liquid magma are briefly considered, the environmental aspects of heat mining are discussed, and finally the remaining problems and probable economics of the process are reviewed.

The potential readership of this book will range between two extremes. On the one hand, there is the general class of reader who may be assumed to have received a broadly scientific training and who is both interested in, and concerned with, the great social and economic problem of how the world is to be supplied in the next and subsequent centuries with the energy it will so desperately need if civilization is to be preserved. This class of reader should have no difficulty in absorbing the whole of Part I (Chapters 1 to 7) and Chapters 11 to 15 of Part II *in toto*, all of which are mainly commonsensical and include only the simplest forms of mathematical analysis that are essential to a proper understanding of the underlying philosophy and the physical processes involved. The same reader may be tempted to skim rather superficially through Chapters 8, 9, and 10, and to skip over the detailed theoretical analyses. On the other hand, there is the class of reader of greater mathematical virtuosity, with a more intense interest in the details of heat mining. This class of reader will not be satisfied with many of the unsupported allegations or arguments concerning phenomena that are not self-evident in Part I and will need to examine the various mathematical and theoretical analyses and summaries of fieldwork presented in Chapters 8, 9, and 10, to be convinced.

One of the authors recently met a sceptic who had little faith in the notion of heat mining, and considered research work in this field to be a waste of money.

He argued that each of the few countries now engaged in this activity was motivated by a spirit of 'one-upmanship' lest they be thought to lag behind the others, rather than by a true commitment to their professed objective. The reader is assured that this cynical view is totally misconceived. The small but dedicated teams of research workers, certainly for the most part, are convinced that they are almost within grasping distance of the key to an awe-inspiring problem.

We are deeply indebted to all of the tireless workers at Los Alamos and Rosemanowes who, in the new quest for crustal heat, have provided the basis from which this book has evolved. Many of these people directly helped and encouraged us during the preparation of the manuscript by providing timely information and reviews. It would be impossible to name them all, but we would like to express our particular gratitude to M.C. Smith, J. Whetten, G. Nunz, M. Fehler, J. Albright, D. Brown, H. Murphy, R. Duffield, R. Brownlee, R. Potter, C.O. Grigsby, R. Pettitt, B. Dennis, R.L. Aamodt, H.N. Fisher, R. Hendron, P. Franke, R. Benson, J.C. Rowley and the late A. Blair of the Los Alamos National Laboratory; to A.S. Batchelor, C. Pearson, R. Pine, P. Ledingham and K.A. Kwakwa of the Camborne School of Mines, now of Geo Science Ltd; to J.D. Garnish of the Energy Technology Support Unit of the UK Department of Energy; to F. Rummel, O. Kappelmeyer, and H. Kepler of the Federal Republic of Germany; to doctoral students B.A. Robinson, R. Rauenzahn, M. Wilkinson, and J. Ferguson of the Massachusetts Institute of Technology and to Profs K. Aki, M.G. Simmons, and W.F. Brace.

Without the supreme commitment of our typists and word processors, Judy Hale, Linda MacDonald, Anne Sollitto, and Sharon Gray, it would not have been possible to produce the countless drafts of this manuscript.

Unit conversion table

	B.t.u.	Quads	Calories	Kilogram-calories	kWht	MWht
B.t.u.	1	10^{-15}	252	0.252	2.93×10^{-4}	2.93×10^{-7}
Quads	10^{15}	1	2.52×10^{17}	2.52×10^{14}	2.93×10^{11}	2.93×10^{8}
Cal.	3.969×10^{-3}	3.969×10^{-18}	1	10^{-3}	1.163×10^{-6}	1.163×10^{-9}
kgcal	3.969	3.969×10^{-15}	1000	1	1.163×10^{-3}	1.163×10^{-6}
kWht	3412.5	3.4125×10^{-12}	8.6×10^{5}	860	1	10^{-3}
MWht	3.4125×10^{6}	3.4125×10^{-9}	8.6×10^{8}	8.6×10^{5}	1000	1
MWyt	2.989×10^{10}	2.989×10^{-5}	7.535×10^{12}	7.535×10^{9}	8.76×10^{6}	8760
bbl	5.5×10^{6}	5.5×10^{-9}	1.386×10^{9}	1.386×10^{6}	1611.5	1.6115
t.o.e.	4.04×10^{7}	4.04×10^{-8}	1.018×10^{10}	1.018×10^{7}	1.184×10^{4}	11.84
t.c.e.	2.778×10^{7}	2.778×10^{-8}	7×10^{9}	7×10^{6}	8139.5	8.1395
J	9.48×10^{-4}	9.48×10^{-19}	0.2389	2.389×10^{-4}	2.778×10^{-7}	2.778×10^{-10}
TJ	9.48×10^{8}	9.48×10^{-7}	2.389×10^{11}	2.389×10^{8}	2.778×10^{5}	277.8
EJ	9.48×10^{14}	0.948	2.389×10^{17}	2.389×10^{14}	2.778×10^{11}	2.778×10^{8}

MWyt	Barrels bbl	Tons oil equivalent	Tons coal equivalent	Joules J	TJ	EJ
3.345×10^{-11}	1.818×10^{-7}	2.475×10^{-8}	3.6×10^{-8}	1055	1.055×10^{-9}	1.055×10^{-15}
3.345×10^{4}	1.818×10^{8}	2.475×10^{7}	3.6×10^{7}	1.055×10^{18}	1.055×10^{6}	1.055
1.327×10^{-13}	7.215×10^{-10}	9.823×10^{-4}	1.429×10^{-10}	4.186	4.186×10^{-12}	4.186×10^{-18}
1.327×10^{-10}	7.215×10^{-7}	9.823×10^{-8}	1.429×10^{-7}	4186	4.186×10^{-9}	4.186×10^{-15}
1.142×10^{-7}	6.205×10^{-4}	8.448×10^{-5}	1.229×10^{-4}	3.6×10^{6}	3.6×10^{-6}	3.6×10^{-12}
1.142×10^{-4}	0.6205	8.448×10^{-2}	0.1229	3.6×10^{9}	3.6×10^{-3}	3.6×10^{-9}
1	5436	740.1	1076.2	3.153×10^{13}	31.53	3.153×10^{-5}
1.84×10^{-4}	1	0.1361	0.1979	5.8×10^{9}	5.8×10^{-3}	5.8×10^{-9}
1.351×10^{-3}	7.3451	1	1.4542	4.261×10^{10}	4.261×10^{-2}	4.261×10^{-8}
9.2916×10^{-4}	5.051	0.6877	1	2.93×10^{10}	2.93×10^{-2}	2.93×10^{-8}
3.1712×10^{-14}	1.724×10^{-10}	2.347×10^{-11}	3.413×10^{-11}	1	10^{-12}	10^{-18}
3.1712×10^{-2}	172.4	23.47	34.13	10^{12}	1	10^{-6}
3.1712×10^{4}	1.724×10^{8}	2.347×10^{7}	3.413×10^{7}	10^{18}	10^{6}	1

To convert the first column units to other units, multiply by the factors shown e.g. 1 kWht = 860 kgcal.

Assumed calorific values:
coal – 7000 cal/g
oil – 10180 cal/g

To our wives, Guendolen and Sue, without whose infinite patience and understanding neither our marriages nor this book would have survived.

Part I

Basic concepts and principles

Chapter 1

Introduction

1.1 What is heat mining?

As the term suggests, 'heat mining' is simply the extraction of heat or thermal energy from underground. Any kind of geothermal exploitation could therefore be regarded as a form of heat mining. Up to now such activities have commonly involved also the extraction of hot *fluids* from the earth's crust; that is to say, the removal not only of heat but also of matter from belowground. Usually it is the heat alone that is prized, with the fluid serving only as a heat transport medium possessing little or no value itself. The fluid may even cause environmental embarrassment by requiring proper disposal after some of the heat has been removed. Sometimes however, the fluid itself may have intrinsic value, by virtue of its mineral content. This may arise from its alleged 'crenotherapeutic' merit in balneology, or it may be due to the commercial worth of the extractable minerals it contains.

Today most geothermal operations, although embodying the concept of heat mining, are hybrid activities involving the mining of both heat and fluids. However, there have been a few geothermal developments where heat, and heat alone, without accompanying fluids, has been extracted from underground. Certain closed cycle heat pumps have used subterranean heat as a source to be transferred to a circulating working fluid that is thermodynamically upgraded to a higher temperature without lifting any of the underground warm waters themselves. In places where hot, lateral flowing ground-water is accessible at shallow depths, as at Klamath Falls, Oregon, submerged downhole heat-exchangers have been used; so that here again no subterranean fluids are themselves extracted from below – only their heat content. In Iceland and France, extensive district heating installations derive their energy from hot underground waters that are simply 'borrowed' for the purpose, being reinjected belowground after surrendering their thermal energy: no *net* fluid extraction being involved. All such cases can truly be looked upon as examples of pure heat mining; but in the majority of geothermal sites that have been developed to date, complete reinjection of fluids does not occur, so in a strict sense the fluid itself is mined but only its heat content is used.

1

As will be shown later, the heat resources of the earth's crust are so vast that it would be unreasonably pessimistic to suggest that they can never be extracted for the use of man except from natural hydrothermal fields, which are comparatively rare phenomena. These resources, however, are not for the most part associated with hot fluids, but are contained in the rock itself as *hot dry rock* (hereinafter abbreviated as 'HDR'). It is the heat content alone of this HDR that will probably be the principal goal of future heat mining operations, and it is in this context that the term 'heat mining' is used in this book. A note of warning is necessary in using the term HDR: the word 'dry' must not be taken literally. Although it may sometimes mean exactly what it says, it is also applied to hot rock zones of such low natural permeability that they are incapable of producing hot fluid of commercial or practical significance without artificial stimulation. In short, they may sometimes be 'wet' but unproductive.

By way of historical note, the first formal mention of heat mining that we could find is contained in a 1904 paper by Sir Charles Parsons where all the essential features of an HDR system are described. He even called it 'thermal mining' and said it would take 85 years to develop a working system, and as we will see, he was not wrong by much!

In addition to HDR there is another source of crustal heat that could make a very significant contribution to the world's energy needs if we could successfully tap it at acceptable cost; namely, the heat contained in magma bodies that are known to exist at various depths in certain tectonically active parts of the world. Although the technology for extracting this heat would differ greatly from that for exploiting HDR, this too would be a form of heat mining.

There is a particular merit in using 'heat mining' as a descriptive term. When we mine fuels or minerals from the earth's crust we are conscious that we are raiding a capital store of non-renewable resources. Even if it is contended that after millions of years further fossil fuels will have been formed by the same processes as those which gave us our coalfields and oilfields, this can be of no value to the human race because of the probable ephemerism of its existence in the scheme of evolution. For practical purposes fuels and minerals are non-renewable. There is a common public delusion that geothermal energy, like solar radiation or wind power, is renewable. This is not so. Again it may be argued that if we could wait long enough, any heat extracted from underground would ultimately be restored from the combined effects of crustal radioactivity and of conducted heat from the mantle and innermost recesses of the earth; but here also the time scale would be such that for practical human purposes earth heat is non-renewable. The term 'mining' serves as a useful reminder of that fact.

1.2 A technology not yet perfected

In writing about a technology that has not yet become a commercial reality, we, the authors of this book, are very conscious of laying ourselves open to a charge of peddling 'science fiction' for the sake of entertainment. On the contrary, we wish to draw attention not only to the vast quantity of heat that resides in the earth's crust, but also to the practical feasibility of gaining access to some of it.

In the 1930s, prophesies of splitting the atom and of travelling in outer space were greeted by all but a very few with amused tolerance and regarded as the fanciful diversions of brilliant but over-imaginative minds; but in less than 25 years both of these dreams became a practical reality. In both cases – nuclear fission and space travel – people were talking and writing about future achievements before the necessary technology had been perfected. We shall here be discussing the potential of heat mining as an alternative energy source from a similar perspective of an unperfected technology.

There is, however, a big difference between heat mining as a technology and the other two examples cited, in that its development as an energy source has little or no connection with strategic defense issues. The strategic elements of space travel and nuclear fission have been at least partially used to justify the tremendous sums of money and effort that have been expended in developing these technologies. In stark contrast, the realization of heat mining will not only be directly applicable to tangible human needs but will also almost certainly require far less expenditure and fewer technological breakthroughs than were needed for the achievement either of nuclear fission or of space travel.

There are, however, a number of non-technical factors that may delay the coming of heat mining. Because heat mining has to be considered in the context of all international programs for developing new energy resources, it is subject to fluctuations in research funding levels which are unfortunately closely tied to political, economic, and social pressures. For long term energy prospects like HDR, the funding situation for projects is frequently worsened because of 'more pressing' short term needs or because of the impact of such phenomena as a temporary downturn in the price of a barrel of crude oil. This short-sightedness and resulting oscillatory funding behavior makes it difficult to provide a stable, well organized program for developing the technology. The importance of these issues to heat mining is real, but unfortunately beyond the scope of this book.

The focus of this book is to evaluate what must be achieved technically for the heat mining concept to work in practice and to illustrate how its development, if successful, would provide access to a large, well-distributed energy resource that should be rapidly beneficial with negligible adverse environmental side effects.

Chapter 2

The need for a vast
new source of energy

Even primitive humans needed energy – apart from that contained in their food to sustain their muscular efforts. With the slow advance of civilization their descendents consumed it in ever-increasing quantities for cooking, for warming their dwellings, for grinding their grain, for forging and smelting, for pumping water and for machines of growing complexity. With the dawn of the Industrial Revolution humanity's appetite for energy started to accelerate rapidly. More and more energy was needed for factories, for transport, for food production, for warmth, for waging war and for all the sophisticated embellishments that have both enriched and befouled modern life. The amount of energy consumption *per capita* is sometimes regarded as a rough measure – not of peoples' happiness – but of their material well-being.

Some idea of the rate of growth of the world's total energy demand during the twentieth century can be seen from Table 2.1 and from Fig. 2.1 and 2.2. As shown, the world's total energy demand rose more than twelve-fold in the first 80 years of the century, despite the occurrence of two devastating world wars and despite the dramatic setback caused by the much smaller, but highly significant, Middle Eastern War of 7 October 1973, which led to the Arab oil embargo.

The figures of the 7th column of Table 2.1 emphasize the explosive growth that took place in the 1950s and '60s; partly caused, no doubt, by the reconstruction that followed the immediate post-war years of shortages. The contrast between the mean growth rates in those two decades and that of the first half-century shows that growth had, until 1970, been hyper-exponential. Clearly this could not have lasted indefinitely: indeed, had the mean growth of the 1960s persisted at 5.31% p.a. without further increase (i.e. at simple exponential growth) the demand by 1980 would have been about 11 800 Mtce p.a., or nearly 28% more than in fact it was. As might have been expected the world demand for energy continued to grow after 1970 though at a slightly falling rate, until 1973, when the Middle East crisis brought about a sharp check to the pattern of growth. As can be seen from Fig. 2.1, the growth pattern has recently shown hesitancy, as though some inherent resilience was struggling with repressive forces which have sometimes even been so strong as to induce

4

Table 2.1. The world's historic primary energy needs

Year	Total world energy for year			World population	Energy per capita	Mean compound annual growth rate			
						World energy	Energy per capita	Period	Ref
	Mtce	Quads	EJ	10⁶	kg.ce				
(1)	(2)	(3)	(4)	(5)	(6)	(7)	(8)	(9)	
1900	753	20.9	22.1	1572	479				1
1925	1520	42.2	44.5	1931	787	2.46%	1.39%	(1900–1950)	1
1950	2542	70.6	74.5	2662	955				3
1955	3281	91.1	96.1	2901	1131	5.15%	3.15%	(1950–1960)	2
1960	4200	116.7	123.1	3226	1302				3
1965	5363	149.0	157.1	3528	1520	5.31%	2.99%	(1960–1970)	2
1970	7045	195.7	206.4	4030	1748				3
1971	7374	204.8	216.1	4106	1796				2
1972	7683	213.4	225.1	4153	1850	4.89%	2.94%	(1970–1973)	2
1973	8130	225.8	235.3	4263	1970				2
1974	8199	227.8	240.2	4366	1878				2
1975	8087	224.7	236.9	4431	1825				3
1976	8573	238.2	251.2	4442	1930				2
1977	8859	246.1	259.6	4552	1946				2
1978	9034	251.0	264.7	4583	1971	1.05%	−0.5%	(1973–1982)	2
1979	9470	263.1	277.5	4709	2011				2
1980	9257	257.2	271.2	4836	1914				3
1981	9048	251.4	265.1	4867	1859				3
1982	8933	248.1	261.7	4895	1825				3

References
(1) World Energy Conference (WEC) Survey of Energy Resources 1974.
(2) UN Yearbook of World Energy Statistics, 1980. Table 1.
(3) UN Yearbook of World Energy Statistics, 1982. Table 1.

Notes
(a) At the time of going to press, the 1982 energy statistics were the latest available.
(b) Population figures deducted from columns (2) and (6).
(c) Units and Conversions:
   Mtce = 1 million tons coal equivalent at calorific value of 7000 cal/g
   kg.ce = 1 kilogram coal equivalent at same calorific value
   1 quad = 10¹⁵ Btu = 0.252 kg cal = 36 Mtce
   1 EJ (Exajoule) = 10¹⁸ J = 0.948 quads = 34.13 Mtce

negative growth. The broken line shown as an extension of the world energy total demand curve beyond 1973 represents an extrapolation at a growth rate of 4.89% p.a. – i.e. at the prevailing trend of the early 1970s. Already by 1982 the demand had fallen short of that extrapolation by about 29%. It seems probable that the political tensions engendered by the Middle East War of 1973, combined with the 'muscle-power' developed since then by the OPEC countries, and the economic troubles that have beset the industrial world largely as a result of these influences will conspire to cause a further falling away of the world energy demand curve from the trend prevailing in the early 1970s,

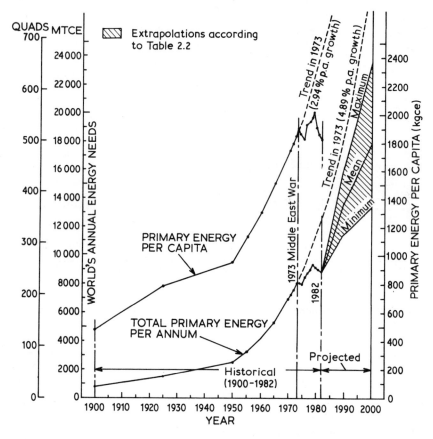

Fig. 2.1 Actual and projected word primary energy growth pattern (1900–2000).

as has been prophesied elsewhere (Armstead (1978) Fig. 79, p. 328) – perhaps more rapidly than had then been expected.

Extrapolation of energy demands into the future is an exercise that has to be performed by governments, private utilities, the oil and coal industries and other organizations that are responsible for providing the means of meeting these expected demands. Until 1973 energy forecasting was regarded, if not as a precise art, at least as one involving only a small element of speculation. For although growth rates were never constant they changed but slowly (except at times of wars and their aftermath). But, as can be seen from Fig. 2.1, future energy projections have become highly speculative since 1973, when all semblance of a regular growth pattern vanished.

Many economists maintain that a relationship can be discerned between gross national product (GNP) and energy consumption, and that the art of forecasting energy requirements can be aided if future economic trends can be foreseen. It is true that if 'historical' statistics of GNP and of energy

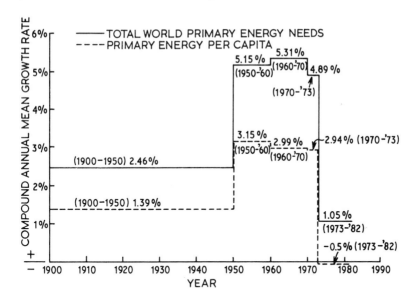

Fig. 2.2 Mean annual compound growth rates of total world needs of primary energy per capita.

consumption were plotted on logarithmic graph paper there would often appear to be some superficial evidence of a relationship of this kind, but it is easy to read too much into such graphs, which have the property of smoothing out wide irregularities. Even where it is possible to detect a broad trend in the GNP/energy ratio for one particular case, it is quite impossible to apply the same trend to another case; for the ratio deduced for an industrialized country would certainly differ widely from that deduced for a country with a mainly agricultural economy, since industry is generally far more energy-intensive than agriculture. It might be argued that the *world* GNP should bear a fairly steady relationship with the world energy consumption, but since 1973 there have been too many local instabilities for any clear pattern to emerge.

The probability referred to above, that the world energy demand curve will fall further away from the upward trend curve prevailing in the early 1970s, is reinforced by various forecasts that have been made by certain prestigious authorities on the basis of careful economic and demographic studies. The only thing that can be said with confidence about any long-term quantified forecast is that it will almost certainly be wrong! There are too many unforeseeable factors that could arise to confound even the most well reasoned prognostications of world energy consumption. Who, even in 1972, could have foreseen the 1973 perturbations that suddenly interrupted the past smoothness of the energy curves shown in Fig. 2.1? Favorable and adverse conditions or events such as economic recoveries, trade recessions, wars, cataclysms, 'bonanzas' of newly discovered fuel deposits as yet unsuspected; all these and other unknowable influences could either raise or depress the world energy consumption curve. However, since some attempts at forecasting are necessary, the following carefully considered figures shown in Table 2.2 are quoted in the hope of seeing what sort of future energy pattern is foreseen by certain highly responsible bodies.

Table 2.2. Various forecasts of future world energy needs

Authority		Annual world energy needs Mtce (quads)		
		1990	2000	2020
Energy Research Group, Cavendish	(Min.)	13 583(377)*	17 713(492)	28 634(795)
Laboratory, Cambridge University (WEC, 1978, p. 235)	(Max.)	16 143(448)*	23 515(653)	45 153(1254)
Workshop on Alternative Energy	(Min.)	—	15 381(427)	—
Strategies (Basile, 1977)	(Max.)	—	19 821(551)	—
Mean of estimates prepared by: Exxon (1980) Tenneco (1980) Oil and Gas Journal (1981) National Geographic Magazine (1981) World Energy Conference (1974, 1980)	(Mean)	11 390(316)	13 340(371)	—
Mean values		13 705(381)	17 954(499)	36 894(1025)

*1990 figures obtained by exponential interpolation of WEC figures for 1985 and 2000.

We are of the opinion that 2020 is too far ahead for any reliance to be placed on the quoted figures, though they may be noted with interest; but the forecasts for 1990 and 2000 have been shown on Fig. 2.1 as a shaded band bounded by the maximum and minimum prognostications, and with the mean value shown as a line. This clearly shows the wide margin of uncertainty to which even the opinions of acknowledged 'experts' are prone. The amounts by which the forecasts fall short of the extrapolated 1970/73 trend curve are as follows:

	1990	2000
Maxima	11.8%	20.3%
Means	25.1%	39.2%
Minima	37.8%	54.9%

What accounted for the high energy growth during the first 73 years of this century? Partly, of course, it was due to the rise in world population, which increased by a factor of 2.7 during that period. But 2.7 is only a fraction of the 10.8-fold increase in the total world energy consumption during the same 73 years. The other reason for the high growth was the steady rise in the energy consumption *per capita* resulting from higher living standards and increasing industrialization. The energy consumption *per capita* in 1973 was about four times what it was in 1900. The upper curve of Fig. 2.1 clearly shows the spectacular increase in *per capita* energy consumption in the 23 years following the mid-century until the severe shock of the 1973 crisis caused a diversion from the prevailing trend of the early 1970s. (There are of course wide differences in *per capita* energy consumption as between one country and another – e.g. in 1973 the figures were 11 494 for the USA, 4621 for the USSR and only 9 for Burundi, in kg.ce; but this book is concerned with the world rather than with the distribution of wealth between nations).

As Third World countries develop, the average energy consumption *per capita* could well rise causing an even higher rate of increase in the total world energy consumption. The average forecast of 17 954 Mtce (499 quads) for the year 2000 represents a mean compound growth in world energy consumption of nearly 4% p.a. from 1982. By contrast, the world population growth rate appears to have reached a peak of about 2.7% p.a. during the 1960s, since when it has shown signs of falling: it was about 1.8% p.a. during the 1970s. Whether this falling trend will be sustained remains to be seen; but if the quoted energy forecasts are even approximately right, a steady rise in *per capita* energy consumption would seem to be likely – at least for some time.

The statistics published by the World Energy Conference in 1974 and 1980 showed a decline of 1.33% during those six years in the estimated proven world reserves of fossil fuels. Although this decline may not look very serious, it nevertheless represents a reversal of the steadily rising estimates of previous years. Furthermore, because over 95% of our current energy usage is supplied by fossil fuels, the broad implication of Fig. 2.1 is that worldwide fossil fuel consumption will rise for many years; so unless this can be matched by some very major new discoveries of proven deposits, the reserves will decline at a growing pace.

The perturbations in the world energy market that were first felt in 1973, and the signs of which can be seen so clearly from Fig. 2.1, have started a process of rising prices and approaching scarcities of fossil fuels that has already dislocated many national economies and created severe international tensions. This is a process that can only grow more acute as resources become more and more depleted. Rising prices are symptomatic not only of growing scarcity but also of more expensive extraction of less accessible reserves – e.g. off-shore oil production; increasing drilling depths now required to tap gas and oil reserves; the need to turn to less economically attractive fuel resources such as oil shales, tar sands and biomass; the production of synthetic fuels from liquefied or gasified coal; recourse to tertiary oil recovery in order to squeeze the utmost from existing oil formations.

In addition to fossil fuels there are of course *fissile* nuclear fuels that could help ease the situation; but these fuels too are by no means unlimited in supply, nor are they evenly distributed throughout the world. Furthermore, there are other factors that have inhibited the development of nuclear power. Although controlled nuclear fission has opened up a large new source of clean energy, it has also introduced a negative element of fear into the world. Worldwide concern over the safety of nuclear reactors and the treatment of reactor waste products are real technical and social issues. Unfortunately, their impact is disproportionately inflated by the emotion associated with a pervasive fear of radioactivity by many people as well as by a technically incorrect linking of power reactors *per se* to the development of nuclear armaments. If we discount the highly emotive popular opposition to nuclear power, fissile fuels could, and probably will, give the world a useful 'stay of execution' of several decades – especially if breeder reactors are developed.

Fuel resources of all kinds, fossil and fissile, are finite and must therefore steadily dwindle as they are consumed. Continued reliance upon them can only lead to a deteriorating situation. Matters will not be helped by world population increases which can only raise the pressure of demand. The occurence of short term energy oversupplies caused by conservation measures and trade recessions, such as have occurred in the 1981–86 period, are but surface ripples on a smooth long term trend. Conservation, rigidly enforced, could serve as a short term time-gaining palliative; but enforcement would be difficult. Conservation does not nor would it offer a permanent solution to the threat, sooner or later, of an energy famine. In the long term, if we are to survive without agonizing readjustments to our way of life, it is of vital importance that we gain access to new energy sources far larger than all the fossil and fissile fuels hitherto used or likely ever to be discovered.

Although energy growth rates will probably be less rapid than before 1973, there is a strong concensus of opinion that growth will continue at quite substantial rates at least until the end of the century and very possibly well into the 21st century, as shown in Table 2.2. If the forecasts in that table are even approximately realistic, the implications are distinctly alarming, as may be seen from Table 2.3. The total primary energy consumed by the world since the beginning of the century until 1982 was about 266 650 Mtce or 7410 quads. This figure may conveniently be designated as a factor 1.0. The amounts of new primary energy that must be found to satisfy the prognostications quoted in Table 2.2 are shown in Mtce and quads and as multiples of this factor in Table 2-3.

Even at the most moderate of the forecast needs shown in Table 2.2, we shall have to find about three-quarters as much (77%) new primary energy in the last 18 years of this century as the whole amount consumed in the preceding 82 years: according to the other forecasts we shall need substantially more. Although we have rather discounted, as too speculative, the long term forecast of the Cavendish Group to the year 2020, we have nevertheless included it in Table 2.3 as a matter of interest, but have shown only the mean of their prognostication. If correct, we would have to find almost three times as much new primary energy in 38 years as the total consumed during the first 82 years

Table 2.3. Additional primary energy required to satisfy the expected needs of the world

Total primary energy consumed by the world from 1900–1982		Range	Additional energy required								
			From 1982–1990			From 1982–2000			From 1982–2020		
Mtce	quads		Mtce	quads	R	Mtce	quads	R	Mtce	quads	R
266 650	7407	Minimum	82 128	2281	0.31	206 498	5736	0.77			
		Mean	91 602	2545	0.34	251 077	6974	0.94	786 340	21 843	2.95
		Maximum	101 125	2809	0.38	300 818	8356	1.13			

$R = \dfrac{Mtce}{266\,650} = \dfrac{quads}{7407}$; i.e. the ratio of additional energy required to total energy consumed from 1900–1982

of this century. Where are we to find such huge quantities of energy? It is true that there is still an abundance of energy available – but not of *cheap* energy. As mentioned earlier in this chapter, greater reliance must increasingly be placed upon costlier lower-grade energy sources. The danger is not that energy supplies will suddenly grind to a halt: somehow or other we can probably scrape up enough to satisfy our more essential needs for several decades to come, but (in the absence of a new break-through) only at the cost of growing economic stresses and greater pollution. In the 1960s the energy component of the cost of manufactured goods accounted, on average, for something like 1% to 5% of their total cost, in the more highly industrialized countries (Goetz, 1973). Such an economically insignificant component was conducive to wastage. But for several years this percentage has been rising rapidly, and it will continue to rise until the economic stresses have been released by the availability of some new vast source of energy that will reduce our dangerous dependence upon fossil and fissile fuels.

For the first 73 years of this century, we were living in a fool's paradise of inexpensive fossil energy. Then came disillusionment. Within seven years the price of energy had risen over ten-fold, and the economic and political effects of this have been acute. We believe the dramatic collapse of oil prices in the mid-1980s to be a short-term phenomenon attributable to 'market correction' aggravated by political pressures. In the longer term, energy prices must inevitably rise again in terms of 'real' money after discounting general inflation: that is to say, they will rise more rapidly than the prices of labor and materials. This will not necessarily be a continuous process. The upward trend may well again be interrupted by further temporary falls caused by active energy conservation programs, new discoveries, price wars, trade recessions and political pressures; all partially induced by the excessive rises in energy costs that have preceded them.

Despite popular opposition to nuclear power, we think it inevitable that increasing reliance will be placed upon it. Tragic though the Chernobyl disaster was, it will at least have one beneficial result in that safety measures will almost certainly henceforth be so rigidly enforced that such a calamity is unlikely to recur. Without the support of nuclear energy, which even in 1984 was supplying more than one-eighth of the world's electricity, the pressures of demand on fossil fuels would very soon become intolerable.

Eventually, access to a huge alternative energy resource on economically and environmentally acceptable terms is necessary to ensure worldwide prosperity and political stability.

Chapter 3

Energy options

As cited in Chapter 2, our present fossil fuel resources are incapable of sustaining our needs for more than perhaps a few tens of decades unaided by new sources of energy and by drastic conservation measures. Thus, it is well to evaluate other options. For convenience, they will be divided into two categories – *non-renewable* and *renewable*.

3.1 Terminology and definitions

Before attempting to assess these resources it is necessary to distinguish between *reserves* and *resources*.

A *reserve* is generally accepted to mean the quantity of some non-renewable commodity – be it a mineral or energy – that is known to exist, has been identified and approximately quantified, is recoverable with today's technology, and is commercially worth harvesting at today's prices. Although there should be little that is speculative about a reserve in theory, its evaluation is unlikely to be very accurate in practice. In the first place, the assessment of all fuels and geothermal reserves depends on the judgments of geologists, reservoir engineers, and other earth scientists, even the best considered opinions of whom are apt to involve an element of intelligent guesswork. Then again, the use of the word 'today' twice in the above definition of a 'reserve' shows that any quantification can be valid only at the time of assessment: thereafter it may be increased by further discoveries or reduced by further extractions. Furthermore, advancing technology may at any time increase a reserve by making accessible what was formerly regarded as inaccessible; while changing market prices could either increase or reduce it. For example, petroleum is now being extracted and consumed more rapidly than new discoveries are being made, so that petroleum reserves are shrinking. Less economic reserves such as off-shore oil as well as new technologies became feasible with rising prices. Higher prices also enable improved recovery factors to be economically applied to 'in place' oil deposits with tertiary methods as well as to unconventional sources of fossil fuel from oil shales and tar sands.

A *resource* implies the total quantity of any energy commodity that is believed to exist in all the world – identified, inferred and undiscovered but

Table 3.1. Approximate worldwide reserves and resources of non-renewable energy contained in the earth's crust, excluding HDR (in quads of thermal energy).

Energy form (1)	Identified reserves (2)	Additional resources (inferred and undiscovered) (3)	Total potentially recoverable resources (Column 2 plus Column 3) (4)	Resource base (5)	Authentication (6)
FOSSIL FUELS	29 900	170 600	200 500	358 900	WEC (1980) p. 5 used for columns (2) and (5); Linden (1980) used for column (3)
FISSILE FUELS:					
Uranium (non-breeder)	964	2 111	3 075	4 300	see Note (ii)
Uranium (breeder)	72 230	159 170	231 400	344 000	
TOTAL FOSSIL AND FISSILE FUELS	103 094	331 881	434 975	707 200	
GEOTHERMAL:					
Hydrothermal (>150°C to 3 km depth)	2 500	30 500	33 000	130 000	see Note (iv)
Geopressured (including methane)*	—	135 000	135 000	540 000	see Note (v)
TOTAL GEOTHERMAL	2 500	165 500	168 000	670 000	
TOTAL NON-RENEWABLE ENERGY	105 594	497 381	602 975	1 377 200	

*Here treated as a geothermal resource although to some extent it can be regarded as a fossil fuel owing to its methane content.

suspected of probably existing – and which may become technically recoverable within the foreseeable future regardless of cost. 'Resources' include, and are likely to be much larger than, their corresponding 'reserves': quantitative assessment of resources will always be highly speculative. Such words as 'believed', 'inferred', 'suspected', 'may', and 'foreseeable', are all imbued with vagueness. The omission of cost as a factor influencing the size of a resource may seem strange; but its explanation lies in the fact that growing scarcities and pressures of need can force up prices more or less indefinitely.

A still wider concept than 'resource' is 'resource base', which may be defined as the estimated gross quantity of any energy commodity in all the world, regardless of its recoverability.

While it may be possible to evaluate energy reserves within reasonable limits of error, it is very difficult – indeed, virtually impossible – to quote reliable assessments of the earth's potentially recoverable energy resources, which include that which has not yet even been discovered, but is only suspected of existing; all such estimates should be treated with the utmost caution and as 'order of magnitude' assessments only. Matters are not improved by the many discrepancies to be found between the figures, both of reserves and resources, published by various international and other organizations – e.g. The World Energy Conference, oil companies, coal authorities, various petroleum institutes, departments within national governments, etc. – and by individual authors of books and papers. Furthermore, statistics of all kinds relating to energy are constantly being revised owing to the delays and difficulties of collecting up-to-date information from all over the world.

3.2 Non-renewable energy resources

If any degree of substance is to be given to what would otherwise be mere vague arguments *in vacuo*, it is necessary that some form of inventory be presented. After examining several publications the authors offer in Table 3.1 a very tentative evaluation of the non-renewable energy content of the earth's crust, with the exception of the heat contained in HDR which will be discussed in depth later. The geothermal component included pertains only to those resources that contain hot fluids – steam, water or mixtures thereof in their natural state: it excludes HDR and magma.

The following points concerning Table 3.1 may be noted:

(i) The figures of reserves are somewhat speculative: those of additional resources and resource bases are highly speculative. All should be treated with the utmost caution; but they may serve to provide some degree of perspective to the overall energy problem if taken in conjunction with the figures of Table 2.1.

(ii) Breeder technology has not yet been fully established as commercially (as distinct from technically) feasible, although regarded so by many. As no breeder estimate was given in WEC (1980), the figure here used is based on Linden's (1980) estimate for the total non-breeder potential. This ratio does not differ widely from the 80 factor used by the WEC in their 1974 Survey of Energy Resources. Breeder reactors could clearly be of immense help in extending the time available for developing the technology for exploiting entirely new energy sources.

(iii) The totals for fossil fuels include oil shales and tar sands – even under 'reserves'. This is because the exploitation of some of these deposits is already becoming more economically attractive.

(iv) Surveying for hydrothermal reserves is a far more recent activity than the search for fossil fuels, or even for nuclear fuels. The degree of uncertainty in estimating the world's total hydrothermal resources and particularly the reserves is therefore even greater than in the case of fuels. The way we have approached the estimation of hydrothermal resources and reserves is to use the guidelines established by White and Williams (1975) for the US. They estimated that about 25% of the enthalpy contained in fluids within the reservoir could be recovered at the wellhead. Rowley's (p.145), 1982 figure of 1.1×10^6 MW-centuries (or 3290 quads) for the worldwide electric energy potential from reservoirs at depths to 3 km at temperatures above 150°C was used to calculate an upper limit for the total recoverable resource. With an overall average conversion efficiency of 10%, this amounts to 32 900 quads of heat energy at the wellhead. Using the 25% figure this would mean that the resource base is approximately 130 000 quads. White and Williams (1975) used approximately 7.6% for the reserve portion of the total hydrothermal resource for the US. Using that percentage of 32 900 quads, about 2500 quads could be considered as the worldwide hydrothermal reserve. The figures supplied by Boldizsár (1980) suggest that the combined hydrothermal reserve for the USA, USSR, Italy, Iceland, New Zealand, and Hungary was equivalent to 1260 quads. Thus to be consistent with our earlier estimate, the Boldizsár figure represents about one-half of the world's total. This seems not unreasonable. However, Table 3.1 would seem to suggest that hydrothermal energy could never play a very major role in the context of world energy supplies, though it could be of great local importance and also in short term time-winning strategy. Its estimated reserves (as shown in Table 3.1) would seem to account for about 8% of the total fuel reserves including fossil fuel and nuclear fission energy (without breeders). Nevertheless, the figures for hydrothermal energy quoted in Table 3.1 are very conservative, for they take no account of low grade heat at less than 150°C which could be suitable for direct applications for industrial and farming purposes, and which is believed by some authorities to be much more plentiful than the high grade resources.

(v) The figures for geopressured energy were extrapolated arbitrarily from US figures for geopressured estimates from US Geological Survey Circular 790 (Muffler and Guffanti, 1978) at 2.5 times the US figure of 216 000 quads for the resource base. The resource was then estimated at 25% of the resource base. This estimate is admittedly very uncertain but with lack of any other data seems reasonable. No reserves are included at this time, as their commercial exploitation has not yet been established.

The fact that between 96 and 97% of the world's energy needs are still derived from fossil fuels, the reserves of which are now known to be consumed more rapidly than new deposits are being discovered, underlines our frightening dependence upon such fuels – a dependence that is likely to persist for many years to come. Moreover, nearly 70% of the fuels now used are in the form of liquid petroleum distillates or natural gas (in terms of heat content); hence the great sensitivity of world energy supplies to the political conditions prevailing in the Middle East and in other regions that produce these fuels. Because of the world's enormous appetite for fossil fuels, there are risks of international chaos when fuel supplies are even temporarily interrupted. The danger lies not only in the strategic control of fossil fuel supplies by a small number of petroleum producing countries but also by the eventual depletion of these resources that will occur. Hydrocarbons have great value not only as fuels but also as chemical feedstocks for many important industries – e.g. plastics, synthetic rubber, fertilizers, textiles, paints.

It is of interest to examine the question 'For how many years would our present reserves and resources of non-renewable energy last at the present level of demand?' There is of course no simple answer to this question, but it is possible to arrive at an order of magnitude if the figures of Table 3.1 are provisionally accepted as correct – a most improbable assumption, but in the

Table 3.2. Approximate duration of non-renewable energy at the 1982 level of demand of 248.1 quads

	Identified reserves	Total potentially recoverable resources
FOSSIL FUELS	120 years	808 years
FISSILE FUELS:		
uranium (non-breeder)	4 years	12 years
uranium (breeder)	291 years	932 years
TOTAL FOSSIL AND FISSILE FUELS	415 years	1752 years
GEOTHERMAL:		
hydrothermal	10 years	133 years
geopressured (incl. methane)*	— years	544 years
Totals	425 years	2429 years

(All figures to the nearest year).
*Here treated as a geothermal resource although to some extent it can be regarded as a fossil fuel owing to its methane content.

absence of reliable evidence to the contrary the only reasonable one to make. Comparison may be made between these figures and the estimated total world demand for energy in 1982. Hence, at the 1982 level of demand, the reserves and resources of Table 3.1 would theoretically last for the following periods shown in Table 3.2. The apparent precision of the figures in this table should not be regarded as too realistic. They are simply consistent with the figures shown in Tables 2.1 and 3.1 and are of course susceptible to any errors contained in these earlier tables.

It is important that these figures should not be misinterpreted. For example, it is not implied that the uranium reserves will in fact last for only four years if none of the inferred and undiscovered uranium should ever materialize. But if the known reserves of uranium were the *sole* source of energy available, then they would suffice to satisfy the world for only four years at the 1982 level of demand.

At first Table 3.2 does not look too bad, for it would suggest that we have over four centuries in which to eke out our reserves even without relying at all upon any further resources; but this would be a most misleading conclusion for the following reasons:

(1) It assumes that Table 3.1 is reasonably accurate; and while this just could be so it would be unwise not to allow for the possibility that it is an over-estimate. It is true that there has been a past tendency for the published figures of reserves to continue to rise as new discoveries have been made, but for some years this has not been true of petroleum: new discoveries are now failing to keep pace with consumption.

(2) It assumes that nuclear energy will be developed to the greatest possible extent. In view of the very powerful and vociferous sector of the public who are implacably opposed to this form of energy, it is quite possible that nuclear power will remain largely undeveloped until too late. Without fissile fuels we would be left with only about 130 years from fossil fuels and geothermal energy if reserves alone are considered, and 1485 years if additional potentially recoverable non-uranium resources are also included.

(3) Even these figures do not seem very alarming, but it must be remembered that in Table 3.2 no growth has been assumed in the world's annual energy needs. Although future growth is likely to be slower than in recent decades, it seems probable that there will be at least *some* growth for a while. Even a devastating war would not necessarily curtail growth, though it would sadly divert our remaining non-renewable energy resources from constructive to destructive purposes – probably including the destruction of much of the energy resources themselves by the sinking of tankers and the burning of oil storage tanks by military action. The effects of even quite a modest sustained annual growth rate upon the duration periods of resources can be very marked. A 2% compound annual growth rate, for example, would shrink the figures of Table 3.2 to those of Table 3.3. As with Table 3.2,

estimates of years may seem unrealistically precise; but they are arithmetically consistent with what has preceded them. In this case it is not permissible to *add* the durability periods of the different energy forms, as was possible in Table 3.2, which represented 'no-growth' conditions. Some readers may find this fact, as well as the extent of curtailments, surprising. (See Armstead, 1980 for explanation).

Table 3.3. Approximate duration periods (to the nearest year) of non-renewable reserves and resources assuming 2% annual growth

Resources	Reserves	Total potentially recoverable resources
Fossil fuels alone	61 years	143 years
Fossil and fissile fuels together	113 years	181 years
All non-renewable energy	114 years	197 years

(4) Last, but not least, all these estimations assume that our demands could be satisfied right up to the last moment when our non-renewable energy resources become exhausted, just as though we had a 'tank' of energy that could be drained to the last drop at any desired rate. In practice, of course, it would not be possible to draw on our resources in this way, for they will become increasingly difficult to acquire with growing scarcities, rocketing prices and less accessibility. Hence the figures of Table 3.3 are somewhat academic in that even a 2% annual growth rate could not be sustained until the point of exhaustion. Scarcity and price dislocations are already beginning to be felt; and if we can survive even into the first or second decades of the 21st century by relying upon fossil fuels and accessible geothermal energy we shall indeed be fortunate. Disturbances could of course be postponed for a while by making great efforts to conserve what energy we have; for example, by adopting large combined heat and power projects.

All this underlines the message of Chapter 2: it is vital that new and large energy sources be tapped without delay. The options that are open to us for achieving this with our present technology are confined to the various renewable energy sources.

3.3 Renewable energy sources

With renewable energy there is no need to distinguish between 'reserves' and 'resources'. Reserves have no meaning, for there is no finite stock of renewable energy on which we can draw: it is all virtually infinite, and may be regarded simply as 'resources'. Moreover, with non-renewable energy, which by definition is finite in quantity, it was appropriate that estimates should be quoted in energy units (e.g. MWyr): with renewable energy the 'years' are, for all practical purposes, infinite, and all we can do is to attempt to estimate this in units of power (e.g. MW) both average and maximum or of energy available per annum.

Although all renewable energy sources owe their origin, wholly or in part, to the sun, it is useful to treat them in separate categories and to reserve the term 'solar energy' for that which can be derived from direct solar radiation. Although fossil fuels also derived their energy from the sun eons ago, it would be absurd to treat them other than as non-renewable.

3.3.1   Hydro power

The first renewable form of energy that comes to mind is hydro power. It can be, and is, exploited by well established technologies; it is clean; it is reasonably dependable; it can provide secondary benefits such as flood control, irrigation, navigation, and recreation; and it is often cheap. The ultimate theoretical world hydro power potential has been estimated (WEC, 1980, Table 4.7) at about 159.4 million TJ p.a., equivalent to 5.05 million MW continuous power, of which only 44% is technically feasible. By 1980 about 16.5% of this feasible potential had already been developed, 5.6% was under construction and a further 10.9% was planned, making a total of about one-third of the ultimate potential. Hydropower is now providing almost a quarter of the world's electricity. If the remaining two-thirds of the world's feasible hydro potential could be developed – and difficulties of site location could well make this impracticable – the additional energy obtainable would be about 1.48 million MWe continuous. This can be expressed as about 44 quads (1593 Mtce) p.a., which is only 17% of the world's total primary energy production in 1980. This figure, however, does not do justice to the true value of hydro power, because the UN energy statistics credit hydro power only with the heat equivalent of its electricity production. What is of significance is that 44 quads (1593 Mtce) of electrical energy, if *thermally* generated, would require about 125 quads (4500 Mtce) of fuel energy, which is almost half of the present total annual world primary energy produced. It is clear, however, that although hydro power could make a very significant contribution to the world's energy problem it could never suffice as a permanent solution. There simply is not enough of it to supply more than a fraction of what we must have.

3.3.2   Solar energy

The amount of solar radiation that reaches the earth's surface averages about 4 kWh/m^2/day (WEC, 1980, p.300). The land surface area of the world is about 148 million km^2, but after setting aside the requirements of farming, forestry, amenities, etc. and allowing for rugged and inaccessible terrain, there would remain only about 25 million km² available for the interception of solar radiation. At 4 kWh/m²/day it would still theoretically suffice to collect about 125 000 quads of energy per annum from the sun. The world is now consuming about 250 quads p.a. of primary energy – a mere 0.2% of that figure. At first sight it would therefore seem that solar energy could provide the answer to our quest for a vast energy source as an alternative to fuels. However, the difficulties are many and formidable.

In the first place much of the available interception areas (e.g. tropical deserts) would be ill adapted to the areas of greatest energy demands (e.g. the large industrial cities in northern latitudes. Then it must be recognized that solar energy is very diffuse, so that huge areas of interception are needed for quite moderate quantities of energy. To produce 1 MWe of power at 22% efficiency would require a collection area of about 27 270 m², though in a tropical desert this figure might be reduced to about 18 180 m². Another difficulty is the low utilization factor arising from the hours of darkness, the obliquity of incidence and the fortuitous obstruction by clouds over much of the land surfaces of the earth. This means that full availability of a required rate of energy supply can be obtained only by providing either 100% standby in some other energy form or very large heat storage facilities – both of which would make very costly additions to an already costly method of energy interception. Furthermore, solar energy is unsuited to transportation, which is a demand that may account for about one-quarter of the total energy used by industrialized countries and about 10% of the world's total energy consumption. It is true that demonstration solar vehicles have been made, but the notion cannot be regarded as commercially feasible for heavy transportation. A partial answer to the transportation question, and also to that of energy storage, could be provided by the use of synthetic fuels made from biomass (WEC, 1980, pp. 328–340, and Crabbe and McBride, 1978, p. 27), or perhaps of hydrogen from electrolysis activated by photovoltaics.

The capital cost per maximum kilowatt of delivered power is at present very high for solar energy installations, even for domestic water heating systems energized by roof panels; and of course, much higher per mean kilowatt. With an increasing scale of development these costs would undoubtedly fall, but are likely to remain comparatively high; and it must be remembered that there is always a limit to capital availability.

So far it has here been tacitly assumed that only solar radiation falling upon the land masses would be of use as an energy source, but it has been demonstrated that ocean temperature gradients caused by solar radiation in tropical waters can be used to activate a thermodynamic cycle to produce motive power from a low pressure turbine (WEC, 1980, pp. 317–328, and Crabbe and McBride 1978, pp. 133–134). This notion is known as OTEC (Ocean Thermal Energy Conversion). In 1930, a 22 kW installation was operated for two years off Cuba by Georges Claude, using a 14°C differential (Crabbe and McBride, 1978 p. 134). Because of corrosive troubles and very high capital costs this experiment was not a commercial success. Thought is now being given to the use of non-corrosive working fluids such as ammonia or propane in a closed cycle, but obviously such projects could never provide more than a tiny fraction of the world's energy needs. OTEC has inherent economic barriers. Its thermodynamic efficiency must be very low (even the maximum Carnot efficiency is only about 5%), equipment is large on a 'per kilowatt' basis and therefore costly, and, as with a hydro project, major costs must be expended before any power can be produced.

In the realms of futuristic possibilities – not necessarily to be dismissed out of hand – it has been suggested that artifical satellites could be placed in geosynchronous orbit and kept in continuous sunlight. Solar energy would be converted on the satellite into electricity by means of solar cells or a thermodynamic cycle, and power would be beamed to the earth by lasers or microwaves so as to provide continuous power (Glaser, 1983). In this way several of the limitations referred to above would be overcome, but the costs would be very high.

In conclusion, solar energy does not lack the requisite abundance but it has certain inherent limitations that would seem to preclude its ability to supply more than a useful and significant part of our total energy needs. At present its contribution is minute, but it is likely to become increasingly important as a supplier of low grade heat for domestic hot water and space heating. For power generation it is doubtful whether it could compete economically with other methods for a very long time to come. However, if substantial technological breakthroughs occur, satellite collectors and photovoltaic converters may one day become a practical proposition; but it would be unwise to assume that solar energy could ever become the vast new energy source that we are seeking.

The reader is referred to the following references for further information on solar energy: WEC (1980), pp. 297–311; Crabbe and McBride (1978) pp. 159–165; UN (1961), Vols. 4, 5 and 6.

3.3.3 Wind power

At present, the use of wind power is mainly confined to small boats and isolated farmsteads. The middle of the 20th century saw some improvements in the aerodynamic design of windmills, with considerable success; so that conversion efficiencies were more or less doubled from the 10 to 20% range to the 20 to 40% range (Sharman, 1975, p. 55 and WEC, 1978, p. 147). The total power of the winds throughout the world has been estimated at 2×10^{10} kW (Crabbe and McBride, 1978, p. 191) but this figure can only be a 'guestimate' in view of the paucity of data from many parts of the world. Nor is it clear whether it claims to be the maximum or the mean extractable power, or whether it allows for conversion efficiency. Even if it is accepted that its potential is very large, wind power is subject to similar inherent disadvantages as solar energy in the matters of diffuseness, low utilization factors and the need for a standby plant or large energy storage facilities. There are also technical problems arising from the fact that the power developed by the wind is proportional to the cube of the wind speed, which causes the power generated to be very capricious and ill matched to the patterns of demand: it also necessitates very robust structures to resist the huge forces induced by hurricanes. The land area required to provide a given amount of power will usually be from four to five times that required for solar-electric generation (WEC, 1978, p. 147).

Environmentally, windmills are at a disadvantage both esthetically (when arranged in large clusters, especially along hill tops) and as hazards to aircraft and obstructions to farming operations. At present windmills cost from two to four times more than conventional generators of equivalent rating (WEC, 1978, p. 148).

Despite these limitations several quite impressive wind power devices and systems have been installed or are being planned, but unit sizes have not yet exceeded the 1 – 2 MW range. An ambitious scheme is being debated in the USA to install about 300 000 wind-driven turbines across the Great Plains to give a collective capacity of 189 000 MW – an average of 630 kW per unit (Crabbe and McBride, 1978, p. 192). Such a system could supply about 0.6% of the electricity now used by the USA for every 1% of annual capacity factor obtained. Thus for 20% annual capacity factor the contribution would be about 12% of the country's electricity, but less than 2% of its total energy needs. However, as with hydropower, this 2% contribution is equivalent to perhaps about 5.5% if regarded as the potential saving of fuel that would have to be burned if the contribution were withheld. If this project is realized it would supply an impressive amount of energy, but whether it could be economically justified seems doubtful. Other areas of sufficiently low population density within reasonable range of very large energy demands would be difficult to find.

Thus wind power could perhaps make a significant contribution to the world's energy needs, but does not look very promising as a possible major alternative to fuels.

The reader is referred to the following references for further information on wind power: WEC (1980), pp. 311–321; Crabbe and McBride (1978) pp., 190–194; UN (1961), Vol. 7.

3.3.4 Wave power

Wave power, like wind power, is a byproduct of solar energy. Waves are raised by the winds, sometimes to heights of several metres in heavy storms, and can travel for hundreds of kilometres across oceans with little loss of energy until they break on the coasts. The total world wave power has been estimated at about 2.7 TW (WEC, 1978, p. 148), but the interpretation of this figure is uncertain. The most optimistic meaning would be that it represents the average simultaneous summation of wave power in all the oceans of the world.

The most practical way of converting wave power into a useful form is by means of electricity generation. Overall efficiencies ranging from 10% to 40% are claimed to be practicable according to the wave intensity with an average of about 20% (WEC, 1978, p. 148). It should therefore be theoretically possible to generate up to about 540 GW (average) from all the waves in the world. This could yield 4730 TWh p.a., or about 6% of the world's total annual primary energy consumption. Obviously only a small fraction of this gross potential could ever be recovered in practice.

Many ingenious devices have been invented for converting wave motion into electricity, but the disadvantages of very large scale development of this energy source are great. The available power is spasmodic and unpredictable. Long periods of calm weather would necessitate 100% standby plant relying on some other energy source, and this would reduce the value of wave power to that of a fuel saver only: capital expenditure would be duplicated. The availability factor of wave power is not great. The necessary conversion devices for large scale development would have to be spread over some thousands of kilometres, thus creating navigation hazards. The moorings of off-shore devices would be subjected to very heavy buffeting in stormy waters and would therefore have to be extremely robust: a broken mooring could be a serious hazard to other wave power devices and to navigation. Electric, hydraulic or pneumatic energy transmission from ocean emplacements to the shore would be extremely expensive and vulnerable. Equipment would have to work in hostile and corrosive environments, and maintenance would be difficult, expensive and often impossible during long periods of bad weather.

For all these reasons it seems improbable that wave power will ever play a very significant part in providing the world with its huge energy requirements.

The reader is referred to the following references for further information on wave power: WEC (1980), pp. 321–324;, Crabbe and McBride (1978), pp. 188–190.

3.3.5 Tidal power

Although the recovery of tidal energy, which is mostly derived from the moon, could be quite impressive and of great local significance at a few sites it would be more or less negligible in the world perspective. The feasibility of a tidal project rests on two primary conditions being met. First, the mean vertical tidal range of water levels should be high; secondly a large volume of water must be capable of being impounded by means of one or more barrage(s) of short length, in fairly shallow water, and with good foundation conditions. There are very few places in the world that can satisfy these conditions, and it is significant that although no new technology is required only two sites have been exploited to date.

At La Rance in Northern France (Eléctricité de France, 1961), a basin of 22 km² (184 million useful m³) is impounded by a barrage of little more than

700 m in length. The maximum and mean tidal ranges are 13.5 and 11 m. The installed capacity of 240 MWe, which can produce 544 GWh p.a. (25.9% annual capacity factor). The dam contains 24 bulb type turbines of 10 MWe capacity each. These are reversible, and the scheme can be used for pumped storage, if required. La Rance is the only large scale tidal power plant now operating in the world; and yet it contributes only about one-five hundredth part of France's present electricity needs. Though of doubtful economic worth when installed in 1967, it is probably now paying its way since fuel prices have risen so abruptly.

The only other tidal power installation of appreciable size in operation is a small experimental one at Kislaya Inlet, near Murmansk, USSR: its electrical output is insignificant. There is also a small experimental tidal unit in the Bay of Fundy, Canada for turbine testing purposes.

The World Energy Conference (1980) lists only 25 'identified' sites in all the world that would meet the necessary requirements (though there are known to be other sites of interest, not included in the list, such as Inchon in S. Korea, Secure Bay in Australia, Solway Firth and Menai Straits in Britain, and perhaps a few others). The WEC (1980) also estimates that their 25 sites could produce only about 1% of the world's present total electricity production, and the proportion would, of course, fall if electrical consumption rises.

The highest tidal range known in the world is in the Bay of Fundy (16 m maximum, 10.7m average), while the Severn Estuary in Britain follows next (14 m maximum, 9.8 m average). The Bay of Fundy has the greatest tidal energy potential of all sites in the world, and yet the Canadians have not yet thought it worth exploiting except on a very small scale (see above). Likewise the Severn tidal barrage has been discussed repeatedly in the British Parliament and postponed again and again for further investigation (while the capital costs have steadily mounted). Estimates suggest that the Severn could supply Britain with about 6% of its present electricity needs: if it had been implemented in the 1920s when it was first seriously considered, it could have supplied almost half the country's need at that time).

Although it is true that both the Fundy and Severn projects are yet again being reconsidered, why is it that they have been repeatedly shelved? It is probably because tidal exploitation has the following disadvantages which have so far been thought to outweigh the advantages:

(1) Tidal power is available at low capacity factor only (about 25% annual) owing to the time of waiting while tidal differentals accumulate, and owing to the times of completion of the filling and emptying operations when the available head is too small to be of use; with the result that power can be generated only in intermittent 'bursts'. This is conducive to high investment costs per kilowatt for turbines and per kWh produced.

(2) Not only is the availability intermittent, but the time of incidence changes by 50 minutes every day. These two factors make it impossible to match the supply to the demand; hence there is need for storage, or to rely on tidal

power only as a fuel saver in parallel operation with more conventional sources in a larger base load system. Storage can sometimes be provided by pumping, as at La Rance, or it could be in the form of compressed air for use with peaking gas turbines.

(3) The output varies widely between spring and neap tides, thus further complicating the integration of tidal power with thermal and/or hydro systems.

(4) A tidal project cannot be built piecemeal, except for the turbo sets which form only a small part of the total costs. It is a matter of all or nothing where the civil engineering works are concerned. This, of course, is largely true of hydro power also; but as tidal projects generally involve much higher capital costs, the point acquires greater importance. It means that unless there is an immediate power market ready to receive the full potential of a tidal project, capital charges on investment would accumulate to large values while the scheme is only part loaded.

(5) Construction times are apt to be very long owing to the large scale of the civil engineering works. This involves very high interest during construction.

(6) Certain environmental effects (Shaw, 1977) can follow the construction of a tidal power installation, such as:
   (a) navigation problems due to silting, changes in the local tidal range and wave pattern resulting from the presence of the barrage, changes in currents;
   (b) ecological changes affecting fisheries and bird life;
   (c) sewerage and land drainage problems due to flooding;
   (d) changes in the salinity distribution.

On the other hand, there are certain positive advantages in tidal power, so:

(1) Despite the variations referred to under items (1), (2) and (3) above, the pattern of power generation is at least precisely predictable – unlike the cases of solar, wind and wave power.

(2) Sometimes recreational facilities can be created as a result of the building of a barrage – e.g. boating, swimming.

(3) It may be possible to construct a highway across the barrage, thereby making obsolete inconvenient ferries or long inland detours.

To summarize, tidal power seems to be a doubtful proposition even now, except perhaps at a very few sites. With further increases in fuel costs (in terms of 'real money') it could become increasingly economic, and in view of the long lead times for construction *all* possible tidal sites should be thoroughly investigated without delay. But although tidal power could sometimes be of great local importance in a very few cases it could never be relied upon to supply a significant share of the world's energy needs.

The reader is referred to the following references for further information on tidal power: WEC (1980), pp. 324–327; Crabbe and McBride (1978), pp. 177–178.

3.3.6 Renewable energy sources: summary

Not one of the five renewable energy sources that have here been discussed could fulfil both the necessary criteria of abundance and quality for full reliance to be placed upon them for supplying world energy needs even at present levels of consumption. Limitations of quality, especially irregular availability, and consideration of costs are such that even the five renewable sources collectively would not suffice to free us from dependability – at least in part – upon non-renewable energy sources.

This certainly does not mean that the renewable energy sources are valueless. Far from it: each should be developed as rapidly as technology and funds permit. In certain locations where 'base load' energy cannot be supplied, renewable energy has a major role to play. This would be particularly applicable in developing countries. In addition, exploitation of renewable sources when feasible could win us valuable time in which to pursue and perfect the necessary technology for harnessing other undeveloped resources that are more compatible with longer term, base load demands even at small power levels.

3.4 Remaining options

As was seen in Tables 3.2 and 3.3, the identified reserves of non-renewable energy excluding the dubious uranium breeder potential could satisfy the world's needs at the 1982 level of demand for 124 years if growth were zero, but for only 63 years if growth were a mere 2% p.a. (In both cases the periods would run from 1982). If reliance has to be placed on these reserves alone, the situation would indeed be alarming; for approaching exhaustion would create havoc in the form of exorbitant prices and scarcities long before the periods ran out. Moreover, fossil fuels have to meet the claims not only of energy but also of industrial feedstocks. Although it would be unreasonably pessimistic to ignore all the additional and unverified resources, it would nevertheless be unwise to place too much reliance upon them, partly because the figures are highly speculative but more particularly because such a large proportion of them depend upon the commercial realization of breeder reactors and of geopressured thermal fields – neither of which is by any means assured of a practical future, and both of which might well be prohibitively expensive. It has been shown that the renewable energy sources (especially hydro, and perhaps solar energy) could be of very great help in enabling us to eke out our non-renewable resources for a longer time; but it would seem doubtful whether, even collectively, they could ever take the place of fuels when we can no longer depend upon the latter.

The major criteria that have to be met by any new energy source include:
(1) potential for large-scale base load development.
(2) confidence in the availability and magnitude of the resource (it must be large enough to satisfy foreseeable demand to provide the necessary security for economic growth and stability);
(3) The resource itself must be easily transportable *or* be well distributed throughout the world.

There really are only two resources that meet all these criteria: thermonuclear fusion and heat mining. Before moving on to a discussion of these resources, two other criteria need to be discussed: *storage* and *transportation.*

3.4.1   The storage and transport of energy

Incidental to finding new energy sources are the questions of energy storage and transport. Perhaps the worst failure in quality of renewable energy sources lies in their irregular availability for matching demands. If energy cannot be used when and where it is available, then it must either be wasted or expensive facilities must be provided in order to reconcile supply with demand. The 'when' can be solved by providing storage; the 'where' by providing means of transport.

(a)   *The storage of energy*

Energy can be stored in various ways:

(1) *Electrically* in secondary batteries or *electro-magnetically* in super-conducting loops: both very costly on a large scale.
(2) *Hydraulically*, usually in the form of pumped storage. This can be economic on a large scale, but it involves heavy capital outlay, the flooding of land, and the loss of about 30% of the stored energy.
(3) *Thermally* by means of sensible heat (e.g. hot water in insulated tanks or in underground aquifers) or by means of latent heat (i.e. phase change from solid to liquid and vice versa, or from liquid to vapor and vice versa). Solid/liquid phase change is the more compact, as it avoids voluminous vapors.
(4) *Physico-chemical storage.* This can be achieved by the hydration of salts, reversible chemical reactions, vapor absorption, or electrolysis.
(5) *Pneumatically* by means of compressed air. For large scale use this would imply underground storage. The compressed air would be valuable in association with gas turbines for meeting peak loads.
(6) *As hydrogen*, produced electrolytically from power for which no immediate use for other purposes can be found, or by photolysis from sunlight. (The simultaneous production of oxygen could be a useful by-product.) Hydrogen is not, of course, an energy source, but it is a very convenient

form of energy storage. Where a supply of energy temporarily exceeds the demand, the surplus energy could be converted into hydrogen by electrolysis or other suitable means, transported to distribution points and consumed as and when required. Except for the fact that hydrogen production is a capital-intensive process, the more widespread production and use of this gas would enable greater use to be made of renewable energy sources that can frequently offer a temporary surplus of energy over immediate requirements.

(b) *The transport of energy*

Energy can be transported in various ways:

(1) *Electrically* by means of high voltage transmission lines;
(2) *In pipelines*, where the energy is in the form of liquid or gaseous fuels or hot water;
(3) *In vessels or vehicles* (ships, rail or road vehicles, or even aircraft) where energy is in the form of fossil, fissile or synthetic fuels.

(Electro-magnetic wave transmission, as would be appropriate for solar energy if beamed to earth from an artificial satellite, need not be included as we are here concerned only with terrestial transport.)

All the renewable sources of energy are more or less restricted to electrical transport, as would be thermonuclear fusion. HDR energy, like hydrothermal energy, would sometimes be transported electrically and sometimes (in the form of hot water for direct applications) in pipelines, and would thus be less amenable to being transported than fuels. This limitation must be recognized. On the other hand, if heat mining in non-thermal areas should ever become commercially attractive, then this disadvantage would largely be offset by the almost universal availability of crustal heat, which would more or less dispense with the need for it being transported altogether.

3.4.2 Transportation

It is now necessary to consider transportation in general – not of energy, but of people and goods. One of the most intractable problems of energy strategy is the fact that to a large extent transportation is directly dependent upon fuels, for the reason that fuels are themselves transportable and can therefore be carried with the various forms of transportation vehicle – the train, the plane, the ship or the road vehicle. This dependence can to some extent be circumvented by electrification, so that those energy forms that must be produced and converted at the locations where they occur in nature can be consumed at fairly remote places. However, although electrification can often serve the train, and rarely the road vehicle, it is of no use for the plane or the ship.

One possible solution, or partial solution, to this problem lies in the use of hydrogen, which could thus serve the dual interests of storage and transportation. It is a non-pollutive fuel, particularly well suited to the combustion engine, and therefore convenient as a substitute for petroleum products. Another potential solution could lie in the production of synthetic fuels from biomass in conjunction with heat produced from any other form of energy. Whilst HDR could perhaps provide such a source of heat, *magmatic* energy might have greater potentialities in this respect, as will be seen in Chapter 12.

3.4.3 Controlled thermonuclear fusion

Whereas nuclear fission is the splitting of heavy atomic nuclei into lighter components with a small net resultant loss of mass which is converted into energy, nuclear fusion is the building up of heavier nuclei by the coagulation of very light ones; for example, deuterium and tritium can be fused to form helium. The fusion process also results in a net loss of mass and the consequent release of energy. Thermonuclear fusion is the process whereby the sun and other stars sustain the emission of huge quantities of energy for immensely long times. Unlike nuclear fission, the undesired radioactive waste products should be smaller in quantity and easier to deal with from thermonuclear fusion. The fusion process has already been accomplished, but only in an uncontrolled way. This 'achievement' has taken the form of the devastating hydrogen bomb, in which a huge quantity of energy is released in a split second. For the process to have use as a commercial energy source, it must be controlled so that the rate of energy release is adjustable to our requirements and sustained for as long as we wish. Up to now we have not succeeded in commercially producing a controlled sustained reaction.

Although fusion can be initiated at a temperature of about 40 million K it requires about 100 million K for it to be sustained continuously, and then only at very high ion densities that endure for sufficient time for the fusion process to become self-maintained. In the laboratory the initiation temperature has already been exceeded, and the requisite ion density has been momentarily achieved with the aid of intense magnetic fields or high-powered lasers, but fusion has not yet been sustained for more than a fraction of a second.

Below a certain critical rate of energy input the fusion reaction, even when initiated, is insufficient to offset that input; and as the critical input is very large, it follows that the process becomes energy-profitable only if performed on a huge scale. The vast amount of capital needed to achieve this means that only a few nations or regional groups of nations can afford the necessary research. Experimentation is nevertheless being undertaken in Japan, the USA and the USSR, and also on a collaborative basis in Europe.

Of the raw materials for fusion, deuterium, an isotope of hydrogen, is abundantly contained in the oceans. In fact, if it were fully utilized there would be sufficient deuterium available in the world's oceans to provide 16×10^{30} J of fusion energy, equivalent to 1.5×10^{13} quads – more than fifty billion times the present annual world consumption of primary energy! However, another ingredient, tritium, which is essential to the first generation systems, must be artificially created with the aid of lithium. This imposes another potential concern as to the availability and recoverability of lithium. However, it is believed that there is sufficient lithium available to support the fusion process to satisfy the world's electricity needs for many centuries. (WEC, 1978, pp. 177–178). For the purposes of comparison with other resources, it is frequently assumed that the fusion resource base is defined by the total quantity of deuterium in the oceans.

So far the mastery of controlled thermonuclear fusion has eluded us, but intense research will doubtless continue; for despite the huge cost involved, the goal is so splendid that a sustained effort is warranted.

3.4.4 Heat mining

This book deals with the second remaining option which also has huge potential. Even though it is non-renewable, it should be plentiful enough in terms of future human 'history,' as will be shown in Chapter 4.

Access to neither of these two vast energy sources has yet been commercially mastered. The implication is that both are the subjects of considerable international activity; however this is far from being true. HDR research and development is in the hands of perhaps fifty professional engineers and scientists worldwide with an annual investment of less than twenty million dollars, while fusion research has hundreds of people involved and an annual investment of the order of one billion dollars. If we are to achieve long-term energy self-sufficiency, scientific activity must be intensified in heat mining as well as for any other options that may appear promising. Some small scale success has already been attained with the proof of the heat mining concept, and there are grounds for optimism that controlled thermonuclear fusion may one day become a practicable and economic proposition. The fact that this book deals only with heat mining does not imply that the authors undervalue the possibilities of fusion (or for that matter any other large-scale options that may yet become worthy of closer study). Apart from the practical technology for mastering heat mining and thermonuclear fusion, economic evaluations for them are on completely different bases.

At present, heat mining is reasonably close to being a technical reality – at least within its *partial* potential in comparison to nuclear fusion which still requires major technological breakthroughs to reach a proof of concept. However, it is important to pursue the possibilities of both these new energy sources so that each can serve as an insurance for the possible failure of the other. The obstacles are still formidable; but when we pause to consider how the miracles of nuclear fission and of space flight were both accomplished within very few years under the concentrated pressure of effort and determination, it would seem that there are good chances of harnessing both of these new energy sources and of making them commercially attractive.

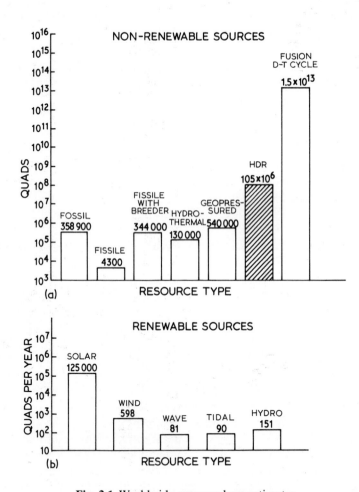

Fig. 3.1 Worldwide resource base estimates.

Note: the estimation of the HDR resource is fully explained in Chapter 4, Section 10. It excludes the energy content of crustal and ocean-based magma (see Chapter 12) which could increase it by a small percentage.

3.5 Summary of worldwide resource base estimates

The estimated resource bases of the various non-renewable and renewable energy resources are shown in Figs. 3.1 a and b respectively, in the form of bar-graphs drawn to logarithmic vertical scales. Of the non-renewable resources, HDR is outstandingly the most abundant with the exception of thermonuclear fusion, despite the very cautious basis on which it has been estimated (as will later be shown in Section 4.10). HDR is almost 300 times as abundant as the total of fossil fuels. The fact that HDR falls so far short, quantitatively, of thermonuclear fusion should not be allowed to obscure the fact that HDR is already far closer to commercial realization, requiring very much less in the way of 'quantum jumps' of technical achievement than does the conquest of thermonuclear fusion. It is also far less limited in potential application because it is more uniformly distributed than other hydrothermal resources which are highly localized. In addition, we have also chosen to use generous estimates of other resources so as not to favor HDR. It is true that both fissile (without breeder technology) and hydrothermal energy are both already established commercially; but their combined resource bases amount to only about one eight hundredth part of that of HDR. In racing parlance, HDR might well be described as 'the favorite' of the non-renewable energy resources of the future.

Coming now to the renewable resources as depicted in Fig. 3.1 (b), it is virtually impossible to make direct comparison with the non-renewable resources; for they differ in nature from the latter in much the same way as income differs from capital. They too are impressive in magnitude, but are all subject to the limitations that have been described in Section 2 of this chapter. In different degrees they are all of value but, for the reasons explained in the text, are handicapped by various factors that detract from their economic competitiveness to become the ultimate solution to the world's energy problems.

An obvious element of caution must be applied to all resource base estimates, not only because of their inherent uncertainty but also because they are all, to various degrees, unattainable figments. The HDR resource base estimate as a measure of total energy in place, however, is probably the most soundly and cautiously based of all – soundly, in the sense that it depends only on the heat content of a specified volume of rock as will be justified in Chapter 4, and cautiously because we have limited our estimate to attainable depths and minimum useful temperatures. It should be noted that only a very small portion (about 0.2%) of the HDR resource base is believed to consist of very hot, molten magma in locations where it exists near the surface (see Chapter 12 for details). What remains is the key question of just how much of this massive HDR resource base can be mined economically.

Chapter 4

The approximate quantification
of crustal heat resources

4.1   Basis of estimation

The heat content of the mantle and cores of the earth is probably between 3000
and 4000 times that of the crust, yet the crust alone contains enough heat to
satisfy a large part of the world's probable energy needs for a very long time. As
it is improbable that we shall be able to penetrate down to the Moho, which
defines the boundary between the 'rigid' crust and the more plastic mantle,
within the foreseeable future (except in a few places where the crust is
exceptionally thin) it would be realistic to assess the resource base of crustal
heat in terms of our penetration ability. Obviously deep heat would have the
advantage of high grade (i.e. temperature), and this would have attractions for
certain applications such as high efficiency power generation; but apart from
the problem of accessibility there would be formidable difficulties of chemistry
and of containment in the handling of a fluid in contact with rock at very high
temperatures. It is for these reasons that it makes practical sense to limit our
initial target to the recovery of crustal heat at fairly modest temperatures and
depths. Our present drilling capabilities are to depths of the order of 10 km for
vertical wells and about 6 km for directionally deviated wells. These depth
limitations as well as high temperature considerations will help to define the
HDR resource base. It is also proposed for the present to consider only the heat
underlying the land masses; for even if, as it well may, it should ever become
practicable to recover off-shore crustal heat, access beneath the land surfaces
would obviously be cheaper (except perhaps in the comparatively rare
instances where use could conceivably be made of abandoned off-shore oil or
gas wells).

In certain parts of the crust abnormally high temperatures are found at
shallow depths. Such places are evinced by very high heat flow rates and
thermal gradients and they often occur at or near the tectonic plate boundaries
at rift or subduction zones of high seismic activity, where hydrothermal fields
or volcanoes sometimes occur. To avoid over-estimating the accessible crustal
heat, all such abnormally hot areas will initially be ignored, and for the purpose

of approximately assessing the crustal heat resources of the earth it will at first be assumed that all the land areas are 'non-thermal' – i.e. having temperature gradients near the surface of not more than 25°C/km of depth.

There are two possible ways of expressing crustal heat quantitatively. The more cautious is to express it as 'specific' crustal heat, or as so much per square kilometre of land area per degree Celsius of average crustal cooling as applied to a vertical column beneath that land area. This involves making assumptions as to the physical properties of the rock and the depth of the column. Alternatively, crustal heat may be expressed as the estimated total heat content of a vertical crustal column beneath each square kilometre of land area, as reckoned above some postulated base temperature. This involves not only making the same assumptions as for the estimation of specific crustal heat, but also assuming some temperature/depth distribution curve from the ground surface to the Moho. The temperature-depth relationship will in turn depend upon the distribution of radioactivity among the crustal rocks and on the temperature at the Moho. Both methods of expression involve speculation: the second method involves a greater degree of speculation than the first. The most that can be reasonably attempted will be 'order-of-magnitude' estimates.

Although the geological 'mix' of rocks at different depths will vary from place to place – and this would affect both the average specific heat and the average density of a crustal column beneath any one area – it is deemed sufficient for our immediate purpose to assume homogeneity. The following constants will be applied to all areas:

Mean crustal rock density $= 2700$ kg/m^3 (2.7 g/cm^3)
Mean crustal specific heat $=$ 839 J/kg K (0.2 cal/g°C)

4.2 Specific crustal heat

Consider a vertical column of crust beneath a land area of 1 km^2 extending to a depth of d km. Its volume will be $d \times 10^9$ m^3 and its mass at a density of 2700 kg/m^3 will be $2.7d \times 10^{12}$ kg. With a specific heat of 839 J/kgK, this mass of rock, in cooling through 1°C (or 1K) would release 2.265 $d \times 10^{15}$J. This specific crustal heat may be expressed in any of the following forms:

$$\left.\begin{array}{l} 71.83\ d\ \text{MWyt} \\ \text{or } (5.41 \times 10^{11})\ d\ \text{kgcal} \\ \text{or } 2265\ d\ \text{TJ} \\ \text{or } 77\,310\ d\ \text{tce} \\ \text{or } 0.00215\ d\ \text{quad} \end{array}\right\} \begin{array}{l} \text{per km}^2 \text{ of ground surface area} \\ \text{per °C of average cooling.} \end{array}$$

(1 tce $=$ one tonne coal equivalent on a basis of an assumed calorific value of 7000 cal/g).

These figures should of course be regarded as approximations only.

The average thickness of the earth's crust beneath the land masses is assumed to be about 35 km. Substituting $d = 35$ km in the above expressions, the specific crustal heat *for the whole crust* beneath the land masses would be approximately as follows:

$$\left.\begin{array}{l} 2500 \text{ MWyt} \\ 1.89 \times 10^{13} \text{ kgcal} \\ 79\,000 \text{ TJ} \\ 2.7 \text{ Mtce} \\ 0.0752 \text{ quads} \end{array}\right\} \begin{array}{l} \text{per km}^2 \text{ of ground surface area} \\ \text{per } °C \text{ of average cooling.} \end{array}$$

It is unrealistic to regard these figures as the ultimate resource base as we cannot penetrate the crust to its full thickness of approximately 35 km down to the Moho. However, if we equate the depth d appearing in the first set of values given above to the practically attainable depths that can be reached with current drilling technology, then it is a simple matter to estimate the resource base accordingly. As mentioned before, a value of about 10 km (or 30 000 ft) could reasonably be assigned to d for vertical wells, in which case the resource base could be taken as follows:

To a depth of 10 km:

$$\left.\begin{array}{l} 718.3 \text{ MWyt} \\ 5.41 \times 10^{12} \text{ kgcal} \\ 22\,650 \text{ TJ} \\ 773\,100 \text{ tce} \\ 0.0215 \text{ quads} \end{array}\right\} \begin{array}{l} \text{per km}^2 \text{ of ground surface area} \\ \text{per } °C \text{ of average cooling} \end{array}$$

For directionally deviated wells, d will be somewhat less than 10 km. Assuming $d=6$ km (or 18 000 ft), the corresponding resource base figures are lowered:

To a depth of 6 km:

$$\left.\begin{array}{l} 431 \text{ MWyt} \\ 3.25 \times 10^{12} \text{ kgcal} \\ 13\,590 \text{ TJ} \\ 463\,860 \text{ tce} \\ 0.0129 \text{ quads} \end{array}\right\} \begin{array}{l} \text{per km}^2 \text{ of ground surface area} \\ \text{per } °C \text{ of average cooling} \end{array}$$

In the course of time, when technological advances may be expected to improve our penetration ability to greater depths than 10 km, these figures can be raised proportionately.

An advantage of expressing the resource base as a 'specific' figure is that it can apply equally to a non-thermal or to a hyper-thermal location. A hyper-thermal zone would of course contain more total heat than a non-thermal zone between the surface and a specified depth; but we are here talking of 'degrees of average cooling' reckoned from the original, natural average crustal temperature at the particular location, regardless of whether that temperature is normal or exceptionally high.

4.3 Total crustal heat

Adopting now the alternative basis of estimation mentioned in Section 4.1, it is first necessary to anticipate a conclusion that will be reached later (in Chapter 6); namely, that a not improbable temperature/depth distribution curve in a typical non-thermal part of the crust would be somewhat as shown in Fig. 4.1. This shows a mean surface temperature of 15°C, thermal gradients of 25°C/km near the surface and about 11°C/km near the Moho, a crustal thickness of 35 km and a temperature of 600°C at the Moho. The arguments by which this curve has been deduced will be explained in Chapter 6, but for the present the curve may be accepted as a reasonable approximation to fact. The actual temperature/depth distribution will of course vary quite widely from place to place and will show irregularities, but for the purpose of making an approximate resource assessment the smooth curve in Fig. 4.1 should suffice.

The area SRM in that figure is a measure of the total heat content of the crust reckoned above 15°C – a figure that is generally accepted as the approximate world mean surface temperature. Its area is 11 613 km-degrees (km°C) which, for a crustal depth of 35 km, implies a mean crust temperature of 11 613/35, or 331.8°C above the base temperature of 15°C – i.e. 346.8°C, say 347°C. The total crustal heat beneath the land masses may therefore be taken at

35 000 m (depth) $\times 10^6$ (m^2/km^2) $\times 2700$ (kg/m^3) $\times 331.8$°C $\times 839$ (J/kg K)

or 2.631×10^{19} J/km^2 of ground surface. This huge figure is of academic interest

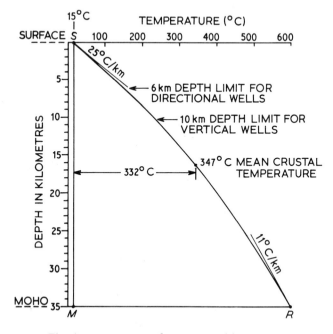

Fig. 4.1 Assessment of gross crustal heat content.

only as it is unlikely that we will ever succeed in extracting more than a tiny fraction of it.

The term 'kilometre-degree' has been used above as a unit of area in a temperature/depth curve: it can also be used as a crude, but convenient, unit of heat. It is equivalent to the heat that would be released by 1 km³ of crustal rock if cooled through 1°C. Clearly this can be no more than a convenience and an approximation, as the specific heat and density of different crustal layers will vary from place to place; but the degree of approximation would be of the same order as that involved in the estimation of 'specific crustal heat' (Section 4.2). For the rough assessment of heat mining possibilities the unit has its uses. Expressed in other energy units the km°C will be those quoted near the beginning of Section 4.2 on the assumption that $d = 1$ km, namely:

$$1 \text{ km}°\text{C} = \begin{cases} 71.83 \text{ MWyt} \\ 5.41 \times 10^{11} \text{ kgcal} \\ 2265 \text{ TJ} \\ 77\,310 \text{ tce} \\ 0.00215 \text{ quads} \end{cases}$$

4.4   Restricted resource base within reach

There are two primary limitations that must be applied before consideration is given to the permissible degree of heat removal:

(1) *Attainable depth of penetration.* The effect of this limitation is to raise the level of the base line of Fig. 4.1 to a depth that is regarded as attainable with the available penetration technology.
(2) *Minimum useful rock temperature.* Although practical uses can be found even for tepid waters in cold climates, the total quantity of heat that can be used commercially at low temperatures would never be sufficient to justify heat mining operations for the sake of such low grade heat alone. There will always be a useful minimum temperature below which heat mining could never be a commercial proposition. This minimum useful temperature will not be the same in all parts of the world: it will depend upon the climate. This will be discussed in more detail in Chapter 5.

The effects of these two primary limitations on the heat resource base can be very marked, as can be seen from Fig. 4.2, in which the *restricted* heat resource base for an attainable depth d and for a minimum useful rock temperature t is shown shaded. The size of this shaded area is very sensitive to the values of d and t and will form only a small part of the gross crustal heat content represented by the area SRM.

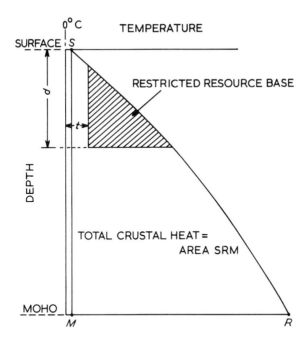

Fig. 4.2 Restricted heat resource base as defined by the attainable depth *d* and by the minimum useful rock temperature *t*.

Figure 4.3 shows the heat market spectrum for the USA in terms of the utilization temperature (up to 300°C): it shows that there is a big market for space-heating. It would be wise to assume that the minimum commercially useful *delivered* temperature that would justify heat mining operations in the USA is 60°C; and it would not be unreasonable to make the same assumption for all other countries having a substantial space-heating market. Lower temperature uses for agriculture, aquaculture, de-icing, etc. could then be regarded as a useful by-product, but not as a *justification* for heat mining operations.

In many tropical and sub-tropical countries, however, the space-heating market is negligible or non-existent. This absent market would include not only district heating but also part of the energy requirements for farming (in colder countries) such as the heating of greenhouses and farm buildings, soil heating, etc. In these warmer countries the heat market would mainly be for power generation and for industrial purposes, for which the minimum commercially useful temperature for non-electric applications might be around 100°C at the users' premises.

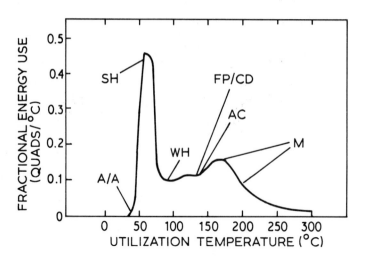

Fig. 4.3 Heat market spectrum for the USA

Key: A/A   agriculture/aquaculture   M   manufacturing (chemicals, pulp and
     AC   air conditioning   paper, petroleum)
     CD   clothes drying   SH   space heating
     FP   food processing   WH   water heating

Fractional energy use distribution in the USA as a function of utilization temperature.
The area under the curve from 0 to 300°C equals 28 quads of process and space heat per
annum and is equivalent to the US demand up to 300°C. (Reproduced, with permission,
from Tester, 1982a.)

There are thus two extreme types of country, or region, as follows:

Category I: Countries having substantial space-heating markets, for which the
             useful minimum delivered temperature would be about 60°C.
Category II: Relatively hot countries, having no or negligible space-heating
             markets, for which the minimum useful delivered temperature
             would be about 100°C.

It will later be shown that the heat mining process will consist of fracturing
the rock, forcing water through artificial fractures to pick up the heat by
conduction, and bringing the heated water to the surface. If heat at a specified
minimum temperature is required at the users' premises, an adequate
allowance must be made for inevitable temperature drops through the rock to
the circulated water and also in suitable heat exchangers at the surface in which
the extracted rock heat is transferred to the process fluid. If 25°C be regarded as
sufficient for these temperature drops, then the minimum commercially useful
rock temperatures would be:

For Category I countries 60 + 25, or 85°C.
For Category II countries 100 + 25, or 125°C.

The choice of 25°C is purely illustrative. We recognize that the actual temperature drop will depend on a number of factors such as the rock temperature, the flow rate up the production well, and the period over which the reservoir has been exploited.

The values of the restricted heat resource base for these two categories of country are illustrated and quantified (approximately) in Fig. 4.4 for 6 and 10 km attainable depths. In Fig. 4.5 the restricted heat resource base for both categories is shown for other attainable depths. It will be seen that the minimum depths at which heat mining could start to become worthwhile in non-thermal areas are about 3 km and 4.8 km for categories I and II countries, respectively. The following deductions may be drawn from Figs. 4.4 and 4.5:

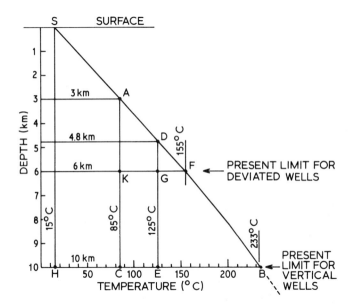

Fig. 4.4 Restricted heat resource base for different categories of country reckoned to 6 km and 10 km depths.

Restricted resource base per km²	Category I			Category II		
	Countries with substantial space-heating markets			Hot countries		
		km°C	Quads		km°C	Quads
To 6 km depth	AFK	112	0.241	DFG	20.2	0.043
To 10 km depth	ABC	533	1.146	DBE	289	0.621

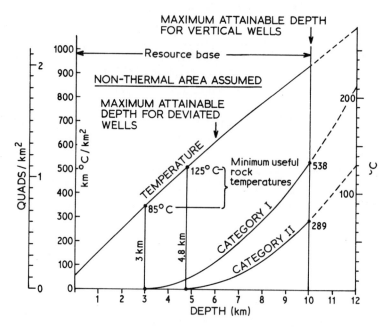

Fig. 4.5 Restricted crustal heat resource base in terms of attainable depth.
Category I: countries with a substantial space-heating market
Category II: relatively hot countries

(1) As the attainable depth increases, the proportional gain in restricted heat rises rapidly and the attainable temperature also rises. This emphasizes the importance of improving the economics of penetration technology.
(2) The proportional difference between the restricted heat resource bases for the two categories of countries is very large at depths of around 5 km, but decreases rapidly with increasing depth.
(3) Category II countries are unlikely to be interested in heat mining in non-thermal areas unless it is economic to penetrate to depths of at least 6 or 7 km.

The quantified values ascribed to the restricted heat resource base, as shown in Figs. 4.4 and 4.5, should be fairly conservative, for two reasons. First, as already mentioned, heat below the prescribed minimum temperatures is not all worthless, especially in the colder countries. Secondly, it is possible to make use of rock heat even after its average temperature has dropped to the prescribed minimum levels, by reducing the water circulation rate or by increasing the rock heat transfer surface during a later restimulation. This would reduce the drawdown and enable heat still to be delivered at the prescribed minimum temperatures at the points of use, though in reduced quantity. It is of course not possible to divide the territories of the world precisely into one or the other of the two categories defined above, but approximately half the world, both in terms of population and of land area, does experience winters that are hard enough to provide a space-heating market; while the other half is, broadly, 'hot'. Nevertheless, there will be many parts of the world of an intermediate category.

It may be noted that at 10 km depth a rock temperature of about 233°C could be expected in non-thermal areas, thus providing a temperature drop of 148°C in cold countries and 108°C in hot countries before the rock becomes locally cooled to the prescribed minimum useful temperatures.

4.5 Restricted crustal heat resource base in terms of world energy needs and of fossil fuel reserves

It has been stated in Table 2.1 that the world energy needs in 1982 were 248.1 quads. If grade could be disregarded, if there were no environmental restrictions upon the permissible extent of crustal cooling, and if our penetration ability is taken at 10 km, then this energy could have been supplied by mining the heat to the fullest practicable extent beneath an area intermediate between the following extremes:

$248.1/(533 \times 0.00215)$, or 217 km^2 (Category I)
$248.1/(289 \times 0.00215)$, or 399 km^2 (Category II)

This gives an average of 308 km^2 for 10 km penetration ability. ✓

Extending this argument to an assumed penetration ability for directionally deviated wells of 6 km, the requisite area would lie between the following extremes:

$248.1/(112 \times 0.00215)$ or 1030 km^2 (Category I)
$248.1/(20.2 \times 0.00215)$ or 5713 km^2 (Category II)

This gives an average of 3370 km^2 for 6 km penetration ability.

The total land area of the world amounts to about 148 million km^2. If it be assumed that about one-third of this area is inaccessible on account of rugged terrain or remoteness from human habitation, there would still be about 100 million km^2 available for heat mining. At first sight the recoverable energy from this area, if heat mining should become a commerical reality, would apparently suffice to supply the entire world with its energy needs at the 1982 level of demand for the following periods:

$10^8/308$, or 325 000 years if the attainable depth is 10 km
or $10^8/3370$, or 29 700 years if the attainable depth is only 6 km

The arguments whereby these areas and times have been deduced are, of course, very weak. It is not possible to ignore grade, and there would be environmental restrictions upon excessive crustal cooling. It would also almost certainly be impracticable to spread heat mining operations evenly over two thirds of the world's land surfaces. Nevertheless, the deduced figures do indeed suggest that in crustal heat we may have found the huge source of energy we need. For even if our rash assumptions about grade, environmental constraints and exploited area were wrong by a factor of ten or more, we would still be talking of a resource equivalent to some thousands of times greater than the entire world's energy needs in 1982.

It would be both wise and cautious to express the 'life' of the restricted crustal heat resource base (as a multiple of the world's total energy needs in 1982) in terms of the permissible extent of crustal cooling and of the land area exploited, so:

(1) For 10 km attainable depth

$$\frac{0.00215 \text{ quads/km}^\circ\text{C} \times 10 \text{ km}}{248.1 \text{ quads p.a.}} = 8.67 \times 10^{-5} \text{ years per } ^\circ\text{C of}$$
average cooling of the upper 10 km of the crust per km^2 of exploited area.

(2) For 6 km attainable depth

$$0.6 \times 8.67 \times 10^{-5} = 5.2 \times 10^{-5} \text{ years per } ^\circ\text{C of average cooling of}$$
the crust per km^2 of exploited area.

Perhaps a better perspective of the crustal heat resource may be gained by comparing it with the estimated identified world reserves of fossil fuels as shown in Table 3.1, namely 29 900 quads. Since 1 km°C is equivalent to 0.00215 quads (see Section 4.3 above), the fossil fuel reserves may be expressed as 29 900/0.00215, or 13.9×10^6 km°C. If heat mining were permissible to the extent of 1°C average cooling of the upper 10 km of the crust, then the fossil fuel reserves would be equivalent to heat mining beneath an area of 1.39×10^6 km^2, or only about 1% of the earth's land surface area. This clearly illustrates the immense size of the crustal heat resource (unless it can be shown that 1°C average cooling of the upper 10 km of the crust is excessive).

Earlier it has been said that it should be possible to reach rock at about 233°C at 10 km depth in non-thermal areas. After applying an allowance of 25°C for temperature drops incurred from the core of the rock to the point of application, this means that temperatures of up to about 208°C should be available for use. A glance at Fig. 4.3 shows that a large fraction of the heat market in Category I countries could be satisfied with heat at this temperature, but the pattern of incidence from a mined zone corresponding to the area ABC in Fig. 4.4 would differ greatly from that of Fig. 4.3. This would substantially reduce the contribution that could be made from heat mining to 10 km depth in non-thermal areas. In Category II countries the discrepancy between the patterns of incidence of supply and demand would be greater. Thus with present drilling technology, heat mining could never satisfy the whole of the world's heat requirements unless very great use were made of electricity generation at rather low efficiency for reconversion into higher grade heat; and this would almost certainly never be economic. The question of matching the pattern of heat availability with that of demand will be examined more closely in Chapter 5. For heat mining to be economic it will be necessary mainly to exploit areas with higher than normal gradients first, until our penetration capabilities enable us at acceptable cost to reach depths of about 10 km, at which rock temperatures around 230°C will be available even in non-thermal areas.

It is not yet possible to decide whether the permissible extent of crustal cooling will limit practical heat mining to avoid environmental ill effects, but eventually, given no other limitations such as heat extraction rates, a maximum level of crustal cooling would enter the picture. Conservatively speaking, at least 1°C of average crustal cooling (for the full crustal depth) should be permissible; so it would seem that crustal heat should be able to supply a very large share of the world's energy needs for many centuries. This point is examined again in Chapter 5.

4.6   Recapitulation

Before proceeding further, it is useful to summarize some of the various numerical deductions made in this chapter up to this point. This has been done in Table 4.1. In order to stress the smallness of the fraction of total crustal heat to which we are likely to have access (at least for a very long time), heat content figures right down to the Moho have also been shown in this table.

4.7   Resource in thermal areas

Up to this point only non-thermal areas have been considered having temperature gradients averaging 25°C/km at the surface. As the greater part of the earth's land areas is non-thermal, it is both reasonable and cautious to have ignored the hyperthermal regions of the world for the purpose of assessing the ultimate potentialities of crustal heat. Nevertheless, there are many places where surface thermal gradients of considerably more than 25°C/km are to be found, and where the immediate commerical prospects of heat mining would consequently be far better than in more 'normal' areas. Obviously, it is these hyper-thermal areas that will first be exploited; for the technical problems will be fewer and the returns on investment will be potentially greater. Whereas it was possible to deduce average crustal conditions for non-thermal areas, as shown in Figs. 4.1, 4.4, and 4.5, and in the deduced values of the specific, total and restricted crustal heat resources, no such standardization can be made for thermal areas, as each will have its own individual characteristics of temperature/depth relationship and of crustal thickness.

Although more heat will be accessible in a thermal than in a non-thermal area, this does not necessarily imply that the crust in a thermal area contains

Table 4.1. Summarized energy potentialities of crustal heat in non-thermal areas

	Energy concentration per km² of surface					
	$km°C/km^2$	TJ/km^2	$Quads/km^2$	$kg\ cal/km^2$	tce/km^2	$MWyt/km^2$
SPECIFIC CRUSTAL HEAT: (i.e. energy per °C of average crustal cooling to the specified depth)						
For attainable penetration depth of d km.	d	$2265d$	$0.00215d$	$5.41 \times 10^{11}d$	$77310d$	$71.83d$
For attainable penetration depth of 10 km.	10	22650	0.0215	5.41×10^{12}	773100	718.3
For attainable penetration depth of 6 km.	6	13590	0.0129	3.246×10^{12}	463900^5	431
TOTAL CRUSTAL HEAT: (reckoned above a mean surface temperature of 15°C)						
Energy in crust between surface and Moho	11613	2.63×10^7	24.968	6.283×10^{15}	8.978×10^8	834162
RESTRICTED HEAT RESOURCES: (i.e. crustal heat at adequate temperatures above the specified depth)						
Category I countries, having substantial space heating markets (above 85°C):						
For attainable penetration depth of 10 km	533	1.207×10^6	1.146	2.884×10^{14}	4.121×10^7	38285
For attainable penetration depth of 6 km	112	2.537×10^5	0.241	6.059×10^{13}	8.659×10^6	8045
Category II countries, having no space heating markets (above 125°C):						
For attainable penetration depth of 10 km	289	6.546×10^5	0.621	1.563×10^{14}	2.234×10^7	20759
For attainable penetration depth of 6 km	20	4.575×10^4	0.043	1.093×10^{13}	1.562×10^6	1451

more absolute heat than in a non-thermal area, because the crust could well be thinner. Thermal areas can be caused by abnormally high concentrations of radioactive rocks, by abnormally thin local crust, by abnormally high magmatic temperature just below the Moho, or by combinations of these three basic causes. (It is true that a thermal area can also be caused by the nearby presence of a submerged pocket of magma contained within the crust, but such occurrences need not be considered at this stage.)

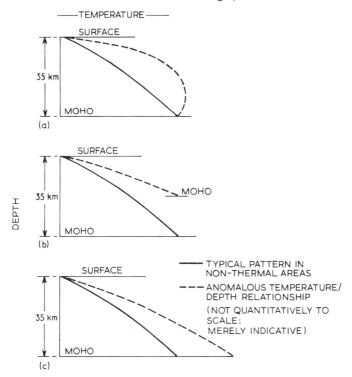

Fig. 4.6 Possible contributory causes of high thermal gradients in the upper layers of the crust. (a) Abnormally high degree of radioactivity. (Moho temperature unaffected) (b) Thin crust. (Moho temperature unaffected) (c) Abnormally hot mantle beneath the Moho.

Possible patterns of temperature/depth distributions for these three possible conditions are shown in Fig. 4.6. It would not be profitable to analyse such curves, for they could be of infinite variety; but the point to be particularly noted is that in each case the temperature gradients in the upper layers of crust are higher than in non-thermal areas. The extent to which a thermal area in a Category I country offers heat mining advantage over a non-thermal area can be seen from Fig. 4.7, in which temperature/depth curves are shown in the left hand diagram for the upper 10 km of crust for a non-thermal area (in accordance with Fig. 4.1) and for a thermal area having temperature gradients

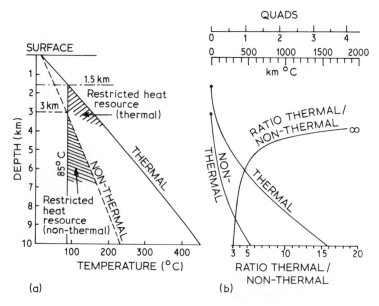

Fig. 4.7 Comparison of minimum useful depths and restricted heat resources as between a non-thermal and an arbitrary thermal area. The arbitrary thermal area is assumed to have twice the thermal gradients of a typical non-thermal area in the upper 10 km of the crust – i.e. 50°C/km just below the surface. (a) Temperature/depth curves. (b) Restricted heat resource.

twice as great – i.e. 50°C/km near the surface, and decreasing slowly with depth. In this example the advantages of the thermal area over the non thermal area are very marked, so:

	Non-thermal area	Thermal area	Ratio of thermal to non-thermal area
Depth at which a minimum useful rock temperature of 85°C is reached	3.0 km	1.5 km	0.5
Temperature reached at 10 km depth	230°C	450°C	2.0
Restricted heat resource to 10 km depth for a 1 km² area	533 km°C (1.15 quads)	1550 km°C (3.33 quads)	2.9

Although it is not possible to study 'typical' thermal areas, as it is with non-thermal areas, estimates have been made of the occurrence of thermal areas in the USA (LANL, 1982, Table IX, p. 98) and in Europe (Cermak and Rybach, 1979). For the USA the gross resource base of thermal areas totalling 1 495 098 km² and having thermal gradients ranging from 30 to over 60°C/km, was estimated in 1980 at 5 972 329 quads to 10 km depth (reckoned above 15°C). This is equivalent to an average energy density of 4 quads per km². In typical non-thermal areas the corresponding figure (area SBH in Fig. 4.4) is 2.425 quads/km². Those areas designated as 'thermal' in the US Survey should therefore contain about 4/2.425, or 1.65 times as much total heat/km² down to 10 km depth as non-thermal areas; but the useful, or *restricted* heat resource base of the thermal areas would be of far greater relative value per km² by comparison with non-thermal areas. If, for example, the total heat content of the thermal area down to 10 km depth were simply due to average temperature gradients consistently 1.65 times as high as in non-thermal areas – a not unreasonable assumption – then the comparison between the two would be as shown in Fig. 4.8 and Table 4.2, from which it will be noted that the restricted heat in the thermal area would on average be 2.6 times as great as in a non-thermal area. Not only this, but the minimum useful temperature would be reached at only 60% of the depth required in non-thermal areas, and at 10 km depth it should be possible to reach a temperature of 375°C as against 233°C in non-thermal areas.

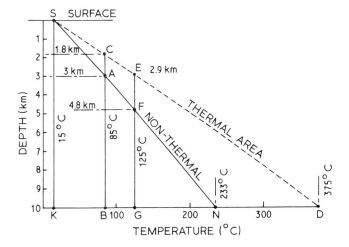

Fig. 4.8 Comparison between a fairly representative thermal area in the USA and a non-thermal area.

It will be noted from Table 4.2 that the restricted heat in a typical US thermal area down to 10 km depth is 2.63 quads/km². Hence the total restricted heat resource base for hyperthermal areas in the USA will be 2.63 × 1 495 098 or 3.93 × 10⁶ quads. This is about 58 000 times as much as the total consumption

Table 4.2. Comparison between thermal and non-thermal areas (cf. Fig. 4.1)

	Temperature gradients		Temperature at 10 km depth °C	Total heat to 10 km depth above 15°C km°C (quads) per km²	Category I countries		Category II countries	
	At surface °C/km	At depth °C/km			Depth at which 85°C is reached km	Restricted heat to 10 km depth km°C (quads) per km²	Depth at which 25°C is reached km	Restricted heat to 10 km depth km°C (quads) per km²
Non-thermal areas	25	19	233	Area SNK 1128 (2.425)	3.0	Area ANB 533 (1.15)	4.8	Area FNG 289 (0.62)
Thermal areas	41.25	31.35	375	Area SDK 1860 (4.0)	1.8	Area CDB 1223 (2.63)	2.9	Area EDG 917 (1.97)
Ratio Thermal/ non-thermal	1.65	1.65	1.61	1.65	0.6	2.3	0.6	3.2

Average advantage of thermal to non-thermal area in restricted heat to 10 km depth $\left.\right\}$ $\dfrac{1223+917}{533+289}=2.6$

of primary energy in the USA in 1980 (about 68 quads). For the removal of all this restricted heat resource the average degree of cooling in the upper 10 km of the crust would be 1223 km°C/10 km, or 122.3°C. Hence even 1°C of average crustal cooling in this upper layer in the thermal areas of the USA would produce over 470 times the 1980 national total consumption of primary energy – i.e. 58 000/122.3. This would be equivalent to cooling the whole of the crustal column by only 10/35, or 0.29°C.

From all this it seems clear that in the USA it will be unnecessary to consider heat mining in non-thermal areas within the foreseeable future, as the hyperthermal areas alone in that country offer such huge potentialities from only the upper 10 km of the crust. These thermal areas amount to 1 495 000/9 205 000, or more than 16% of the land in the USA, and although part of these areas will be only slightly hyperthermal with temperature gradients not greatly exceeding 30°C/km, others will have much higher gradients than the mean of about 41°C/km (as shown in Fig. 4.8).

However, it cannot be claimed that the USA is typical of the world in general; for that country is exceptionally well endowed with HDR. There may be other countries in Latin America, the Western Pacific and elsewhere that are even more greatly favored by nature, but there are also many countries that are far less fortunate. Moreover, many of these countries will belong to Category II (as defined in Section 4.4 above) in which the restricted heat resource is reduced considerably by the absence of a space-heating market. Rowley (1982, Appendix C) has estimated that about 17×10^6 km² of the world's land surfaces are 'thermal', having temperature gradients and heat flows substantially above normal. This is about 11.5% of the world's total land area – a considerably smaller proportion than the 16% in the USA. If these world thermal areas were of the same average quality as those in the USA, their restricted heat resource would be about $17 \times 10^6 \times 2.63$, or 44.7×10^6 quads. If the minimum useful rock temperature were raised from 85°C to 125°C, as would be necessary for Category II countries, the restricted heat resource would be reduced from 2.63 to about 2.0 quads/km², and the total restricted heat resource would be $(2.0/2.63) \times 44.7 \times 10^6$ or 34.0×10^6 quads. It would seem to be a reasonable deduction that the world value of the restricted heat resource in HDR in thermal areas would lie between 44.7 and 34.0 millions of quads, say about 39 million quads, which is more than 150000 times as great as the total world consumption of primary energy in 1982.

Although these estimates are based upon very imprecise assumptions, and although they ignore the consideration that must be given to the grade of the available heat (which would be less of a problem in thermal than in non-thermal areas), the deduced figures imply such a vast quantity of available heat that there can be little doubt that HDR in thermal areas alone could provide far more heat from the upper 10 km of the crust than all the world's resources of fossil fuels – almost 200 times as much – could we but gain access to and recover that heat at acceptable cost. Why then do non-thermal areas need to be considered at all for heat mining? Without doubt early heat mining operations will only be undertaken in places of high temperature gradient; but in some countries, particularly those that are too small to embrace a wide variety of crustal conditions, there will be no hyperthermal areas at all – or at least only a few with gradients only slightly exceeding 25°C/km. Then again, if penetration costs can be sufficiently reduced, it might one day be more economic to mine heat locally in a non-thermal area where it is needed in vast quantities than to do this in a remote thermal area and incur the trouble and expense of long distance energy transmission.

Many countries having poor resources of easily accessible HDR will also have very modest energy demands in terms of their size or population, and will not therefore necessarily be worse endowed (at any rate until their economies have been greatly developed) than some other countries having much bigger HDR resources but with much higher energy needs. Nevertheless, as with all the other resources of the earth, it cannot be expected that the rich potentialities of HDR are in any way 'fairly' distributed among nations.

4.8   Inadequacies of grade and applicability

Although in terms of heat units the energy stored in the earth's crust is clearly gigantic, it must be recognized that it could never entirely displace liquid, gaseous, and solid fossil fuels. Even if heat mining were confined only to thermal areas, of abnormally high gradient, rock temperatures exceeding about 375°C would seldom be encountered for a very long time (until we can penetrate below 10 km depth at acceptable cost) so about 350°C would be about the limit of application temperature after allowing for unavoidable temperature drops between the HDR and the point of use. This assumes technical feasibility in gaining access to these formations. The limit of 350°C would exclude direct use for certain industries such as metals production and the manufacture of glass and ceramics; it would also exclude the generation of electricity at the high thermal efficiencies (e.g. about 40%) which modern power stations can achieve. Another energy use that could not be directly satisfied by crustal heat is transportation (by road, rail, sea and air), which is now almost entirely dependent upon liquid fuels, which possess the requisite mobility.

At present about 25 to 30% of the world's primary energy consumption is used for electricity generation, and about 10% for transportation. Thus at first sight it might be thought that heat mining could never hope to capture more than about 60 to 65% of the energy market of the world. However, if crustal heat is sufficiently cheap and abundant it would be quite acceptable to generate electricity at a considerable sacrifice in thermal efficiency. Figure 4.9 shows the approximate relationship between available temperature and the thermal efficiency of electricity generation. Although a large modern power station can generate at efficiencies around 40%, the world average generation efficiency is more likely to be about 30% or less. With rock at 350°C it should be possible in thermal areas of good quality to generate at about 29.5% efficiency with little sacrifice in using HDR in place of fuel; but where rock temperatures of only 250 or 200°C can be reached the generation efficiencies would fall to about 22.5% to 17.5% respectively, and this would involve a considerable increase in heat consumption. However, with abundant electricity, even if produced rather inefficiently, it would be possible to supply many of the very high temperature industrial processes by using electric furnaces or other appropriate means. As to transportation, an industry that is showing some promise is the manufacture of methanol from low grade heat in conjunction with various forms of biomass. It would also be possible to make hydrogen by electrolysis with abundant electricity, and this too would be very useful as a transportation fuel. Railway electrification is another practice that could be intensified.

Thus the apparent inadequacies of HDR heat are more apparent than real, and most of them could be overcome in time. Meanwhile, the main value of heat mining would be to alleviate the pressure of demand upon fuels, which could then be eked out over a much longer future, and also used as industrial feedstocks.

4.9 General comments on the resource estimates

(1) All the figures of resources derived in this chapter, reckoned on various bases, should be regarded only as 'order-of-magnitude' estimates. Nevertheless, they should be no less reliable than the published estimates of other non-renewable energy resources, for they probably involve less speculation about unseen evidence. In the case of fuels, assumptions of 'in place' reserves involve at least as great an element of guesswork as the assumptions made in this chapter as to the density and specific heat of the crustal rocks and the temperature/depth distribution. In the case of crustal heat, once the basic assumptions have been made, the estimates are amenable to simple mathematical calculation.

(2) In considering the proportion of the land surfaces that could one day be subjected to heat mining operations it would not be necessary to exclude all desert lands or other zones of low or nil population, because the transportation of energy, either electrically or in some cases in the form of alcohol or hydrogen, would render them exploitable. Moreover, if the world's population ever reaches the eight billion mark (as has been suggested by demographers as the maximum that the earth could ever support) many areas now empty will doubtless become inhabited; and these areas could be made more hospitable by means of the crustal heat extracted.

(3) All the estimates in this chapter are based on *stored* crustal heat alone, and give no credit for the further infeed of conducted heat from the mantle or for the heat generated by radioactive decay within the crust, as these energy renewals would usually be far too slow to be of significance. In addition, the relatively small contribution of magma has been ignored (see Chapter 12).

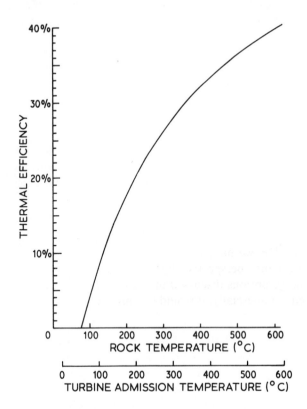

Fig. 4.9 Approximate influence of temperature on electricity generating efficiency. Assumptions: (1) Condenser pressure= 4 in. Hg (13.5 kPa) (2) Actual efficiency/Carnot efficiency=64.4% (3) Turbine admission temperature is 25°C below rock temperature *Note:* The Carnot efficiency is estimated from the turbine inlet and condensing temperatures, with no allowance for auxiliary power.

(4) Frequent reference has been made to *average* crustal cooling. This of course implies that local cooling is likely to be many times as great.

(5) Even though penetration through the whole crust is most unlikely to become possible, figures have nevertheless been quoted in Table 4.1 for specific crustal heat in terms of the average cooling of the whole crust. This is not entirely irrelevant; for both the recovery period and the distribution of stresses caused by the heat mining process are influenced by the crustal rocks both above and below the immediate region of heat extraction.

(6) The thermal areas of the world will undoubtedly be exploited long before the non-thermal, as they offer fewer problems and greater rewards. These areas are fairly easily detected, as minimal prospecting is required to identify areas of adequate thermal gradient and acceptable geology. So much geophysical survey

work has already been done all over the world that we know more or less where to look for favorable sites. Regions of recent volcanic and orogenic activity are likely to exhibit high thermal gradients, while propinquity to a known hydrothermal field is an excellent 'recommendation'.

(7) The broad conclusion of this chapter is that crustal heat represents a huge energy resource, far greater than any other non-renewable resource of the earth except for thermonuclear fusion (see Fig. 3.1). The potential rewards from heat mining are so great that a thorough examination of the practical and economic aspects is imperative. This should not be to the exclusion of the pursuit of thermonuclear fusion or other alternatives; it would be wise insurance for all possibilities to be pursued as urgently as possible.

4.10 Comparison of the HDR resource with other non-renewables

It is hoped that what has been said in this chapter will have convinced the reader that in HDR we have a vast store of energy. This bare fact, however, must be brought into perspective by comparing the resource with the other competing energy options discussed in Chapter 3. As heat mining has not yet been practiced commercially, it would be unrealistic to talk of HDR 'reserves', but it is possible to estimate accurately its resource base in several ways. To avoid possible charges of over-optimism or wishful thinking we have decided to base our estimate on the following cautious assumptions:

(1) Only the HDR lying beneath 100 million km^2 of ground area (about two-thirds of the earth's total land area) is considered. The remaining third of the land area is assumed to be inaccessible or too remote from human habitation. The HDR lying beneath the oceans is excluded.
(2) Only the 'restricted resource base' (as defined in Section 4.4) is considered. Thus rock heat at unattainable depth or of inadequate temperature is excluded.
(3) Of the land area considered, one half is assumed to lie within countries of Category I and the other half within countries of Category II (as defined in Section 4.4).

In Section 4.7 it was stated that about 11.5% of the earth's land areas are thought to be thermal. Hence the thermal and non-thermal areas to be considered for a resource base estimate will be 11.5×10^6 and 88.5×10^6 km^2 respectively. Consider first the *non-thermal* areas. By extending the method used in Fig. 4.4 to other depths, the mean crustal heat content for countries of categories I and II would be approximately as follows. Only those figures relating to depths of 10 km or less should be regarded as *resource bases*.

As to *thermal* areas (see next page) the restricted heat to 10 km depth has been assessed in Fig. 4.8 at 2.63 and 1.97 quads/km^2 for Category I and Category II countries respectively – a mean figure of 2.3 quads/km^2 for thermal

Attainable depth (km)	Heat content in non-thermal areas (quads)	
6	1.22×10^7	
10	7.82×10^7	} Resource base
15	2.46×10^8	
20	4.95×10^8	
25	8.16×10^8	
30	1.20×10^9	
35	1.64×10^9	

areas. The worldwide resource base for these areas would therefore be 2.3 × 11.5 × 10^6, or 2.65 × 10^7 quads. It would be unreasonably optimistic to assume that the temperature gradients in thermal areas would continue to be 1.65 times as high as in non-thermal areas all the way to 35 km depth, for this would imply a temperature of 980°C at that depth – a most improbable figure. A very cautious assumption would be that the higher gradients in thermal areas persist only until the temperature reaches 600°C, which would be at a depth of 17.55 km. The implication of this assumption is that the occurrence of thermal areas is simply due to abnormally thin local crust. To avoid over-rating the HDR potential, this assumption will be made.

Without showing the detailed calculations, the heat content of HDR in thermal areas is estimated as follows:

Attainable depth (km)	Heat content in thermal areas (quads)	
6	0.74×10^7	} Resource base
10	2.65×10^7	
15	8.00×10^7	
17.55 or more	1.15×10^8	

The total heat content for HDR will therefore be the sum of the two sets of figures shown above so:

Attainable depth (km)	Total heat content (quads)	
6	1.96×10^7	
10	1.05×10^8	} Resource base
15	3.26×10^8	
20	6.10×10^8	
25	9.31×10^8	
30	1.32×10^9	
35	1.75×10^9	

From our earlier discussions, 10 km is regarded as a realistic depth for vertical wells and 6 km for directionally deviated wells. By any standards the resource bases calculated to these depths may be regarded as conservative. They are shown graphically in Fig. 4.10, together with the estimated heat contents for other depths: for the 10 km depth, the HDR resource base is compared with other competing resources in Fig. 3.1. Even at an attainable depth of 10 km the HDR resource base is greater than that of other non-renewable resources of energy by the following factors:

Energy resource	10 km HDR resource base/alternative resource base
Fossil fuels	292
Fissile fuels	
(non-breeder)	24419
(breeder)	305
Hydrothermal	808
Geopressured	194

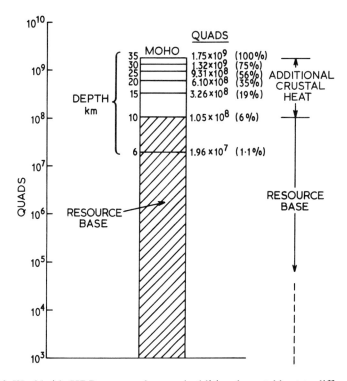

Fig. 4.10 Worldwide HDR resource base and additional crustal heat to different depths as specified in Section 4.10.

Often resource base estimates are apt to be misleading, as the extent of recoverability will vary greatly between one resource and another. In the case of HDR, recoverabililty is dependent upon the permissible extent of cooling, but even if this were limited to only 1°C the HDR resource to 10 km depth would still be 2.15×10^6 quads, which is still far greater than any of the other non-renewable energy resources other than fusion.

There seems little doubt that HDR contains an immense store of recoverable energy that could solve the world's energy problems for a very long time if we could tap it. The remainder of this book will be devoted to the study of ways and means of gaining access to as much as possible of it at economically acceptable cost.

4.11   Questions to be answered

(1) Can we develop the necessary technology for heat mining to become both practical and economic? How much of the restricted or total HDR resource base can be captured and recovered? Of course, this remains the major focus of the book.

(2) To what extent is the temperature spectrum of heat availability matched to human needs? In other words, could we use all the heat available from successful heat mining operations, or would there be a useless residue of unwanted low grade heat? And how much high grade heat, necessary for certain applications would heat mining fail to win for us? These questions are examined in Chapter 5.

(3) Given other practical limitations to heat mining, are the potential environmental hazards of crustal cooling likely to be significant? This question is also considered in Chapter 5.

Chapter 5

Physical constraints and limitations of heat mining

In Chapter 4 an attempt was made to quantify the amount of accessible heat in the crust to various depths, and in Section 4.4 the method was briefly described as to how some of that heat is to be extracted. Nothing, however, has yet been said about how much of it we can hope to extract, how rapidly it would be possible to extract it, or how the availability of the restricted crustal heat is suited to the demand market. Even if we had the technical ability to extract all the restricted crustal heat, certain constraints would prevent us from removing more than a fraction of it, would dictate both minimum and maximum rates of extraction, and would necessitate some degradation of heat. These constraints are imposed by engineering, economic and environmental considerations, some of which will be discussed at length in later chapters. It is well, however, to examine here the broad nature of the constraints and to consider some of them in detail, before proceeding further.

5.1   Engineering and economic constraints

5.1.1   Spectra of heat availability and of heat demand

A glance at Figs. 4.2 and 4.3 will show that there is little compatibility between the spectra of heat demands and of crustal heat availability. To supply the whole of the heat market from crustal heat would inevitably involve degradation of high grade to low grade heat, and this would almost certainly apply very often to restricted markets also. The maximum temperature at which heat could be supplied from heat mining for a specific use (e.g. a power plant or an industry) would of course depend upon the depth to which we are capable of penetrating into the crust, the temperature gradient and the thermal drawdown characteristics of the operating reservoir. Given the present state of drilling technology and costs and expected advances in that technology in the next few decades, exploitation will almost certainly be limited to moderate reservoir temperatures ranging from about 100° to 300°C and probably seldom exceeding about 375°C.

59

As already pointed out in Chapter 4, Section 9, the temperatures attainable in heat mining would be inadequate for direct application in certain industries, and would impose some sacrifice of efficiency in electric power generation, but if mined heat were sufficiently plentiful and cheap these limitations would not necessarily be of great importance; they would also diminish as we learn to penetrate more deeply.

The economic generation of electricity from HDR reservoirs will require sufficiently high rock temperatures to give acceptable (though perhaps rather low) conversion efficiencies (see Fig. 4.9) as well as adequate rates of crustal heat extraction. Furthermore, heat extraction at these rates should be capable of being sustained for·as many years as would be technically and environmentally permissible.

Requirements for lower grade heat can always be obtained by degrading high temperature heat, and to some extent this will always be necessary. As heat degradation is economically wasteful (since low grade heat could be extracted more cheaply at lesser depth if that were all that was needed) it should be limited to a practical minimum.

Figure 5.1 shows three examples of how mined heat availability could be matched to a heat market spectrum. By way of example, the pattern of the demand spectrum has here been taken as that of the USA heat market; so each of the three cases in Fig. 5.1 shows a repetition of Fig. 4.3, but shifted 25°C to the right so as to make reasonable allowance for the inevitable temperature drops from the point at which hot fluid is produced from the reservoir to the points of use. The three heat market curves, as drawn, therefore represent the demands upon the hot rock. Although the shape of each of the three demand curves of Fig. 5.1 has been arbitrarily taken as the same as that of the national USA demand curve, it is emphasized that for a single local heat market – whether a town, a district, or an industrial complex – the demand curve might well be of a very different shape. It will be shown in Chapter 6 that the availability spectrum of mined heat will probably have a shape like that of the area ABCD. If this area be made the same as that which lies beneath the demand curve (plus a small amount of unavoidable low grade waste heat represented by the area ABE), and if the attainable rock temperature is not less than the highest temperature required by the heat market, then the available heat could fully satisfy the whole of the market if enough heat is degraded to supply the 'humps' of the demand curve, as in case (a), Fig. 5.1.

If larger quantities of heat were mined at the same attainable temperature (up to 350°C as drawn) then it would be possible not only to satisfy the whole of the heat market with less degradation, but also to generate some electricity, as in case (b) of Fig. 5.1. In this case there would be rather more low grade heat wasted in the area ABE, and of course also another large quantity of rejected heat from the power plant (here assumed to operate at about 24.5% efficiency in accordance with Fig 4.9).

A third possibility is illustrated as case (c) in Fig. 5.1 in which co-generation

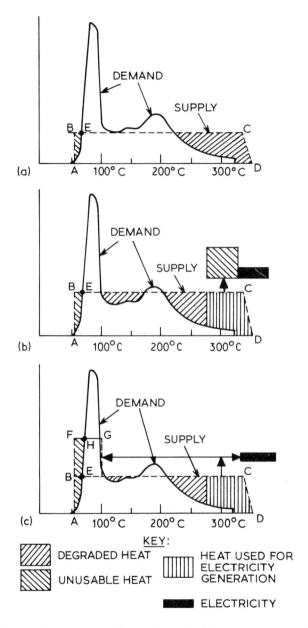

Fig. 5.1 Various ways of matching mined heat to market needs.

Note: A demand pattern resembling that of the USA has been assumed by way of example. (a) Heat market fully satisfied. (b) Heat market fully satisfied plus electricity generation. (c) As for (b), but with co-generation.

enables less heat to be mined than in case (b) while still satisfying the whole of the heat market. In this case a large part of the waste heat from the power plant turbines, assumed here to exhaust at 100°C can be recovered as process heat to supply part of the space-heating load; and although the former area ABE has increased in size to AFH there is far less total heat wastage than in case (b). Rather less electricity than in case (b) would be generated, however, as the efficiency of the back-pressure turbines under the assumed conditions would reduce the cycle efficiency to about 18.5%.

Figure 5.1, although to some extent an over-simplification, serves to illustrate in the simplest terms the broad principles of heat supply and demand reconciliation. The three examples are fairly realisitic in that the maximum assumed temperature of around 350°C should be attainable within 10 km depth in a thermal area of moderate, though above average, temperature gradient (see Fig. 4.8). The vertical height of the supply spectrum ABCD could be adjusted by choosing the appropriate land area beneath which the heat is mined, taking into account the permissible average degree of crustal cooling.

A more realistic example of the possible ways in which energy supply can be matched with demands would be some hypothetical industrial plant such as a large integrated chemical factory or a paper mill, requiring both electricity and process heat of various grades. Such industries frequently need 50 to 100 MWe of electricity and several millions of kilograms per hour of steam and hot water at various temperatures. Figure 5.2 shows three of the many possible options for achieving a balance between energy supply and demand – a simple non-cascaded system, a fully cascaded heat system, and a system combining full cascading with co-generation. Cascading, of course, means the sequential use of heat for two or more processes in declining order of temperature. The energy supply is assumed to be derived from HDR initially at 300°C and capable of sustaining the full needs of a power plant for an economically adequate life without the delivered temperature falling below 250°C at any time. It will be noted that by comparison with case A, cases B and C would enable reductions of 16.8% and 36% to be effected in the quantities of heat extracted from the source. These percentages are of course peculiar to the assumed quantities of electricity and process heat at each of the three grades arbitrarily chosen, and also to the assumed temperatures at which the various processes both take in and reject heat. Nevertheless, the examples clearly show the very great savings that can be made both by cascading and by co-generation. It is even theoretically possible for certain groupings of heat demands and power generation to be conceived that would entirely elimi- nate all wastage except for that arising from the fact that the rejection tempera- tures of the various processes differ from the intake temperatures of the next downstream process; thus almost attaining a 'total energy' complex. Such groupings, however, would be rare in practice.

It is important to understand that even for the simplest 'monochromatic' heat market – a tall rectangle of narrow temperature width – some measure of heat degradation is unavoidable; for if the initial rock temperature were not

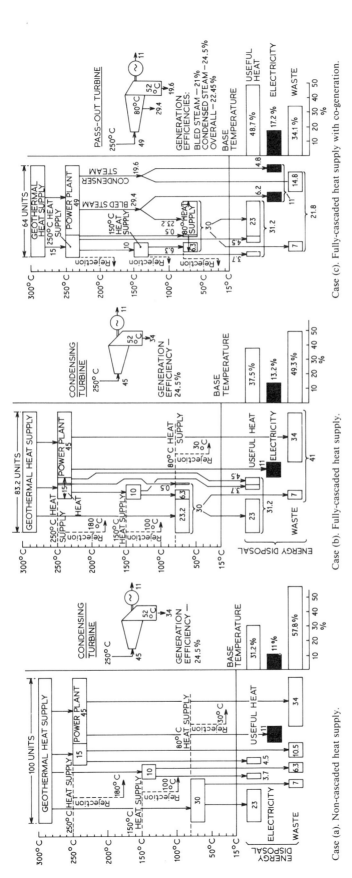

Fig. 5.2 Hypothetical example showing three options for energy distribution from a geothermal source of heat to an industrial plant requiring electricity and also process heat at three different temperatures.

Case (a). Non-cascaded heat supply.

Case (b). Fully-cascaded heat supply.

Case (c). Fully-cascaded heat supply to an industrial plant with co-generation.

Fig. 5.2 Explanatory notes

Assumptions:

(1) In each case the energy requirements of the industrial plant have been arbitrarily taken at the following values:

Electricity	11.0 units	
Process heat at 250°C	4.5 units	
Process heat at 150°C	3.7 units	31.2 units
Process heat at 80°C	23.0 units	
Total	42.2 units	

(2) The *widths* of the energy 'blocks' are proportional to the quantities of energy in each block.
(3) Only the *upper edges* of the energy blocks are of significance in relationship to the temperature scale: the levels of the lower edges are without significance.
(4) Heat energy is reckoned above a base temperature of 15°C.
(5) Process heat for the 250°C, 150°C and 80°C systems is rejected at 180°C, 100°C and 30°C respectively.
(6) No account is here taken of the times of incidence of the various energy demands nor of such energy storage facilities as may be necessary.
(7) The heat shown as 'waste' is that which cannot be used profitably because of the system limitations. The efficiency within each process is not here taken into account: there could well be additional heat losses arising from intrinsic imperfections of the actual processes.

Fig. 5.2 Justification

Case (a) Non-cascaded heat supply – justification

	From HDR	FREE			
		Received	Passed on	Used	Wasted
80°C heat system: rejection at 30°C					
From HDR	30				
Useful heat $= \dfrac{80-30}{80-15} \times 30$ (76.7%)				23	
Waste $= 30 - 23 = 7$ (23%)					7
150°C heat system: rejection at 100°C					
From HDR	10				
Useful heat $= \dfrac{150-100}{150-15} \times 10$ (37%)				3.7	
Waste $= 10 - 3.7$ (63%)					6.3
250°C heat system: rejection at 180°C					
From HDR	15				
Useful heat $= \dfrac{250-180}{250-15} \times 15$ (30%)				4.5	
Waste $= 15 - 4.5$ (70%)					10.5
Carried forward	55			31.2	23.8

Case (a)—*continued*

	From HDR	FREE			
		Received	Passed on	Used	Wasted
Brought forward	55			31.2	238

Electricity: at 250°C and 4 in Hg vacuum.
Fig. 4.9. Efficiency = 24.5%

	From HDR	Received	Passed on	Used	Wasted
From HDR	45				
Electricity = 0.245 × 45				(11)	
Waste = 45 − 11					34
TOTALS	100			31.2 (11)	57.8

Percentages the same.

Case (b) Fully cascaded heat system – justification

	From HDR	FREE			
		Received	Passed on	Used	Waste
250°C heat system					
From HDR	15				
Used (30% as for case (a))				4.5	
Passed on to 150°C system			10		
Passed on to 80°C system			0.5		
150°C heat system					
From 250°C system		10			
Used (37% as for case (a))				3.7	
Passed on to 80°C system					
= 10 − 3.7			6.3		
80°C heat system					
From 250°C system		0.5			
From 150°C system		6.3			
From HDR	23.2				
Total received = 30					
Used (76.7% as for case (a))				23	
Waste (30 − 23)					7
Electricity					
As for case (a)	45			(11)	34
Totals	83.2			31.2 (11)	41
Percentages	100%			37.5% (13.2%)	49.3%

By comparison with case (a):

Reduction in mined heat $= 16.8\% = (100 - 83.2)$

Reduction in *proportion* of waste $= 14.7\% = \left(\dfrac{57.8 - 49.3}{57.8}\right) \times 100$

Case (c) Fully cascaded heat supply with co-generation – justification

	From HDR	FREE			
		Received	Passed on	Used	Waste
250°C heat system					
From HDR	15				
Used (30% as for case (a))				4.5	
Passed on to 150°C system			10		
Passed on to 80°C system			0.5		
150°C heat system					
From 250°C system		10			
Used (37% as for case (a))				3.7	
Passed on to 80°C system					
$=10-3.7$			6.3		
80°C heat system					
From 250°C system		0.5			
From 150°C system		6.3			
From pass-out turbine					
$=(30-6.3-0.5)$		23.2			
Used (76.7% as for case (a))					
Waste (30–23)					7

Electricity
Bled steam at 80°C. $t=250-80=170°C$
Condensed steam at 52°C. $t=250-52=198°C$
Efficiency of condensed steam $=24.5\%$ (case (a))
Approx. efficiency of bled steam

$$=24.5\times\frac{170}{198}=21\%$$

Electricity from bled steam $=E$, and

$\frac{E}{23.2+E}=0.21$ Hence $E-6.2$				(6.2)	
Balance of electricity required $=(11-6.2)$				(4.8)	

Steam needed for this balance $=\dfrac{4.8}{0.245}=19.6$

Quantity of bled steam $=23.2+6.2=29.4$

Total steam to turbine $19.6+29.4=49$

		49			
Waste $=49-(19.6-4.8)$					14.8
Totals	64			31.2	21.8
				(11)	
Percentages	100%			48.7%	34.1%
				17.2%	

By comparison with case (a)
Reduction in mined heat $=36\%=(100-64)$

Reduction in *proportion* of waste $=41\%=\left(\dfrac{57.8-34.1}{57.8}\right)\times 100$

substantially higher than that required for a particular application, there would be no store of heat on which to draw. In any case from a heat mine at a certain adequate depth and temperature there is no way in which lower grade heat can be extracted without degradation: the high grade 'cream' must be removed first. Thus it will be noted that in cases (b) and (c) of Fig. 5.1 the heat required for electricity generation is all degraded from the original rock temperature to that at which the turbines must be supplied – from 275°C in the rock, as drawn. The exhaustion of a heat mine will occur slowly from right to left in Fig 5.1, which may ultimately necessitate the shedding of load from high to low grade consecutively. Although ultimately unavoidable, this loss of load could in theory be mitigated by mining heat at two or more depths to suit different sectors of the demand spectrum, as shown in Fig. 5.3. A deep mine (case (a)) could supply both electricity and the higher grade sector of the heat market, while a shallower (and cheaper) mine could supply the lower grade sector of the heat market (case (b)). As the high grade reservoir becomes gradually depleted of heat it would become necessary to shed power load at first and then heat load, starting at the highest temperature end of the spectrum, until the whole of the high grade heat market has been lost. There would still then remain enough residual heat in the deep reservoir to supplement the shallow reservoir and to provide the lower grade heat market with a second lease of life.

Three, or even more, heat mines at different levels could in theory be added to enlarge this concept, but very possibly the notion of split level heat mining would not in practice be economically supportable, for the attractions of a quicker return on the smaller capital outlay required for a single deep mine with a large degree of heat degradation would probably outweigh the theoretical thermodynamic advantages of mining at multiple levels at a far greater capital outlay. Variants could also be planned making use of co-generation as in cases (b) and (c) of Fig. 5.2. In theory, if it were possible to forecast heat and electricity demands far into the future, it should be possible to plan the consecutive development of heat mines of different potential and at different depths to keep pace with heat market developments for a very long time ahead. There would inevitably be maladjustments in practice, which would necessitate further heat degradation and the establishment of additional heat mining operations from time to time to redress the imbalance caused by faulty demand forecasts. Thus, compromises between excessive investment in over-large mines and the premature exhaustion of under-sized mines will always be required.

5.1.2 Reservoir requirements

Heat mining must necessarily be a large scale undertaking involving, as it does, deep drilling, rock fracturing operations, the installation of heavy pumping equipment and often expensive local heat distribution pipework – all requiring heavy capital investment on which an adequate financial return must be assured. Since the saleable commodity is essentially hot water, the first

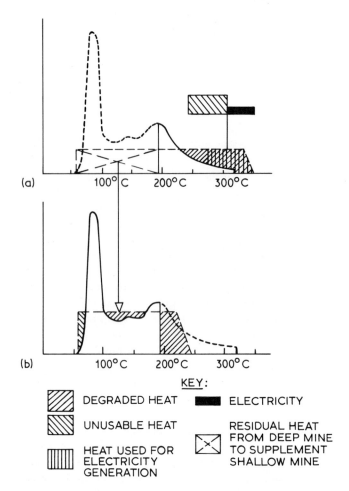

(a)   100°C   200°C   300°C

(b)   100°C   200°C   300°C

KEY:

DEGRADED HEAT   ELECTRICITY

UNUSABLE HEAT   RESIDUAL HEAT
                       FROM DEEP MINE
HEAT USED FOR   TO SUPPLEMENT
ELECTRICITY   SHALLOW MINE
GENERATION

Fig. 5.3 Concept of split level heat mining. (a) Deep heat mine to support power generation and high grade heat market. (b) Shallower heat mine to support low grade heat market.

requirement of a good artificial reservoir is that it should be capable of yielding a high fluid output at a temperature as close as practically possible to that of the natural hot rock within the reservoir. Purely commercial considerations will therefore establish a minimum economic flow of water through the voids and fissures of the reservoir. In addition, physical considerations will establish maximum economic flow above which the pumping costs and temperature drawdown become excessive. Obviously, the economic maximum flow should exceed the economic minimum flow by a wide margin. To meet these economic flow requirements, an artificial reservoir in a mass of HDR must be endowed with the following properties:

(1) A very large contact area between the working fluid and the hot rock.

(2) Adequate conductive communication between the working fluid and a sufficient volume of hot rock to ensure a long life of the 'mine'.

(3) The provision of sufficient void volume within the hot rock mass to ensure that the working fluid exits the reservoir at as high a temperature as possible when circulated at a flow rate well above the economic minimum.

(4) A configuration of voids and fissures that offers a minimum impedance to the required flow of the working fluid, so as to minimize pumping energy while containing the circulating fluid to minimize losses to the non-productive regions outside the boundaries of the active reservoir.

The ways in which it is hoped to attain these properties in an artificial reservoir will be discussed in later chapters.

5.1.3 Recovery time after exploitation

One limitation of heat mining, already mentioned in Section 1.1, is the non-renewability of the extracted heat. A zone cannot be mined indefinitely – at any rate not in the upper parts of the crust that are likely to be of practical interest for quite a long time. As heat is constantly entering the crust from two sources (by conduction from the mantle below, and by radioactive decay from within) the non-renewability of mined heat may cause some surprise to many people. However, to be economic, the rate of heat extraction from a reservoir would usually have to be much faster than the rate at which heat could be replenished from the two sources from which the stored crustal heat originated; hence a continuing run-down of stored heat is inevitable during the mining process. For all practical purposes, heat mined from the upper layers of the crust that are now within commercial reach of the drill would be non-renewable over time scales of interest. This is because the thermal insulating properties of the crust are only affected to a moderate extent by the sort of penetration that we can expect to achieve within the foreseeable future. Recovery periods after the cessation of heat mining operations would therefore be governed mainly by heat flow rates that are comparable with the natural outward heat flow rates that prevailed before those operations began. This natural outward heat flow rate in non-thermal areas is very low, averaging about 1.4 μcal/cm^{2}s, 14 kgcal/km^{2}s, 58.6 mW/m^2, or 1.75×10^{-6} quads/yr km^2; in thermal areas it may be much higher.

It was shown in Section 4.2 that where the thickness of the crust is 35 km the removal of 0.0752 quads (1.89×10^{13} kgcal) would cool a crustal column of 1 km^2 horizontal sectional area by about 1°C. If the insulating properties of the crust were such as to restrict the rate of heat flow through it to 1.75×10^{-6} quads/yr km^2, then the time required to recoup the removed heat would be $0.0752/1.75 \times 10^{-6}$ or about 43 000 years, which for practical purposes may be regarded as infinite. However, the fact that the cooling will have been concentrated within one particular pocket at some depth below the surface means that the recovery heat flow would no longer be 'bottle-necked' through the whole thickness of the crust, so that the recovery time would be somewhat less than 43 000 years. It would nevertheless be so large as to be regarded as essentially infinite. Much of the heat recovery would at first be effected laterally by drawing on the heat stored in the rock surrounding the chilled pocket; but this 'borrowed' heat would have to be repaid before the status quo ante could be restored not only to the cooled column but also to its surroundings. The same would apply to any downward flow of heat from the uncooled rock above the chilled pocket. The ultimate recoupment would be effected partly by increased conduction from the mantle – increased because of the establishment for a long time of a higher temperature gradient from the Moho to the cooled pocket – and partly by virtue of exposure of the cooled pocket to radioactivity from all sides instead of from below only (as at the surface). These factors would combine to shorten the recovery time to less than 43 000 years. A further minor factor that would slightly accelerate recovery would be the temporary reduction in outward heat flow from the surface above the cooled pocket.

The figure of 43 000 years deduced above relates to an average non-thermal area: it is also based on 1 °C of average cooling of a 35 km thick crust. If the degree of cooling were less or more than 1 °C the recovery period would be affected proportionally. It must also be recognized that in thermal areas the outward heat flow may exceed the average of 58.6 mW/m^2 by many times; and this would reduce the recovery time accordingly. Nevertheless, in HDR areas (as distinct from hydrothermal areas where the surface heat flow is greatly increased by fluid convection) heat flows are unlikely to be so great as to reduce recovery times to periods of commercial interest. Even a full recovery time of a century or two would have few economic attractions in terms of 'present worth' applied to future capital savings. Also, if heat mining is to be really profitable, heat extraction should be carried as closely as possible to the permissible limit (see Section 5.2.1 below), and it is quite likely that this may be more than 1 °C of average crustal cooling.

From a practical standpoint there seems to be no alternative but to regard crustal heat as a non-renewable resource if mined at moderate depths. If we consider the exceptional geologic condition where the crust is very thin as at Krafla, Iceland, it is possible to penetrate as far as the Moho. The situation is then dramatically changed; for crustal conductivity would then have virtually no influence on the recovery time. Heat extraction at the Moho, though drawing on stored crustal heat to some extent, would no longer be wholly relying on it. Instead, it would mainly be taking heat directly from the mantle; and as the mantle possesses some measure of very viscous fluidity, it is just possible that the heat supply could be sustained by magmatic convection sufficiently to support heat mining operations, if not indefinitely, at least for very long periods.

Those readers who require mathematical support for the broad conclusions regarding recovery times are referred to Section 10.3, Thermal Performance.

5.2   Environmental concerns

The creation and sustained operation of heat mines causes changes to the earth's condition on a local scale in the vicinity of the exploited volume of rock that *could* have environmental effects. Three areas with potential impact include (1) shrinkage and subsidence caused by heat removal, (2) seismicity induced by thermal and/or hydraulic stresses, and (3) waste heat rejection effects on ecology and climate.

5.2.1 Shrinkage and subsidence: the permissible extent of crustal cooling

It has been shown in Chapter 4 that the cooling of even quite a small section of the crust through a mere 1°C (average) could release an enormous quantity of energy. For example, where the thickness of the crust is (typically) 35 km, an area of only 10 km × 10 km would yield about 2.15 quads if the upper 10 km were cooled through 1°C and 7.52 quads if the whole thickness of the crust were cooled to that extent. These figures respectively amount to 0.87% and 3.0% of the total world energy demand in 1982 disregarding the grade of the energy. They would apply equally to thermal or non-thermal areas.

It is necessary to consider whether even this modest amount of local crustal cooling is likely to cause adverse environmental effects. On the face of it, 1°C of average cooling would seem insignificant; for if this temperature drop were applied to the whole depth of the crust, the heat removal would amount to only 35/11613, or 0.3% of the total heat contained in a crustal column, reckoned above 15°C, as can be seen from Table 4.1. Nevertheless, the question is of sufficient importance to merit closer examination.

Superimposed onto this process of thermal shrinkage are other hydraulic and mechanical effects associated with operating the reservoir to extract heat. This causes the reservoir to behave in a non-linear, coupled manner in response to these stimuli. These hydromechanical effects include the 'swelling' of the reservoir due to transient permeation of fluids into the rock formation comprising and surrounding the active pressurized system, rock failure and opening of natural joints as effective stress levels are altered by the diffusing pressure pulse, and other irreversible changes related to joint or fracture interaction.

All of this complexity makes it virtually impossible to predict behavior *a priori*, but we can simplify matters for assessment purposes by considering the effect of thermal shrinkage separately: When free from internal stress, the linear coefficient of thermal expansion of rocks ranges from about 5×10^{-6} to about 12×10^{-6} per degree Celsius (°C) or Kelvin (K), say about 8×10^{-6} K^{-1} typically. Under conditions of very high compression, this coefficient is reduced considerably. Bullard (1973) has suggested that in the mantle it would be only about 1×10^{-6} K^{-1}. The lithostatic pressure at the Moho in places where the crust is 35 km thick is probably almost 1 kbar, within the mantle it would be more, according to the depth. Even if, for the sake of argument, we assume a conservative value for the mean linear coefficient of expansion of 8×10^{-6} K^{-1} for the crust as a whole, then the effect of cooling the upper 10 km beneath the land areas through 1°C would be to lower the ground surface by about 8 cm. To be more accurate, shrinkage will be a three-dimensional process, so the volumetric change associated with cooling should be estimated. For an unconfined, homogeneous block, three times the linear coefficient is used. Thus the ideal volumetric change for free thermal contraction for a 1°C change in average temperature would be only 2.4×10^{-5} parts in one, or for a cubic kilometer of rock only 2.4×10^4 m³ which is equivalent to a cube about 29 m on a side. Given the complexity of the *in situ* situation and that other hydromechanical effects tend to offset this thermal shrinkage, these estimates are maximum figures: actual shrinkage would probably be far less.

It is also surprising how quite large ground level changes can often be tolerated in actual field situations where subsidence occurs. At Wairakei, New Zealand, the removal of huge volumes of underground waters led to underground subsidence of as much as 7.6 m in one place, after twenty years of geothermal exploitation. As, in this case, the surface fall was spread over several square kilometres, the effects on buildings, roads, pipelines or other structures were either negligible or easily remedied. Consideration has later been given at Wairakei to reinjection. If and when this is practiced further subsidence should be greatly reduced.

The possible effects of shrinkage have been described for a very idealized situation that allowed us to estimate the worst conditions ever likely to arise. Not only have hydromechanical effects been ignored as well as the diminution of the coefficient of thermal expansion under high pressure but also very large crustal sections are assumed to have been cooled through a substantial temperature drop without thermal interaction with the surrounding rock. In practice, the effects of shrinkage are unlikely to be detectable even after substantial heat mining for many years; and even then, they could well be no greater than the effects that are continuously occurring slowly as a result of tectonic forces. For example, there are many parts of the world where the sea has encroached or receded some hundreds of metres within a few centuries.

The heat mining process as practiced would obviously not consist of the uniform cooling of the entire crustal thickness, but is likely to be confined to the extraction of heat from isolated underground pockets of rock, cooled through quite large temperature drops. In Fig. 5.4 is shown a localized crustal pocket A which is to be cooled through N°C. If the mean vertical thickness of this pocket is h metres, and the total thickness of the crust is 35 km, then the mean effective cooling of the crustal column GM containing the exploited zone A would be $N \times h/35\,000$°C. Thus if h were 175 m and N were 200°C, the mean effective crustal cooling would be only 1°C. Concentrated cooling of this nature introduces quite different shrinkage concepts from the idealized cases treated earlier. Before exploitation, the zone A will be in compression to the same extent as the surrounding rock at the same depth, owing to the lithostatic pressure imposed by the over-burden. As the zone A is cooled, it will try to shrink; but rock so confined and in a state of high compression would be unable

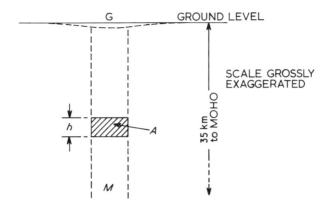

Fig. 5.4 Effect of shrinkage over a cooled deep pocket of rock.

to shrink. Instead, the compressive stress within the zone A will gradually fall and the zone will no longer contribute its full share of carrying the overburden. This share will be transferred to the surrounding rock which, in consequence, will now have to carry a heavier compressive load than formerly. For a homogeneous, isotropic rock, depending on the linear coefficient of expansion and on an effective bulk or Young's modulus of elasticity, it could be that after the zone A has cooled by a certain amount the rock within that zone is first relieved of all compressive stress and later may even be subjected to vertical *tensile* stresses – which could perhaps lead to additional cracking and fissuring, and which, incidentally, could aid the heat mining process because if water can penetrate into these secondary fractures, additional heat could be extracted.

The net result is that all the rock surrounding the exploited zone A would carry increased compressive stresses that attenuate with distance from the crustal column containing A. The ground surface would then tend to become only slightly distorted by the formation of a shallow depression, as shown dotted and greatly exaggerated in Fig. 5.4.

As mentioned earlier, another factor would greatly mitigate, and perhaps even eliminate, surface deformation. The heat mining process as has already been mentioned requires the injection of water at very high pressures into underground fissures within a zone exploited for cooling. This high hydraulic pressure would tend to *inflate* the surrounding rock, and so resist the tendency to collapse as a result of shrinkage. Just as the removal of large quantities of subterranean waters caused subsidence at Wairakei, so should the injection of high pressure water cause the reverse process – a tendency to swell – that could well offset the thermal shrinkage tendency.

For a very long time to come, given finite limitations to demand, heat mining operations are likely to be confined to deep-seated pockets at different depths, spaced at fair distances apart; these local effects would be mitigated by distribution over a much wider area. Only in the very distant future, when exploited rock zones become more or less contiguous over large areas, will these thermal and hydromechanical effects be detectable. By then much useful empirical experience will have been gained as a result of cautiously developing heat mining.

In Chapter 4, we showed that the cooling of the whole crust through 1°C could provide the world with most of its energy needs for many centuries, or even millenia. Given what has been said, shrinkage and subsidence on a world scale are unlikely to be troublesome. It is true that heat mining activities could be fairly highly concentrated in populous countries, so that shrinkage problems could conceivably arise locally. However, the adoption of a wise heat mining strategy should ensure almost complete immunity from shrinkage effects – at any rate for many centuries.

5.2.2. Induced seismicity

Sudden temperature and pressure changes belowground could very possibly give rise to seismic shocks by disturbing the stress pattern within the crust; in fact, earthquakes are themselves caused by stress pattern changes. Exploitation of HDR by heat mining, however, would be a continuous and rather slow process, so that the resultant stress changes would be gradual and not sudden. Rocks possess a degree of 'pseudo-plasticity' due to micro-structure alterations that are on a much larger than molecular scale. Therefore, it is probable that these induced local stresses would have time to redistribute themselves to avoid any build up of stress concentrations that could result in catastrophic failure.

It is expected that thermal stress-cracking will occur at the rock/water interfaces as a result of the contact between hot rock and relatively cool circulating water (see Chapters 8 and 10). These cracks, which could have a beneficial effect on heat mining by reducing the flow resistance to the working fluid and by extending the rock/water contact area, will be individually small and will form one by one – not in a single cataclysmal rupture. They will also be very deep below the ground surface. As each individual crack is first formed, probably no more than a micro-earthquake will occur, detectable only by sensitive seismometers. Once a crack has started it will extend slowly and gently as further cooling takes place. Further cracking may perhaps occur (see Section 5.2.1 above) as a result of gradual shrinkage within a cooled zone of rock if the cooling effect is sufficient to overcome and reverse the lithostatic compressive forces. These cracks (which would be within the body of the cooling pocket of hot rock, and not necessarily at the rock/water interfaces) would also probably be individually small and would form one by one. Hence it would seem that the chances of a large build-up of unrelieved stress that could suddenly lead to failure and cause seismic shocks would be remote. Such limited experience as is yet available would seem to support these expectations.

In addition to thermal-stress induced effects, there is the potential for induced seismicity as water permeates under pressure into the formation surrounding the active reservoir. In this case, hydraulic forces are altering the *in situ* stress conditions. The rate of water lost defines the relative importance of this effect. For a stable fractured reservoir, water loss rates should decrease initially and eventually reach a steady state level. As water diffuses into the formation, microseismicity usually occurs on some level as readjustments in the microstructure occur. The intensity of this microseismicity will be controlled by the formation's ability to redistribute these stress changes. The ideal situation is to have many, more or less continuous, small-scale events, thus avoiding stress build ups.

Field measurements of seismic activity with rock fracturing induced by thermal and/or hydraulic forces typically yield tremors of Richter magnitude −4 to −1 or 7 to 4 orders of magnitude below the level of human sensitivity. However, seismic susceptibility will be very site-specific and difficult to predict in advance of actual measurements. In fact, it may be hard to find two seismologists who agree on a precise estimate before the fact. Extensive seismic monitoring arrays were installed at Fenton Hill and Rosemanowes to characterize any induced seismicity. At Los Alamos and Rosemanowes, where experimental circulation and heat extraction tests from HDR have been in progress for a few years, no seismic tremors of any significance have yet been detected. Although these have so far been only on a small scale relative to proposed commercial operations, they are encouraging.

Swanberg (1975) states that fluid disposal by reinjection will not trigger off earthquakes unless it is carried out in an area where substantial shearing stresses are already present (active faults). He mentions that in literally thousands of fluid injection wells in the USA there have been only two documented instances of earthquakes induced by reinjection. He also gives warning, however, that in tectonically active areas there could be risks of seismic shocks arising from reinjection (which is a process not unlike heat mining). Such areas, however, are comparatively rare; so that by far the greater part of the world should not be susceptible to seismic shocks as a result of heat mining. In any case, reinjection is being practiced in several hydrothermal areas of tectonic activity (e.g. in Japan, El Salvador and California) where no shocks have resulted therefrom though it is understood that some slight seismicity was experienced in Tuscany during reinjection tests. Moreover, tectonically active areas are subject to natural earthquakes, and structures in these areas are consequently designed to withstand such shocks. For a thermally-induced stress, the places most at risk would be the fringes of large heat mined areas, where the unmined outer zone would be resisting the tendency of the mined zone to shrink. Here there could possibly be sufficient build-up of stress for it to be suddenly relieved. For a hydraulically-induced stress, on the other hand, the production zone itself would have the highest risk. Clearly this is a matter where it will be necessary to proceed with caution, with the results of field experiments carefully monitored by seismologists.

5.2.3 Effects on ecology and climate

It will be seen from Fig. 4.8 that in a non-thermal area it would necessary to practice heat mining at depths of not less than 3 km. Even in a thermal area having a temperature gradient of 70°C/km (2.8 times the normal) the minimum depth of interest would be 1 km. In places of still higher thermal gradient it would be possible to reach a minimum useful temperature of 85°C at less than 1 km depth; but in such cases the attraction of attaining higher temperatures would almost certainly encourage exploitation to much greater depths. There is little doubt that heat mining will seldom, if ever, be practiced at less than 1 km below the ground surface. Owing to the low thermal conductivity of rocks, the thermal effects of substantial heat removal at such depths would be virtually undetectable at the surface except as a very small diminution in the rate of outward heat flow from the surface. The top few metres of ground below the surface will be affected to far greater extent by solar radiation and other climatic influences than by the paltry amounts of heat flow ascending from below; and it is within these few metres that the roots of vegetation will be confined, that burrowing life forms will be found, and that irrigation waters are effective. It is worth noting that solar thermal fluxes of 200–1000 W/m² may occur at the earth's surface, as compared with only 60 mW/m² due to the outward flow of crustal heat in non-thermal areas. In hyperthermal areas the outward heat flow would of course be considerably greater, but still insignificant by comparison with solar heat fluxes. Artificial cooling at depth, as a result of heat mining, might neutralize or even reverse the surface geothermal flux, but the effect on the ecology and climate would be puny by comparison with the influence of solar radiation. Heat mining activities at depths of 1 km or more should have no discernible effects whatsoever upon the ecology or the climate above the mined zone, as a result of heat removal from below.

Environmentally, tampering with crustal heat should, of itself, have no harmful side effects whatever. There is, however, a broader question that concerns all forms of energy consumption regardless of whether it originates from fuels, renewable energy sources or from heat mining. This relates to the possible effects of releasing huge quantities of waste heat into the atmosphere after use.

It has already been stated that the world is consuming primary energy at a rate of approximately 250 quads per annum, which is equivalent to a continuous consumption of 8.4×10^6 MWt. Apart from problems of chemical or toxic pollution (which may well be serious but are not germane to the matter under consideration), to what extent is this causing *thermal* pollution? Ultimately, all human energy consumption appears in the form of heat discharged either into the atmosphere or into the waters of earth. Could it be that the huge quantities of heat so discharged are already affecting the world climate?

Perhaps the best way of gaining perspective is to note that solar energy is falling onto the earth at a fairly steady rate of about 178×10^9 MWt. Hence the present rate of human energy consumption is only 8.4/178 000, or less than 1/20 000th part of the solar energy falling on the earth. Although some of this solar energy is reflected off the outer atmosphere, about 70% of it reaches the ground surface and some additional heat is absorbed in the atmosphere as the solar radiation passes through it. Most of the unreflected solar energy is re-radiated into outer space at night, although part is retained by photosynthesis in the formation of organic matter that is the raw material of future fossil fuels. The miniscule addition of present human energy consumption is virtually lost against the vast background of this heat transfer process. The world energy consumption would have to be increased by a huge factor before the rejected heat after use would have any appreciable effect on the climate.

5.2.4 Relevant field experience

Of the various possible environmental side effects of crustal cooling it would seem that none of them – at least for a very long time – poses any serious threat. Only if heat mining were practiced on a gigantic scale, or at very high intensity within a restricted area, would there be any likelihood of local subsidence or of notable induced seismicity. If we proceed with caution and avoid high local extraction rates, no adverse effects need be feared. However, this does not give us *carte blanche*, for it cannot yet be categorically stated what are permissible 'high local extraction rates'. It is not yet possible to draw on direct experience, because heat mining has not yet been practiced except on an experimental scale. However, experience of hydrothermal heat extraction gives us some indirect evidence that is perhaps suggestive:

At Wairakei, New Zealand, the geothermal power station had generated 20 837 GWhe from the date of inception (late 1958) to the end of 1980 (Thain, 1980). The average thermal efficiency, in terms of the heat content of the extracted bore fluid (including the rejected hot water), has been 6.4%. Hence the heat extracted from the field since power generation began has been $20\,837 \times 1\,000 (\mathrm{MWh/GWh})/(8\,760\ (\mathrm{hours/year}) \times 0.064\ (\mathrm{eff.})) = 37\,165\ \mathrm{MWy}$. In the 1950s there had been about seven years of drilling and testing of bores, many of which were allowed to discharge continuously to waste. It would not be unreasonable to add, say, 25% for this unproductive heat that was artificially extracted from the field before electricity production began and also to cover waste steam discharged from vent valves, bores under test and 'rogue' bores. Hence an approximate estimate of the total heat extracted artificially from the Wairakei field up to the end of 1980 would be $1.25 \times 37\,165$, or $46\,456$ MWyt. The Wairakei field extends over an area of about 18 km², so the intensity of artificial heat extraction has been about $46\,456/18$, or $2\,581$ MWyt/km².

In Section 4.2 the specific crustal heat was estimated at $2\,500$ MWyt/km²/°C of average cooling of the whole of the local crust to a depth of 35 km, and at 718.3 MWyt/km²/°C of average cooling of the upper 10 km of the local crust. The total heat extracted from the Wairakei field until the end of 1980 could therefore be considered as the equivalent of the following extents of average crustal cooling:

2581/2500, or about 1°C for the whole thickness of the local crust
or 2581/718.3, or about 3.6°C if all the heat were extracted from the
upper 10 km of the local crust only.

As to the first figure, it is probably an understatement in one sense; for the Wairakei field is believed to owe its existence to a magmatic intrusion, which would mean that the thickness of the crust locally is probably very much less than 35 km. The average temperature drop for the full thickness of the crust would therefore be considerably more than 1°C.

The Wairakei power station has been generating power at high output steadily since 1980, and is expected to continue doing so for several more years. No adverse effects of induced seismicity or of thermal shrinkage have yet been felt there. It might therefore be argued that experience at Wairakei provides evidence to show that average crustal cooling to the extent of at least 1°C for the whole crustal thickness, and at least 3.5°C for the upper 10 km of crust, would be permissible for heat mining. Nevertheless, these figures, though significant, should not be regarded as firm evidence. Heat extraction by thermal conduction (heat mining) is not the same process as heat extraction by fluid removal: the one is confined to a well-defined column of crust, while the other is not. It cannot be claimed that all of the extracted heat at Wairakei has originated from the column of crust vertically beneath the field: some of it may well have been drawn in laterally with fluids migrating from further afield. Moreover, the extracted hot fluids will not yet have been fully replaced by inflowing cold waters from beyond the confines of the field; so the cooling process is not yet complete. If the removal of heat is accompanied by the simultaneous removal of matter, this does not necessarily amount to 'cooling'. However, at Wairakei the rate of fluid recharge has for some years equaled the rate of fluid withdrawal (Hunt, 1970), so that cooling must have been fairly effective for some time.

Despite the need for caution, the experience of Wairakei would seem to suggest that the cooling of a crustal column of any depth should be permissible to the extent of at least 1°C. If so, the resource accessible to heat mining could be immense.

It should be noted that in the statement above to the effect that no adverse effects of thermal shrinkage have been experienced at Wairakei, the accent should be on the word 'thermal'; for there has been considerable ground subsidence there. This, however, is entirely attributable to the removal of very great quantities of fluid from belowground: in no sense is it thermal shrinkage. With heat mining operations no such fluid removal would occur.

It should also be noted that the natural heat flow at Wairakei before exploitation began was many times the world norm of 14 kgcal/km²s, so the net rate of heat removal would apparently be less than the deduced figures would suggest. However, much of this natural heat flow took the form of surface phenomena such as geysers, hot springs, and fumaroles; and most of them have ceased since the field has been exploited. The amount by which the deduced figures would have to be modified on this account would not therefore be very great.

Chapter 6

Graphical representation of heat mining targets

6.1 Temperature distribution through the crust

A very influential factor in the economics of heat mining will be the cost of drilling from the surface to a depth at which the rock is sufficiently hot to be of commercial interest. Drilling costs are very sensitive to the depth of penetration; hence a study of the temperature distribution within the earth's crust is important. This distribution will of course vary from place to place and may vary irregularly, but for convenience and in order to gain some perspective of the problems involved it is instructive to consider the simplified case of a hypothetical non-thermal region of the earth's crust with a smoothly varying gradient. Even this cannot be free from some degree of speculation, but if reasonable assumptions are made it is possible to discern an approximate pattern of the likely way in which the temperature will vary with increasing depth in non-thermal regions; and from this pattern it will be possible to argue with some degree of confidence from the general to the particular. It is of course true that non-thermal regions will for a very long time be of far less commercial interest than thermal regions, which will undoubtedly be exploited first. Nevertheless, non-thermal regions are simpler to analyse; and as their study reveals the most pessimistic of conditions, there is much to be said for considering them first.

6.1.1. Heat flows

At any point P in the earth's crust, let:

q_m be the heat flow due to thermal conduction from the hot mantle;
$q_r{}^*$ be the heat flow due to internal heat generation from radioactive decay of certain crustal elements;
q be the total heat flow ($= q_m + q_r{}^*$).

The $+$ sign will be used to denote upward flow, and the $-$ sign to denote downward flow.

All three values will vary from place to place according to the chemical and physical composition of the crustal rocks, the thickness of the crust from the surface to the Moho (which, in places of magmatic intrusion, may be comparatively thin) and the temperatures at the Moho directly beneath the point P. For the purpose of making a general study it is therefore necessary to make certain simplifying hypotheses:

(1) The radioactive rocks are assumed to be evenly distributed throughout the thickness of the crust.
(2) The crust is assumed to be getting neither hotter nor cooler – it is in a state of thermal equilibrium, with no change in the quantity of *stored* heat within the crust. This assumption relates to the crust only: it does not necessarily apply to the earth's interior beneath the Moho.

The first of these assumptions is almost certainly not true, while the second is not necessarily universally true. Nevertheless, for the purpose of the general analysis which follows the effects of any inaccuracies in these assumptions would be of minor importance.

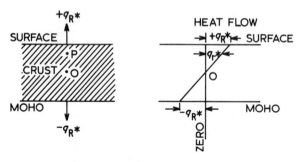

Fig. 6.1 Approximate pattern of radioactive heat flow in crust under assumed conditions.

Figure 6.1 shows an imaginary section of 'disembodied' crust conceived in isolation, with no mantle or other matter beneath it. In the absence of a mantle, q_m would be zero and the total flow q would simply be equal to q_r^*, attaining a maximum value of $+q_R^*$ at the surface. Owing to hypothesis (1) it would follow, from symmetry, that there would also be a downward heat flow, $-q_R^*$, at the Moho. At the mid-point O between the surface and the Moho there would be no net vertical heat flow at all; and the variation in heat flow from point to point through the crust would be linear, as shown. Even after the 'replacement' of the mantle, the radioactive element of heat flow would still act in this manner. In accordance with hypothesis (2) the rate of generation of radioactive heat would be the same as that of the combined outward heat losses at the surface and into the mantle – i.e. $2q_R^*$.

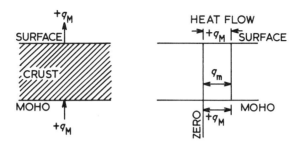

Fig. 6.2 Pattern of conductive heat from the mantle through the crust.

Figure 6.2 shows the pattern of conducted heat from the mantle, upwards through the crust. Owing to hypothesis (2) the heat q_M entering the crust from the mantle must be equal to that escaping from the surface, so that $q_m = q_M$ at all points throughout the crust, as shown. It is now possible to deduce the pattern of the total heat flow, q which at all points must be the sum of q_r^* and q_m. This can be derived from the simple addition of the graphs of Figs. 6.1 and 6.2; allowing for the different scales that apply to the two graphs. According to Bullard (1973) the proportion of surface heat flow in non-thermal areas due to thermal conduction from the mantle varies from place to place on the continents over a range of about one third to two thirds of the total surface heat flow; the balance being due to radioactivity in the crustal rocks. Three cases will therefore be taken to illustrate the range of heat flow patterns through the crust, and these are shown in Fig. 6.3, in which the following symbols are used for the total heat flows at the upper and lower limits of the crust:

q_{HE} = *net* heat flow entering the crust at the Moho
q_{HL} = *total* heat flow leaving the crust at the surface.

The three cases are as follows:

Case (a) $q_M = 2/3\ q_{HL}$ and $q_R^* = 1/3\ q_{HL}$
Case (b) $q_M = 1/2\ q_{HL}$ and $q_R^* = 1/2\ q_{HL}$
Case (c) $q_M = 1/3\ q_{HL}$ and $q_R^* = 2/3\ q_{HL}$

As q_M, q_R^* and q_{HL} will all vary from place to place, both absolutely and relatively, it will here be assumed that q_{HL} is constant. Bullard (1973) states that the world average value for q_{HL} is about 59 mW/m^2 (1.4 μcal/cm^{2}s), and this figure will be taken as typical for non-thermal areas. Figure 6.3 discloses some curious results. In case (a) only one-third as much net heat enters the crust at the Moho as leaves it at the surface. In case (b) no net heat enters the crust at all from below; the upward conductive heat from the mantle exactly balancing the downward radioactive heat. In case (c) there is a net negative or downward heat flow at the Moho equal to one-third of the heat escaping from the surface. The implications are as follows:

(1) Under the conditions of case (a) the mantle would be continuously cooling.
(2) Under the conditions of case (b) the mantle would, in effect, be perfectly lagged: its temperature would remain constant.
(3) Under the conditions of case (c) the interior of the earth would actually be growing hotter.

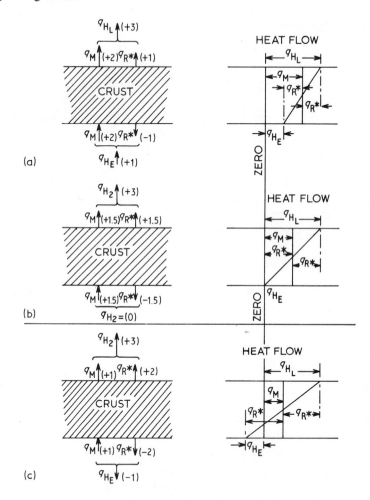

Fig. 6.3 Range of net heat patterns through the earth's crust. ( ) around figures refer to magnitude of heat flow vectors.

These observations are made on the assumption that there is no appreciable amount of radioactivity within the mantle and cores of the earth. In different parts of the crust any of these three broad patterns of heat flow could be present; but of course in hyperthermal areas the heat flow pattern would be altogether different. However, our first aim is to discover 'typical' conditions in non-thermal areas. Whether, on a worldwide basis, the net effect of the different crustal heat flow patterns is such that the interior of the earth is now cooling or whether it is still becoming hotter, at the present level of radioactivity in the crust, cannot be stated with certainty.

6.1.2 Feasible patterns of normal temperature variation with depth in non-thermal regions

In order to deduce a feasible temperature/depth relationship it is necessary to make certain simplifying assumptions. These may conveniently be taken as follows:

(1) The temperature gradient at the surface of a 'typical' non-thermal area is 25°C/km.
(2) The average crustal thickness in non-thermal areas is 35 km.
(3) The average temperature at the Moho is 600°C. Bullard (1973) gives a probable range of 500 to 700°C: hence the assumed figure of 600°C.
(4) The heat flow pattern is in accordance with case (b) of Fig. 6.3. This is simply to avoid the extreme conditions of cases (a) and (c).
(5) The mean surface temperature is 15°C.
(6) The surface heat flow, q_{HL} is 1.4 μcal/cm^{2}s, or 0.059 W/m^2.

None of these assumptions will be precisely correct at all parts of the earth's crust, but all are reasonable for non-thermal areas.

A sustained temperature gradient of 25°C/km through a 35 km thick crust would give a temperature at the Moho of [15 + (25 × 35)], or 890°C, which is at variance with assumption (3). Given these assumptions, clearly the temperature gradient must decline with increasing depth.

By assuming thermal conduction to be the dominant mechanism for heat transport, the heat flow at any point in the crust will be proportional to the product of the geothermal temperature gradient and the local thermal conductivity of the rock.

According to Fourier's law of heat conduction, the rate of heat flow per unit area of surface is given by q:

$$q = -\lambda_r \nabla T = -\lambda_r \, dT/dZ \qquad (6.1)$$

where ... ∇T = geothermal temperature gradient = dT/dZ, °C/km (K/m)
q = upward (+) heat flow per unit area of surface, W/m^2 (cal/cm^{2}s)
λ_r = thermal conductivity, W/m K or W/m°C
Z = depth m(km)

In fact, if ∇T, q and λ_r are all expressed in mutually consistent units – preferably in terms of watts, metres, seconds and degrees Kelvin or Celsius – then it would be true to say that $\nabla T = -q/\lambda_r$ without any conversion factors required. According to Assumptions (1) and (6), ∇T is -2.5×10^{-2} K/m and

q is 59mW/m² (1.4 × 10⁻⁶ cal/cm²s) at the surface. Hence... λ_r = 5.9 × 10⁻²/2.5 × 10⁻² = 2.34 W/m K or 0.0056 cal/cm s °C at the surface.

It is clear from Assumption (6) and from case (b) of Fig. 6.3 that

$$q = q_{HL} - A^* \cdot Z \tag{6.2}$$

where A* is a constant in W/m³ proportional to the rate of heat generation primarily caused by radioactive decay of the isotopes of uranium, thorium and potassium contained in the crust and Z is the depth below the surface in metres (m). It is also clear that at the Moho, where Z = 3.5 × 10⁴ m, q is zero.

Therefore,

$$0 = (5.9 \times 10^{-2}) - (3.5 \times 10^4)A^*$$

or

$$A^* = 1.7 \times 10^{-6} \text{ W/m}^3.$$

By way of trial, let it first be assumed that λ_r is constant at all depths. Then

$$\nabla T = \frac{-q}{\lambda_r} = \frac{-q}{2.34}$$

or

$$\nabla T = -0.427q \text{ K/m (or °C/m)} \tag{6.3}$$

But

$$q = q_{HL} - A^* \cdot Z = (5.9 \times 10^{-2}) - (1.7 \times 10^{-6})Z$$

By definition

$$\nabla T = dT/dZ, \text{ hence}$$

$$\frac{dT}{dZ} = -0.427 \left[(5.9 \times 10^{-2}) - (1.7 \times 10^{-6})Z \right] \tag{6.4}$$

$$= -(2.5 \times 10^{-2}) + (7.26 \times 10^{-7})Z$$

and by integrating to depth Z from the surface ($Z=0$) to obtain T:

$$T = T_o + \int_0^Z [+(2.5 \times 10^{-2}) - (7.26 \times 10^{-7})Z]dZ$$

where T_o is the surface temperature, which, by Assumption (5) is 15°C.

Hence

$$T = 15 + (2.5 \times 10^{-2})Z - (3.6295 \times 10^{-7})Z^2 \tag{6.5}$$

The curvature shown in curve 1 of Fig. 6.4 is a direct result of the quadratic depth dependence depicted by equation (6.5) above, which in turn resulted from the amount of radioactive heat generation being proportional to depth. According to the model, the gradient decreases with increasing depth but unfortunately too fast to match the estimated Moho temperature. Curve (1) of Fig. 6.4 shows that the temperature at the Moho, according to this equation, would be about 450°C instead of 600°C. This suggests that the hypothesis of constant thermal conductivity at all depths cannot be true, or alternatively that the radioactivity cannot be evenly distributed through the crust. For the Moho temperature to be 600°C, without sacrificing any other assumptions, λ_r would have to decrease with increasing depth and temperature so as to give steeper temperature gradients at all depths except at the surface (where it must remain at 25°C/km) and at 35 km depth (where it must be zero).

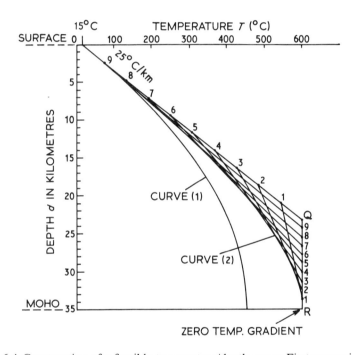

Fig. 6.4 Construction of a feasible temperature/depth curve: First approximation.
Curve 1 constant thermal conductivity
Curve 2 variable thermal conductivity

The thermal conductivity of silica (SiO_2), an important crustal mineral ingredient, may increase or decrease with temperature depending on its structural form. For example, λ_r for quartz or crystalline silica decreases significantly with increasing temperature by a factor of two or more as the temperature rises from 0 to 200°C. λ_r for amorphous or glassy silica, on the other hand, increases by a factor of 1.8 from 0 to 600°C (Birch and Clark, 1940). If the variations in the gradient are to be such that the assumed temperature of 600°C is reached at the Moho, the complex assemblage of minerals comprising the crust must produce an average thermal conductivity that decreases with increasing depth.

In view of the many unknowns and the doubtful nature of some of the basic assumptions made at the beginning of this section it is not possible to deduce rigorously the temperature/depth relationship in a typical non-thermal area. Instead, an arbitrary 'short cut' will have to be applied by making use of a well-known construction method for forming a curve tangential to two straight lines at two defined points. The straight lines are SQ, the tangent to curve (1) of Fig. 6.4 at the surface point S ($T = 15°C$), and QR, the vertical at ($T = 600°C$) to which the required curve must be tangential if no temperature gradient is to occur at the Moho, – i.e. in order to conform with case (b) of Fig. 6.3. The construction is as follows. Divide both SQ and QR into an equal, but arbitrary, number of sectors (ten have been chosen in Fig. 6.4) and number the division points serially as shown. Join up points (1) to (1), (2) to (2), and so on. The lines so formed will present an envelope, tangential to which a smooth curve may easily be drawn. This construction gives curve (2) of Fig. 6.4. While it cannot be justified on strictly scientific grounds, it at least satisfies all the assumptions that have been made: it is also consistent with case (b) of Fig. 6.3. It does, however, require the acceptance of decreasing thermal conductivity with depth. The implications of curve (2) of Fig. 6.4 are shown in Fig. 6.5. The heat flow falls linearly with depth, as required for case (b) of Fig. 6.3, from 59 mW/m² (1.4 μcal/cm²s) at the surface to zero at the Moho. The gradients have been measured off curve (2) of Fig. 6.4, as constructed; while the thermal conductivity has been deduced from the quotient $q/\nabla T$ and expressed in appropriate units. It will be seen that the thermal conductivity, λ_r, would seem to fall more or less linearly from 2.34 W/m K (0.0056 cal/cm s °C) at the surface to 0.84 W/m K (0.002 cal/cm s °C), or about one-third as much, at the Moho.

There would seem to be no escape from the acceptance of something approximating to curve (2) of Fig. 6.4 unless it be admitted that one or more of the six basic assumptions was wrong, or unless the heat flow pattern differs substantially from case (b) of Fig. 6.3. In practice, of course, the temperature/depth distribution curve would almost certainly vary from place to place: it is also unlikely to be perfectly 'smooth' in form, owing to random geological variations at different depths. But in the absence of factual knowledge of the lithology or petrological 'mix' at different depths – and this will never be the same at all places – curve (2) of Fig. 6.4 may reasonably be taken as a feasible, though not accurate, typical temperature/depth relationship for non-thermal areas where the heat flow at the surface is composed of equal parts of radioactive and conductive elements.

So far only case (b) of Fig. 6.3 has been considered in detail because it is a compromise between the extremes of cases (a) and (c). Different temperature/depth relationships would be expected for these other two cases. It is almost certain that for them also it would not be possible to retain the six basic

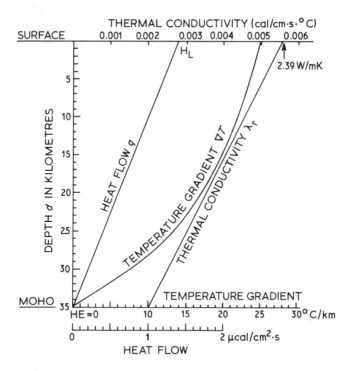

Fig. 6.5 Implications of the variable thermal conductivity model given by curve (2) of Fig. 6.4.

assumptions listed earlier. The one most likely to be jettisoned would be the second – that the crustal thickness is 35 km everywhere. There is of course an infinite number of other possibilities. For example, if the radioactive rocks were unevenly distributed through the crust, with a preponderance of radioactivity in the upper layers, this would imply a diminution of upward heat flow at great depths, which would reduce the discrepancy between curves 1 and 2 of Fig. 6.4 without the need for such drastic reduction of conductivity at increasing depths. However, for the purpose of this simple analysis it would be both laborious and unprofitable to pursue too many speculative hypotheses; so only the simple postulations laid down early in this chapter will be adopted.

Since case (a) represents a preponderance of conducted heat over radioactive heat it would seem likely that such conditions would be realized at places where the crust is thinner than 'normal'. Conversely case (c) represents a preponderance of radioactive heat over conducted heat; and this might be expected where the crust is abnormally thick. Without analysing these cases in detail, the *sort* of temperature/depth distributions that might be expected for cases (a) and (c) would be more or less as illustrated in Fig. 6.6, from which the following points may be noted:

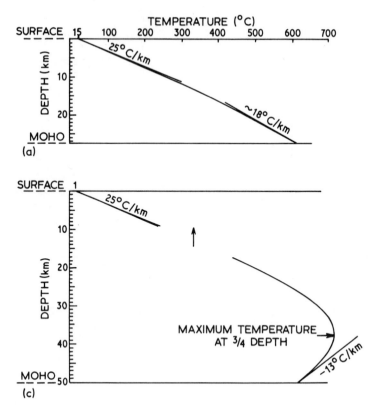

Fig. 6.6 Types of temperature/depth distribution curves to be expected for cases (a) and (c) of Fig. 6.3. (a) Thin crust: Preponderance of conducted heat. (c) Thick crust: Preponderance of radioactive heat.
Note: The second case is designated (c) for consistency with Fig. 6.3. There is no case (b) to be considered here.

CASE (a)

A *positive* temperature gradient persists at the Moho, so the earth would be losing heat at such places as a result of a net outward flow. As the heat flow falls less rapidly with depth than in case (b), the temperature gradient will change less from surface to Moho. In consequence, the prescribed Moho temperature of 600°C will be reached at a shallower depth.

CASE (c)

A *negative* thermal gradient will exist at the Moho, so the earth would be gaining heat at such places as a result of net downward flow at this depth. The maximum attained temperature will exceed 600°C, and this will be at three-quarters depth where the net flow is zero (see case (c), Fig. 6.3).

The depths, gradients and maximum temperature shown in Fig. 6.6 should not be taken as factual, but merely as indicative of possibilities. Quite obviously actual heat flow measurements are not available for the entire crustal thickness to 35 km. In fact, no direct data in the form of rock temperatures and/or rock thermal conductivities exists beyond the maximum depths that have been drilled in the crust. Today, this would be about 10 km or 32 000 ft. Other geophysical evidence, usually in the form of deep seismic soundings or magnetotelluric measurements, is used to model the physical and chemical state of the crust below depths accessible to the drill bit; but substantial uncertainty and controversy exists in predicting the temperature of the Moho as well as the thermal conductivity of the crust below 10 km and this no doubt would have an impact on the temperature–depth relationships cited so far.

It has been the subject of debate (Boldizsar, 1970) as to whether, on average, the mantle and cores of the earth are cooling or growing hotter. Most opinion seems to favor the belief that the earth is slowly cooling; but in past ages when the planet was young the radioactivity in the crustal rocks would have been far greater than now, so that a temperature/depth distribution curve resembling case (c), Fig. 6.6, would have been quite probable – very likely with a lower Moho temperature, higher surface heat flow, and with a more pronounced bulge to the curve. In the course of eons the earth would have heated up to attain a maximum Moho temperature, after which the cooling process would begin – and probably has begun.

If credence is to be given to the theory of a cooling earth, then a 'flatter' temperature/depth curve than curve 2 of Fig. 6.4 would be implied if all the basic assumptions made at the beginning of this section are to be retained except item (4). In place of Assumption (5) it would be necessary to assume a heat flow pattern somewhat intermediate between cases (a) and (b) of Fig. 6.3. The question is 'How much flatter?' Bullard (1973) has suggested a temperature gradient of 12°C/km just *below* the Moho. This, of course, does not imply that the same gradient would be found just *above* the Moho, for although the heat flow would be the same on either side of the Moho boundary the thermal conductivities could be different.

In the presence of so much uncertainty, what may reasonably be assumed in a working model? A fairly plausible hypothesis would be as illustrated in Fig. 6.7, in which the gradient just above the Moho is $+11$°C/km. – implying an almost symmetrical curve having tangents SQ and QR of equal length. Although this is only one of an infinite number of other curves that could be drawn tangential to SQ, representing a surface temperature gradient of 25°C/km, and reaching a temperature of 600°C at Moho, it can in no way be regarded as over-optimistic in so far as deductions of crustal heat content are concerned. If we further postulate that radioactivity accounts for 40% of the total heat flow at the surface (as against 33.3% for case (a) and 50% for case (b)) then the implications of Fig. 6.7 would be as shown in Fig. 6.8. For if q_{HL} were 59 mW/m^2 (as before) then the radioactive component would be 40% of 59, or 24 mW/m^2 and the

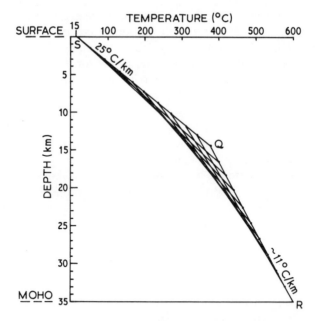

Fig. 6.7 Compromise temperature/depth curve for 'typical' non-thermal area.

conductive component would be 60% of 59, or 35 mW/m² at the surface. Hence the heat flow at the Moho would be 35–24, or 11 mW/m². The gradients shown in Fig. 6.8 have been taken from Fig. 6.7, and the resulting thermal conductivities λ_r at various depths would be as shown in Fig. 6.8.

This arbitrary theory requires, as with curve (2) of Fig. 6.4, that thermal conductivity falls with increasing depth; but the conductivity at the Moho would now be 43.4% of the conductivity at the surface (as against 35.7% in Figs. 6.4 and 6.5). It further suggests that the conductivity in the upper third of crust does not greatly vary.

Two possible 'typical' temperature/depth curves have now been deduced for non-thermal areas – curve 2 of Fig. 6.4 and Fig. 6.7. Neither can be expected to be correct, but either could be regarded as reasonable. However, Fig. 6.7 would appear to be not only the more cautious but also very slightly more probable; for it would indeed be coincidental if just at the present time the earth, on average, were neither gaining nor losing heat – and that is what is implied by curve 2 of Fig. 6.4, and it would, moreover, be a momentary transient condition. The general trend of modern opinion is that the earth is in fact in the process of slowly cooling; hence Fig. 6.7 would seem to be the rather more probable curve in the light of present beliefs. In any case, in so far as total crustal heat is concerned there is not a great deal of difference between the two curves, for that of Fig. 6.7 implies only 6.9% less stored crustal heat than curve 2 of Fig. 6.4; and we are now seeking only 'order-of-magnitude' estimates of heat mining potential.

Apart from these considerations the uncertainties of temperature prediction

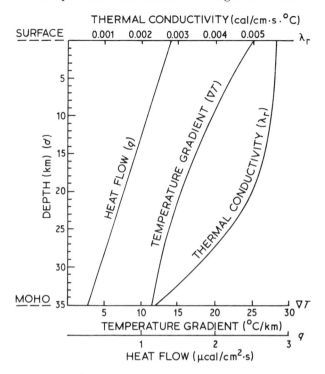

Fig. 6.8 The implications of the temperature/depth curve of Fig. 6.7.

increase with depth. In the upper layers of the crust in which we are likely to be interested for many years (say to 10 km depth) the curve of Fig. 6.7 is not inconsistent with actual measurements. Any inaccuracies are therefore of little practical importance in the short term future.

It is Fig. 6.7 that has been anticipated in Section 4.3 and Figs. 4.1 and 4.3.

6.1.3 Temperature/depth variation in thermal regions

It is emphasized that nothing in this chapter concerns hydrothermal fields, where convection may play an important part in the upward transport of heat from the depths of the crust to the surface. Here we are only concerned with those regions that rely on thermal conductivity for the maintenance of upward heat flow. Thus the HDR resource can be regarded as *conduction-dominated*.

In Section 6.1.2 an attempt has been made to establish an approximate norm for temperature/depth variation in non-thermal regions; that is, in most parts of the land masses. This was possible because typical, or average, conditions were assumed. When we come to examine thermal regions, however, nothing is typical or average. Temperature gradients, crustal thicknesses, Moho temperatures, heat flow patterns and surface heat flows can all depart widely from the norms assumed at the beginning of Section 6.1.2. Any kind of systematic analysis of temperature/depth relationships in thermal regions down to great depths is therefore impracticable, for the possible range of variations in the characteristics of thermal regions is infinite. Nevertheless, certain classes of temperature/depth curves have already been shown qualitatively in Fig. 4.6; and although the behavior of temperature at great depths differs widely between the three classes of curves shown in that figure, they all have two important features in common: the temperature gradients at and near the surface are higher than the norm of 25°C/km assumed for non-thermal regions in Section 6.1.2 and the curvature of the temperature/depth characteristic does not vary greatly within the depths that are likely to be of commercial interest for a very long time – say, up to 10 km. All that need here be noted about temperature/depth relationships in thermal regions is that hot rock is encountered closer to the surface than in non-thermal regions, and that the analysis of non-thermal regions as performed in Section 6.1.2 represents the *worst* conditions likely to be encountered from the commercial aspect.

6.2   Incidence of restricted, or accessible heat in terms of grade

It has been shown in Chapter 4, Section 4 that for the purpose of assessing the restricted quantity of heat that could theoretically be accessible to heat mining operations it is necessary to discount quite a large proportion of the total crustal heat (reckoned above 15°C), owing partly to the limited market for very low grade heat and partly to the need to provide an adequate margin for unavoidable temperature drops from the source of hot rock to the point of use (even though a fraction of this discounted heat may ultimately be put to good use and win a modest bonus). In a non-thermal region this discounted heat is represented by the cross-hatched vertical band SQUM in Fig. 6.9, in which the curve SR is a repetition of the curve deduced in Fig. 6.7. Although it cannot yet be stated with confidence how wide this band of discounted heat should be, it will here be assumed to be 70°C (i.e. $85° - 15°C$) in accordance with what has been said in Chapter 4, Section 4, apropos of countries having a substantial space heating market in winter; in which case the discounted heat would amount to 20.2% of the total crustal heat in a typical non-thermal region if the full thickness of the crust (35 km) is considered. This follows from the figures quoted in Table 4.1 $(11\,613 - 9\,268) \times 100/11\,613 = 20.2\%$. The temperature range of this vertical band of discounted heat is only $70/(600-15) \times 100$, or less than 12% of total temperature range within the crust. The disproportion between 12% and 20.2% shows that there is a preponderance of low grade heat by comparison with high grade heat in the crust; and this can further be seen from Fig. 6.10, which represents the integration of area QRU of Fig. 6.9 from left to right. (Fig. 4.5 was the integration of this curve from top to bottom). The way in which the curve of Fig. 6.10 bends over to the right at the top shows that only a small part of the crustal heat is at very high grade – approaching 600°C.

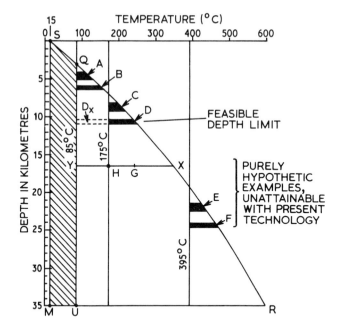

Fig. 6.9 The concept of heat extraction from deep crystal pockets. See note on p. 94.

What has been said above applies to the whole thickness of the crust in a typical non-thermal region. If consideration is confined only to the upper part of the crust that lies within the depth attainable with our present penetration capabilities, then the amount of low grade heat to be discounted will of course form a higher proportion of the total crustal heat to that depth. For example, for the upper 10 km of the crust in a non-thermal region the discounted heat would amount to about 53% of the total.

In a thermal region, not only is the restricted, or accessible, heat greater than in a non-thermal region to the same depth, but the proportion of discounted low grade heat is less. This can be seen from Fig. 4.8, in which the following proportions of discounted heat at 10 km depth would apply:

Non-thermal region $\quad \dfrac{\text{area SABK}}{\text{area SNK}} = \dfrac{1128-533}{1128} = 52.8\%$

Thermal region, as
illustrated $\quad \dfrac{\text{area SCBK}}{\text{area SDK}} = \dfrac{1860-1223}{1860} = 34.2\%$

It must be remembered that even in countries having a substantial space-heating market the minimum useful rock temperature, here assumed to be 85°C, could well have some other value that must be determined by experience; in hot countries it would certainly be about 40°C higher. The figures here

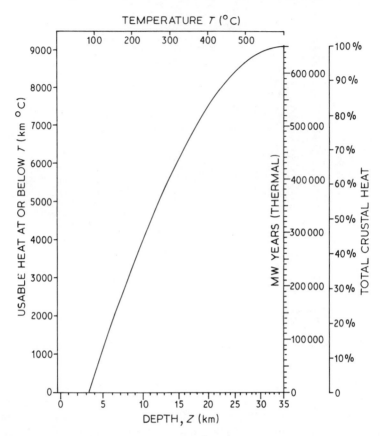

Fig. 6.10 Incidence of restricted crustal heat in terms of grade and depth.
Note: the use of a linear temperature scale requires the use of a non-linear depth scale.

quoted are by way of example only, they do not claim to be absolute truths. It must also be stressed that the restricted, or accessible, heat (as limited by the attainable depth and by the minimum useful rock temperature) is not the same thing as the *recoverable* heat; for the latter will be further limited by the various constraints discussed in Chapter 5.

6.3   The concept of cooled pockets of rock

Although it has been estimated in Chapter 4 that in a typical non-thermal area in a country having a substantial space-heating market the crust contains about 9268 km°C of restricted, or accessible heat per km^2 (see Table 4.1), it is most improbable that we shall ever succeed in winning more than a small fraction of this heat for commercial use, but even a small fraction could provide comparative abundance. In Chapter 4, Section 2, it was estimated that by locally cooling the whole crust through an average temperature drop of 1°C we could probably extract about 0.0752 quads (2500 MWyr(th)) of heat per square kilometre of horizontal area in non-thermal areas. It is of interest to see how this modest degree of cooling can be represented graphically in Fig. 6.9. As the estimated heat yield of 0.0752 quads was based on the assumption of a crust 35 km thick, this quantity of heat can equally be expressed as 35 km°C (see Section 4.3) and shown very simply as an area on Fig. 6.9. Examples of 35 km°C or 0.0752 quads of heat extraction per 1 km² of land area are to be seen in the areas A and B. A represents the cooling of a crustal layer 1 km in thickness through an average temperature drop of 35°C, B represents the cooling of a crustal layer 0.5 km in thickness through an average temperature drop of 70°C. The areas A and B represent low grade heat exploitation for such purposes as space heating or farming in climates with cold winters.

In the arguments that follow, we are forced to consider the notion of mining at depths below 10 km, which we have agreed to be impracticable; but this need not invalidate the logic of the arguments, which have a useful illustrative purpose. In practice, we would either have to be content with a larger degree of crustal cooling than here assumed, or we would have to mine thicker 'seams' of rock, if operations are to be restricted to within the 10 km depth limit. Figure 6.9 should therefore be regarded as to some extent fanciful, but the arguments that follow can be more clearly illustrated if the practical depth limitation of 10 km be temporarily ignored. When mining in a thermal area, the boundary curve SR of Fig. 6.9 would be displaced to the right, and some of the proposals that follow would then fall within practical range.

Clearly any 'pocket' of crust could be chosen such that the product of its thickness and temperature drop is 35 km°C, provided it falls within the shaded area QRU. Each of these areas is only 0.378%, or 1/265 part of the total heat represented by the area QRU. To win a pocket of heat equivalent to 35 km°C it is necessary to compromise between layer thickness and depth. It would probably be easier to confine the heat extraction process to a thin layer rather than a thick one; but to do this in a single pocket having an area of 35 km°C it would be necessary to drill more deeply to a point where a higher temperature drop is available without cooling the rock below 85°C. Alternatively we could exploit two or more thinner pockets at different depths, having a combined area of 35 km°C. With example A, Fig. 6.9, we must exploit rock at depths between about 4.2 and 5.2 km; in the case of B the depth range would be about 5.9 to 6.4 km. The corresponding mean initial rock temperatures (before exploitation) would be 120°C and 155°C respectively.

Although these temperatures would initially be much higher than the minimum of 85°C stipulated in Section 6.2 above, the pockets would be drawn down to an average temperature of 85°C by the end of the extraction process if the full quantity of 35 km°C is to be won. It must however be recognized that when an average rock temperature of 85°C has been attained, temperatures will vary quite widely within the mass of the cooled rock, being lower at the water/rock interfaces and higher at those points within the rock that are furthest away from the nearest water-bearing fissure. Moreover, rock temperatures and water temperatures near the upriser bore will remain hotter for longer than those near the injection bore. The problems of drawdown are far from simple and they will be examined in Chapter 7. For the purpose of interpreting Fig. 6.9, however, it will suffice to think in terms of average temperatures only.

Suppose now that our aim is to use extracted heat not for direct application, as with low grade heat, but for power generation. Although the minimum temperature at which power generation becomes commercially worthwhile cannot be precisely defined (it depends upon too many market factors) we shall here assume, by way of example only, that nothing less than 150°C is acceptable. At this temperature at turbine inlet and with dry saturated steam or an organic vapor, an overall generation efficiency of about 15% should be attainable as shown in Fig. 4.9. In order to produce a secondary fluid at 150°C it would be necessary to add a margin for temperature drops between the hot dry rock source and the turbine inlet. Assuming this to be 25°C, as before, the minimum required initial rock temperature would be 175°C.

To extract 35 km°C of heat for the purpose of generating power at about 15% overall efficiency it would therefore be necessary to cool pockets of rock such as shown at C and D in Fig. 6.9, of 1 km and 0.5 km thickness respectively – or other deeper or thinner layers subject to thermal drawdown constraints (see Chapter 7). In these examples the exploited pockets of hot rock would range in depth from 8.25 to 9.25 km in the case of area C and from 10.4 to 10.9 km in the case of area D. Again, both these areas represent only 1/265th part of the total accessible heat content of the crust; but being at high grade, it is necessary to penetrate more deeply than for areas A and B.

Finally, if we assume that our goal is to generate power at an efficiency of about 32%, which would require dry saturated steam at about 370°C at the turbine inlet, then after adding the same 25°C margin as before for covering temperature drops, it would be necessary to exploit rock at not less than 395°C. Theoretical examples of possible single exploitation pockets for this purpose are illustrated by E and F in that figure, representing layers of 1 km and 0.5 km thickness respectively; but this would have to remain a fantasy, in non-thermal regions, as the possibility of penetrating to 22 km and of working at such high temperatures must be ruled out.

In the examples C, D, E, and F it is clear that much more heat, though of lower grade, could be extracted at the depths of the pockets illustrated and used for other purposes; provided that the constraints on excessive cooling (see Chapter 5) and thermal drawdown (see Chapter 7) permit. For example, the pocket represented by the area D_x in Fig. 6.9 could be fully exploited if the extraction of 45 km°C additional heat can be tolerated. In this case, the average extent of cooling of the crustal column would the be (35 + 45)/35, or 2.29°C.

However, on no account should the available heat be degraded prematurely; that is to say, the area D_x should first be fully exploited for power generation before further heat is extracted for some other purpose, such as process or space heat. For if low grade exploitation were practiced simultaneously with high grade exploitation there would be excessive degradation, possibly with insufficient heat available for the full extent of power generation (in this example) and an excess of heat available for direct application. If two or more different forms of utilization are required, there are two possible courses of action that can be taken. Either the high grade application can be first exploited to the fullest possible extent, followed some years later by the lower grade exploitation; or the different applications may be simultaneously exploited from different pockets at different depths within the same vertical column of crust. For example, let it be assumed that it were possible to penetrate to a depth of 16.5 km in a non-thermal area and that practical constraints limit heat mining to an extraction level of 35 km°C, i.e. 1°C average crustal cooling. The first possible course of action would be represented by a single pocket XY at a depth of 16.5 km and thickness of 132 m. This thickness is derived so

Rock temperature at 16.5 km depth $= 350°C$
Minimum useful rock temperature $= 85°C$
Permissible temperature drop $= 265°C$
Permissible heat extraction level $= 35$ km°C
Thickness of pocket $= 35/265 \times 1000 =$ ca. 132 m

Such a pocket could, for example, be exploited for the following consecutive purposes:

(1) 13.87 km°C (XG), or 990 MWyr(th) for power generation at about 22% efficiency, with 220°C at turbine entry.

(2) 9.24 km°C (GH), or 660 MWyr(th) for power generation at about 15% efficiency, with 150°C at turbine entry.

(3) 11.89 km°C (HY), or 850 MWyr(th) for direct applications at temperatures ranging from 150°C to 60°C at the point of use.

Which totals 35.00 km°C (XY), or 2500 MWyr(th). There could of course be variants of this, with or without co-generation.

The alternative would be to create heat extraction pockets at different depths and to exploit them all simultaneously subject to economic constraints. For example:

(1) One pocket XG at about 16.5 km depth to be used for power generation at about 22% efficiency with turbine entry at 220°C.

(2) One pocket at about 10.6 km depth to be used for power generation at about 15% efficiency with turbine entry at 150°C.

(3) One pocket at about 7.2 km depth to be used for direct application purposes at temperatures ranging from 150°C to 60°C at the point of use.

With this alternative arrangement the residual heat from the deepest pocket (after its exhaustion for its intended purpose) could supplement the heat from the two upper pockets; while the residual heat from the middle pocket (after its exhaustion for its intended purpose) could supplement the heat from the uppermost pocket. All this has been foreshadowed in Chapter 5, Sections 1 and 2. The three pockets could be of the same thickness, 132 m. Alternatively they could be of different thickness if it were desired to obtain a different quantitative balance between the three classes of utilization, provided that the sum total of the extracted heat does not exceed 35 km°C.

Although Fig. 6.9 shows a final cut-off rock temperature of 85°C, below which no exploitation appears to be possible, this is not strictly true for the reasons given in Chapter 4, Section 4, which are here restated in rather different form:

(1) The area SQUM, though here written off as unusable, could in fact yield moderate amounts of useful heat of very low grade for such purposes as fish-breeding or de-icing. As the commercial value of this heat is very small in the perspective of heat mining, this small bonus is best ignored initially as being insignificant.

(2) The vertical line QU suggests that all points within an exploited pocket of rock would reach a temperature of 85°C simultaneously. What is really intended is that the temperature must not be allowed to fall below 85°C at those regions within the exploited pocket that are most remote from the nearest rock/water interface; so that the prescribed temperature drop of 25°C between the rock core and the point of use on the surface is maintained. When cooling has been effected to an extent that these strategic regions within the reservoir do in fact reach 85°C, the mass of rock surrounding them will be appreciably cooler. By restricting the rate of water circulation through the fractured rock the temperature drawdown could be reduced, thus enabling further heat to be extracted at a diminished rate. In effect, therefore, Fig. 6.9 is over-cautious in that the line QU could probably have been drawn a few degrees to the left, so that rather greater temperature ranges would be exploitable. For the sake of simplification, however, this point may be overlooked.

Figure 6.9 has of course been drawn for non-thermal regions, it therefore represents the most unfavorable conditions for heat mining operations – conditions that are not even likely to be tackled for many years, at least until adequate experience has been gained in thermal regions. The same principles that have been described in this chapter can be applied to thermal regions by redrawing Fig. 6.9 with the appropriate steeper gradients in the upper layers of the crust. It would then be found that desired temperatures could be reached at much shallower depths and that much thinner pockets would be capable of yielding the same quantities of heat as in non-thermal regions owing to the greater available temperature drops. Obviously, if heat is thought of in terms of 'kilometer-degrees,' then by raising the degrees, the kilometers (i.e. the pocket thickness) can be reduced.

Chapter 7

The basic principles of heat mining

7.1 Need to create permeability where none exists

Hydrothermal fields are exploitable because they contain a permeable layer of rock at depth capable of containing large volumes of water which serve to collect crustal heat by conduction, and convey it to the surface through wells sunk into the permeable layer, or aquifer, for the purpose. The subterranean hot water thus removed is replenished by the natural inflow of cooler waters from outside the 'field'; and these waters pick up further crustal heat by conduction from the permeable rock within, and bounding, the aquifer. In the course of time, the temperature of the water in the aquifer must inevitably and gradually fall, owing to the ingress of the cooler waters from outside. After extraction of the heat, at the surface, from the up-flowing bore water, the cooled (but often still quite hot) residual water is sometimes re-injected into the aquifer at suitable points with the combined aims of mitigating ground subsidence (see Chapter 5, Section 5.1), conveniently disposing of large quantitites of hot water which (if discharged into natural water courses) might give rise to toxic or thermal pollution, and perhaps of prolonging the life of the geothermal field by tempering the chilling effect of the inflowing cooler waters from outside the field. A highly stylized diagram of an exploited hydrothermal field is shown in Fig. 7.1.

Unfortunately, aquifers at suitable depth are comparatively rare. Most parts of the upper crust are almost impermeable except very close to the surface, where useful temperatures are not encountered. How then is it possible to extract stored heat from deep-seated hot rocks? Of the three possible heat transfer mechanisms – conduction, convection and radiation – it is clear that in virtually solid rock buried deep in the crust neither convection nor radiation could be of use in transferring the heat to the surface. Conduction too would be useless if reliance were placed on the natural rock overlying the deep-seated hot zone, for all rocks are very poor conductors of heat. The only practical way of extracting the heat is to imitate as closely as possible the provisions of nature in hydrothermal fields; that is to say, we must create permeability where none exists, and supply water to circulate through the artificial aquifer thus formed to convect heat away from the rock. Literally, we must mine the heat!

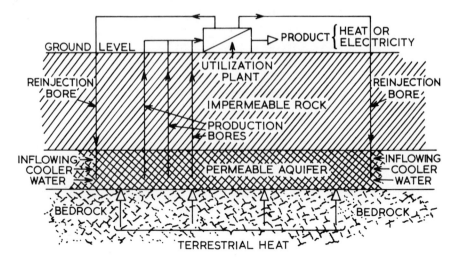

Fig. 7.1 Stylized model of an exploited hydrothermal field.

A natural aquifer is a system of inter-communicating voids, fissures and porous rocks through which water can percolate or flow. To imitate this we must artificially form or stimulate underground voids or fissures, and force water through them in order to collect the crustal heat by conduction from contact with the hot rocks and bring it to the surface. If convection currents within the reservoir can be used to aid the process, then so much the better.

It is conceivable that zones of moderately permeable HDR may occur in nature. If so, they could perhaps be exploited by means of a water drive technique similar to that employed in oil and gas production (Ramey *et al.*, 1973). The possibility will be discussed further in Chapter 10: for the present, attention is focused on HDR of low permabililty.

7.2 Ideal qualities of underground permeable labyrinths

What is required for heat mining is an underground reservoir or a structured *labyrinth*, formed in the heart of a mass of HDR and having the following ideal structural and hydraulic properties, the first four of which have been introduced in Chapter 5, Section 5.1.2.

(1) A very large contact area between the working fluid and the hot rock.
(2) Adequate conductive communication between the working fluid and a sufficient *volume* of hot rock to ensure a long life of the 'mine'.
(3) The provision of sufficient void volume within the hot rock mass to ensure that the working fluid exits the reservoir at as high a temperature as possible when circulated at a flow rate well above the economic minimum.

(4) A configuration of voids and fissures that offers a minimum impedance to the required flow of the working fluid, so as to minimize pumping energy while containing the circulated fluid to minimize losses to the non-productive regions outside the boundaries of the active reservoir.

(5) Wide inequalities of flow resistance through alternative 'parallel' flow paths should be avoided as far as possible, lest parts of the reservoir become over-cooled before other parts have had time to surrender more than a fraction of the heat available.

(6) Every point within an exploited pocket of HDR should be as close as possible to a water/rock interface, so that heat conduction paths shall nowhere be too long. This calls for a high degree of rock fracturing or fragmentation.

It is by no means easy – in fact, probably impossible – to create an underground labyrinth that simultaneously satisfies all these requirements. Properties (3) and (4) are perhaps the most difficult to achieve, because they imply that void space must be created within what was formerly a mass of solid, almost impermeable rock; and the creation of voids can only be achieved by the displacement of rock. At first sight it is not easy to see how rock, firmly embedded in other rock, could ever be displaced from the position it originally occupied. This point will be examined more closely later.

The art of reservoir formation more or less resolves itself into finding the best compromise between costs on the one hand and the achievement of as many as possible of the six desirable properties on the other hand.

7.3   The essence of the heat mining process

Assuming for the present that a suitable underground labyrinth or reservoir has been created in HDR possessing as far as possible the six desirable properties mentioned in Section 7.2 above, there are various possible ways in which bores could be situated in the labyrinth to optimize heat mining. These are shown diagrammatically in Fig. 7.2.

First there is the *open cycle* with cold feed. This is a 'once-through' system whereby cold water is pumped down an injection bore through the reservoir, in which the water extracts heat from the rock with which it is in contact. After being heated, it rises up the recovery bore, reaching the surface hot, and ready to be put to some practical use.

Then there is the *closed cycle* in which the same water is re-circulated repeatedly round a primary loop through the labyrinth, surrendering much of its collected heat to a secondary fluid circuit in a heat exchanger.

Finally there is the idealized *thermo-syphon cycle* in which pumping is dispensed with. For its establishment it is necessary that the difference in liquid density between the hot and cold legs of the wellbore/reservoir system shown in Fig. 7.2(c) must provide a sufficient pressure differential for the system to be

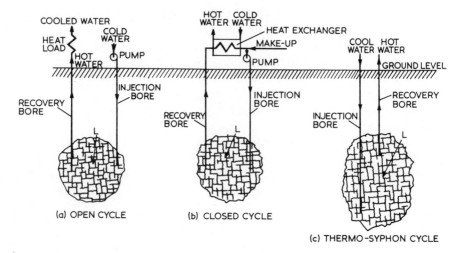

Fig. 7.2 Principle of heat mining.
L=labyrinth of fissures within a mass of impermeable rock.

self-pumped by the buoyant forces between the legs. This cycle would be possible only where a labyrinth has been formed at fairly great depth and if the pressure drop through the reservoir is not too large. The thermo-syphon or buoyantly-driven system could theoretically be used either in an open or closed cycle.

Even where the thermo-syphon effect is insufficient to dispense with pumping altogether, it can sometimes greatly reduce the required pumping effort for the other two cycles. An example of this will be given in Section 7.5 below.

The open cycle has the merit of avoiding the cost of a heat-exchanger: it also has the ability to flush out any solubles that may be present in the HDR, and this could sometimes have the benefit of enlarging very slightly the flow passages through the rock (points (3) and (4) of Section 7.2 above). But for flushing to be effective, the cold feed must be almost pure, otherwise the effect could be just the reverse – the clogging of the labyrinth passages with precipitated solids from the feed. The most practicable cycle on all counts will nearly always be the closed cycle, for the reason that the residual heat contained in the cooled (but still possibly quite warm) water discharged from the heat-exchanger is returned to the labyrinth and not lost; it is therefore likely to be the most thermally efficient of the possible cycles. A watch, however, must be kept on the composition of fluid circulated in the closed cycle. At first, while the more soluble ingredients of the hot rock are being dissolved by the circulating water, in-line water treatment and/or a fairly high rate of purge will probably be needed to prevent excessive concentrations. Later, it will probably be possible to reduce the rate of purge on this account; but as the rock itself is likely to have some degree of solubility it may never be possible to eliminate the need for periodic treatment or purging altogether. Most likely silica (SiO_2) will dissolve continuously from the quartz minerals commonly found in these rocks as long as the hot circulating water is under-saturated. With a closed cycle it may be that a state of silica saturation could be established, thus preventing further dissolution; but this would be conditional on the ability of the cooled fluid at surface level to remain in a state of super-saturation until it is reinjected into the reservoir. If the retention time at surface level is too long for this to be achieved, there would be a risk of precipitation in the heat-exchanger and/or the injection bore unless an adequate purge is maintained. Purging, necessary though it may be, is not conducive to efficient heat extraction as the purge fluid must be replaced by cold makeup.

Although we have assumed that HDR is of low-permeability, it is not absolutely devoid of all natural permeability, there will always be some small measure of permeability that will give rise to finite water loss. With the open cycle, this would mean that the mass flow rate of hot fluid rising up the recovery bore would be fractionally less than the cold water descending the injection bore. With the closed cycle, the same would apply, and in order to keep the primary circuit full, some make-up water would have to be supplied continuously, as shown in Fig. 7.2(b). If the water losses were found to be excessively high, the efficiency of the heat mining process would be lowered drastically. Most lost fluid presumably percolates or diffuses away from the active reservoir region into the outermost, inaccessible regions of the labyrinth or into the formation that contains it. Along these paths, it is mining heat that is not recovered. Water loss always creates a parasitic drag on the system's performance; it increases relative pumping costs, wastes heat, may induce seismicity, and increases operating costs for the makeup water required.

The rate of water loss that can be tolerated will depend on the induced seismic risks and the economics of any particular heat mining operation. In the early experimental work at Fenton Hill, USA, a loss of up to about 10% of injected rate was arbitrarily regarded as 'acceptable' in terms of costs and risk of induced seismicity. For more commercially-mature heat mining systems, this fraction will inevitably be lowered. At present, a figure less than 5% of the injected flow is regarded as a reasonable goal. Even at 5%, the losses are substantial. For a 40 to 80 kg/s circulation, this amounts to an annual loss of 63 to 126×10^6 litres. It may eventually be necessary to reduce losses even further. This may require downhole pumping as discussed in Section 10.5.4. Some small amount of water loss may be necessary to purge the system for chemical reasons.

7.4   Need for pressurized water circulation

The fluid passing into the reservoir through the injection bore must be pressurized for three major reasons. In the first place heat transfer from rock to water is far more efficient than from rock to steam. Secondly, the flow resistance to water through the restricting fissures of the labyrinth is far less for water than for steam. Furthermore, the viscosity of liquid water decreases strongly with rising temperature so it tends to flow preferentially through the hottest rather than the coolest parts of the reservoir. For these reasons it is essential that the water circulating through the reservoir is never allowed to boil. If we were to rely on the hydrostatic head alone, then after deducting the frictional pressure drop caused by water flow, the pressure might fall below the vapor pressure before clearing the labyrinth passages, and boiling would be liable to occur – even in non-thermal areas. In places of high temperature gradient, where HDR is encountered at shallow depths, the tendency to boil would be far greater. It will therefore usually be necessary for the circulating water to be injected at a pressure sufficiently high to ensure that after allowing for the high frictional head loss during its passage through the labyrinth a good margin remains over and above the vapor pressure appropriate to the temperature. This means that some degree of pressurization will be needed. The most vulnerable point in the circuit is likely to be at the outlet of the rising bore, where the temperature (and therefore the vapor pressure) is high but the advantage of hydrostatic pressure is not available for the suppression of boiling, Although the pressure at the pump inlet would be lower, the vapor pressure will there be very low because the fluid has by then been cooled in the heat-exchanger; so there would be no risk of boiling or cavitation at that point. This question of boiling suppression is examined more thoroughly in Section 7.5. Were any fluid other than water used for collecting the heat from the deep rocks, the same arguments would apply – even more so for low boiling point refrigerants.

There is a third reason for pressurizing the circulating water. After a labyrinthine system of fissures has been created within the HDR (by means yet to be described in Chapters 8 and 9) the fissures would tend to reclose, partly or completely, because of the elasticity of the rock and the *in situ* earth stress field. Some degree of dilation is desirable to hold the fissures slightly open so as to mase the passage of the heat-collecting fluid. There is a limit to the amount of dilation possible, as will be shown in Chapter 8, but some dilation pressure is desirable. For this reason also it is necessary to pump the circulating water through the labyrinth at fairly high pressure. Nature will sometimes contrive to aid the dilation process in places where the HDR reservoir is in a state of tectonic shear stress before the artificial fissures have been created; for a slight relative movement of the walls of a fissure may occur at the moment when the fissure forms. As the fissure walls will be somewhat rough and irregular, this relative movement may prevent them from precise mating when they reclose, so that narrow gaps will remain between the walls and the fracture is 'self-propped.' The breaking off of rock fragments while the walls are momentarily separated could also provide a further degree of 'propping' to keep the walls slightly apart. A net positive hydraulic pressure in excess of the reclosing pressure would nevertheless aid the circulation.

This reasoning assumes behavior that would allow reservoir pressurization without unstable fracture growth and consequent excessive water loss, as could be experienced if massive shear failure occurred at low overpressures. In these cases, one must rely on self-propping or artificial propping with downhole pumping to keep working pressures low.

There may be certain geologic environments where a two-phase, steam-water system is acceptable for heat extraction. For example, there are at least three quasi-HDR systems in operation currently at the Geysers field in California where downhole flashing occurs. The reservoirs are in the 230 to 260°C range and produce hot fluids at modestly acceptable flow rates.

7.5   Example to illustrate boiling suppression and thermo-syphon effect

Figure 7.3 shows a stylized diagram of a heat mining installation in a hyperthermal region where the mean temperature gradient is constant at 65°C/km., the injection bore is 4 km deep, the production bore is 3.5 km deep and the mean surface temperature is 15°C. The assumed constancy of the temperature gradient simplifies the calculations which follow and neglects the probable slight curvature of the temperature/depth curve.

Rock temperatures before exploitation:
   at C=(4×65)+15, or 275°C
   at D=(3.5×65)+15, or 242.5°C

Mean rock temperatures:
   from B to C=(15+275)/2, or 145°C
   from D to E=(242.5+15)/2, or 128.75°C

Water temperatures at reinjection point: assume the temperature at A and B to be 40°C

The mean rock temperature surrounding the injection bore is higher than the water within the bore by about 100°C on average. There will be some heat pick-up by the water during its descent from B to C. Assume this to be 10°C. Then the water temperature at C, at the point of entry into the reservoir, will be (40 + 10) or 50°C.

(a)   WATER TEMPERATURE AT ENTRY TO THE PRODUCTION BORE

In its passage through the reservoir from C to D the rock/water temperature differential will initially be (275 − 50), or 225°C. This will cause the water to heat up rapidly; but as it approaches D the differential will fall, partly because the water becomes hotter and partly because the rock becomes less hot at the

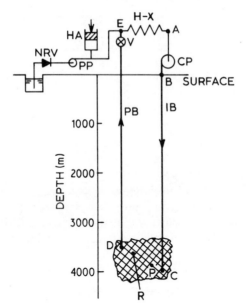

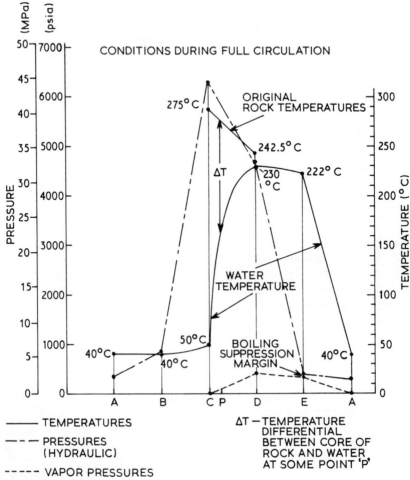

CONDITIONS DURING FULL CIRCULATION

275°C

ORIGINAL ROCK TEMPERATURES

242.5°C

230 °C

222° C

ΔT

WATER TEMPERATURE

BOILING SUPPRESSION MARGIN

50°C

40°C

40°C

40°C

——— TEMPERATURES

– – – PRESSURES (HYDRAULIC)

– · – · – VAPOR PRESSURES

ΔT — TEMPERATURE DIFFERENTIAL BETWEEN CORE OF ROCK AND WATER AT SOME POINT 'P'

shallower depth. Let it be assumed that a rock/water differential of 12.5°C is attained at D. Then the water temperature at D will be (242.5 − 12.5), or 230°C.

(b) WATER TEMPERATURE AT E

In the production bore there will be slightly less mean temperature differential across the bore walls than in the injection bore, and its length is rather less; so an 8°C temperature loss will be assumed during the passage of the water up the bore. The water temperature at E will therefore be (230 − 8), or 222°C. This 8°C drop is somewhat arbitrary. During early production periods and at lower production flows, larger temperature drops would be observed in the wellbore.

(c) TEMPERATURE DROP IN HEAT EXCHANGER AT SURFACE

This must be (222 −40), or 182°C.

(d) WATER TEMPERATURES, SUMMARY:

  at A   40°C
  at B   40°C
  at C   50°C
  at D   230°C
  at E   222°C
These values are plotted in Fig. 7.3.

(e) VAPOR PRESSURES DURING CIRCULATION

These will follow from the water temperatures; they are obtainable from the steam tables, as follows:
  at A   0.007 MPa (1.02 psia)
  at B   0.007 MPa (1.02 psia)
  at C   0.012 MPa (1.74 psia)
  at D   2.80 MPa (406 psia)
  at E   2.41 MPa (349.5 psia)
These values are plotted in Fig. 7.3.

Fig. 7.3 Illustration of heat mining operation.

Key:

NRV	non-return valve	CP	circulating pump
PP	pressurizing pump	IB	injection bore
HA	hydraulic accumulator	R	reservoir
V	throttle valve	PB	production bore
H–X	heat exchanger		

(f) HYDROSTATIC PRESSURES

The mean water temperature from B to C is 45°C, at which the density of water is 990 kg/m³. The static pressure from B to C will therefore be 4000 m × 9.8 m/s² × 990 kg/m³, or 38.8 MPa (5627 psi). The mean water temperature from D to E is 226°C, at which the density of water is 830 kg/m³. The hydrostatic pressure from D to E will therefore be 3500 × 9.8 × 830 or 28.48 MPa (4133 psi). Within the reservoir, from C to D, the mean water temperature will be more than the arithmetic mean of the temperatures at C and D (i.e. 140°C) because of the rapid temperature rise in the lower part of the reservoir. Assume the mean to be 160°C, at which the water density is 905 kg/m³. The hydrostatic pressure from C to D will therefore be 500 × 9.8 × 905, or 4.44 MPa (644 psi).

(g) FLOW FRICTION PRESSURE LOSSES DURING CIRCULATION

Assume these to be as follows:

from B to C 1.10 MPa (160 psi)
from C to D 6.90 MPa (1000 psi)
from D to E 0.97 MPa (140 psi)
from E to A 0.69 MPa (100 psi)

(h) DIFFERENTIAL HYDRAULIC PRESSURES RELATIVE TO THE PRESSURE AT A:

at A   = 0
at E   0+ 0.69   = 0.69   MPa (100 psi)
at D   0.69+28.48+0.97=30.14   MPa (4372 psi)
at C   30.14+ 4.44+6.90=41.48   MPa (6016 psi)
at B   41.48−38.83+1.10= 3.75   MPa (544 psi)

This means that the circulating pump must be capable of supplying differential pressure of 3.75 MPA (554 psi) or more.

(i) ACTUAL HYDRAULIC PRESSURES

If the pressure at A were zero (abs.), the hydraulic pressure at E would be only 0.69 MPa (100 psia), which is far less than the vapor pressure at that point, 2.41 MPa (349.5 psia). To prevent boiling, therefore, it would be necessary to raise the pressures throughout the circuit sufficiently to ensure that the hydraulic pressure at E exceeds the vapor pressure there by an adequate 'boiling suppression margin'. This could be done by means of a pressurizing pump and hydraulic accumulator, connected as shown in Fig. 7.3. Assume that this pump delivers at 2.63 MPa (365 psig, or 381 psia). This would provide a boiling suppression margin of 0.22 MPa (31.9 psi) which should be sufficient. As the pressure difference between E and A is 0.69 MPa, the pressure at A would now become 1.94 MPa (281 psia). The pressure at all other points in the circuit can therefore be deduced by adding 1.94 MPa (281 psi) to the differential pressures calculated above, so:

at A	1.94+	0=	1.94 MPa	(281 psia)
at B	1.94+	3.75=	5.69 MPa	(825 psia)
at C	1.94+	41.48=	43.42 MPa	(6297 psia)
at D	1.94+	30.14=	32.08 MPa	(4653 psia)
at E	1.94+	0.69=	2.63 MPa	(381 psia)

(j) NOTE CONCERNING PUMPS

The circulating pump must be capable of delivering the full required flow against a *differential* pressure of not less than 3.75 MPa (544 psi). The pressurizing pump need have only a small flow capacity, but must have a delivery pressure of not less than 2.52 MPa (365 psig). When the system has been pressurized this pump would have to supply only the system losses: it would serve as the makeup pump. All the surface equipment beyond the non-return valve must be capable of withstanding an internal pressure of not less than 2.52 MPa (365 psig), the pressure at E.

(k) STAGNANT CONDITIONS

Now suppose that the circulating pump fails. If this should happen early in the heat mining operations before there has been appreciable permanent temperature drawdown in the reservoir, the water temperatures at C and D would rise almost to the original rock temperatures, at which the vapor pressures would be 5.95 MPa (863 psia) at C and 3.52 MPa (510 psia) at D. The temperatures at A, B, and E would remain unaffected for some time, but would gradually fall. The hydraulic pressures would also change owing to the removal of the circulating pump head and of the frictional flow heads; they would attain the following values:

At B, E, and A	2.63 MPa/365 psig/381 psia
At C	41.46 MPa/5997 psig/6013 psia
At D	31.12 MPa/4497 psig/4513 psia

As these are all in excess of the local vapor pressures, boiling would be suppressed.

(l) AID PROVIDED BY THERMO-SYPHON ACTION

If the density of water were constant at all the temperatures experienced in the circuit, the hydrostatic pressure in the descending and ascending limbs (including the reservoir in the latter) would balance one another. The circulating pump would then have to provide the friction heads without any thermo-syphon or buoyancy aid, namely 1.10 + 6.90 + 0.97 + 0.69, or about 9.7 MPa (1400 psi). (This neglects friction in the circulating pump itself.) Thus the aid provided by the thermo-syphon effect is therefore 9.66 − 3.75, or about 5.9 MPa (855 psi). This is 60% of the total friction head.

(m) DILATION

This example makes no allowance for any further degree of pressurization that may be needed to distend the flow passages in the reservoir. Some further pressure capacity of the pressurization pump might well be needed for this purpose.

(n) EFFECT OF REDUCED FLOW IMPEDANCE WITHIN THE
      RESERVOIR

It is interesting to examine what would happen if the flow friction through the reservoir induced a much lower pressure drop than the 6.9 MPa (1000 psi) that has here been assumed. If, for example, the frictional pressure drop were only 3.15 MPa (456 psi), the differential pressures relative to A would be as follows:

At A		=0
At E	0+ 0.69	=0.69 MPa (100 psi)
At D	0.69+28.48+0.97	=30.14 MPa (4366 psi)
At C	30.14+ 4.44+3.14	=37.72 MPa (5470 psi)
At B	37.73−38.83+1.10	=0

Thus the pressure differential from A to B would have disappeared, no circulating pump would be required, and the thermo-syphon effect alone would ensure circulation. Any further reduction in frictional pressure loss through the reservoir would necessitate throttling the flow to regulate the rate of thermal drawdown.

(o) GENERAL COMMENTS ON THE EXAMPLE

It is emphasized that many of the figures in the foregoing calculations have been arbitrarily chosen and that others have been approximations. The example should nevertheless serve to illustrate the sorts of conditions that might be expected in a forced circulation system through HDR, and the sorts of precautions that must be taken to suppress boiling. After exploitation has persisted for some time there will be changes in the temperatures at various points in the circuit. The rock surrounding the injection bore will become cooled, while the rock surrounding the production bore will become heated. These effects would improve the efficiency of heat extraction by increasing the rock/water temperature differential near C and by reducing the temperature loss in the rising bore. Another factor for efficiency improvement would be a reduction in the flow impedance through the reservoir due to thermal stress effects, necessitating less pumping effort. This could be achieved by using a variable speed circulating pump, or even by changing the pump to a less powerful one when conditions permit. A factor operating against these improvement tendencies would be the increasing rate of reservoir drawdown at higher mass flows.

7.6   Influence of temperature gradient on boiling suppression.

Figure 7.4 shows the saturation pressure for water plotted against temperature. It will be noted how the vapor pressure rises rapidly with temperature. For example, a temperature rise of 185°C above a surface temperature of 15°C would give a rock temperature of 200°C, at which the vapor pressure is about 1.6 MPa (230 psia). An increase of only 50% in the gradient would raise the rock temperature to (1.5 × 185) + 15, or 292.5°C, at which the vapor pressure is about 7.70 MPa (1120 psia). Thus a mere 50% increase in gradient would result, in this case, in an almost five-fold increase in the vapor pressure corresponding to the rock temperature at the same depth. Although the temperature at the most vulnerable point in the water circuit of a heat mining

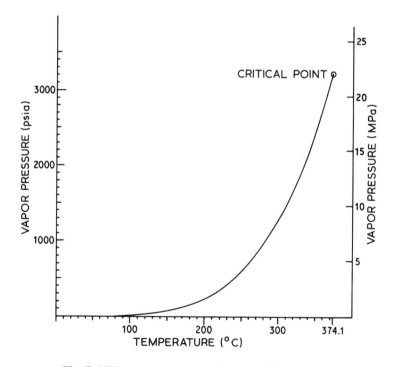

Fig. 7.4 Water vapor pressure in terms of temperature.

operation is not the same as that of the rock at the base of the production bore (e.g. 222°C at E as compared with 242.5°C in the rock at D in the example given in Section 7.5 above) it is nevertheless greatly influenced by it. Hence the necessary degree of pressurization will rise rapidly with the gradient in mined regions. Moreover in a region of high gradient any desired temperature will be reached at a shallower depth than in a region of low gradient; so the hydrostatic pressures involved will be less. Hence there may even be a risk of boiling within the reservoir itself in high gradient regions unless the system is adequately pressurized.

7.7 Temperatures at rock/water interfaces

It will be seen from Fig. 7.3 that substantial and varying differences occur between the original rock temperatures and the temperatures of the water percolating through the reservoir.

It is instructive to study the way in which heat is transferred from the rock to the water, and how the temperature varies from place to place within the body of the rock. This can be understood by referring to Fig. 7.5, which shows a stylized section through a mass of homogeneous hot rock at an initial temperature of T_a and having two parallel plane fissures, each carrying a flow of water $\dot{q}_1$. At first it is necessary to consider only the slab of rock of undetermined depth (measured perpendicular to the paper), neglecting what happens to the extreme right and left of the sketch. As cool water starts to flow through the fissures, the rock temperature at the rock/water interface will fall rapidly. After a time this temperature will have fallen to some value T_b (case I); but within the body of the rock the degree of cooling will be progressively less as the distance from the fissures increases. There will be very high temperature gradients through the 'skin' of rock near the fissures, but the gradient (that is the slope of the temperature curve) will decline with distance from the fissures. The temperatures through the slab of rock will vary somewhat as shown by the curve BCB', with the temperature at C infinitesimally below its original temperature T_a. The shape of the curve BCB' will be such that the shaded area is a measure of the heat extracted since the water began to flow. (This area will be measured in m°C, just as it was shown in Chapter 4 that km°C was also a measure of heat).

Now let it be assumed that the water flow ceases. The thermal gradients through the slab of rock will assure for some time a continuing outflow of heat toward the fissure walls at the cost of heat stored in the core of the rock. Equilibrium will be attained when the uniform temperature throughout the rock is T_d, such that the area AA'D'D is equal to the shaded area. The fissure wall temperature will by then have risen considerably, while the core temperature will have fallen only slightly.

After heat mining has been practiced for a very long time, perhaps several years, the whole of the rock slab will have become cooled, so that the

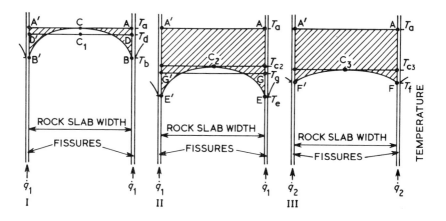

Fig. 7.5 Pattern of temperature distribution within a slab of hot rock between water-bearing fissures.

I Early
II Later
III As for II but with reduced water flow $\dot{q}_2 < \dot{q}_1$

temperature at the core and at the fissure walls will have fallen to T_{c2} and T_e respectively (case II). The total heat extracted from the rock slab will now be proportional to the area AA′E′C₂E. If the water flow were then interrupted, the temperatures throughout the rock slab would gradually approach an equilibrium value of T_g such that the area AA′G′G is the same as AA′E′C₂E. But now suppose that the water flow is not altogether stopped, but merely reduced from $\dot{q}_1$ to $\dot{q}_2$ (case III). The effect would be that the fissure walls would warm up from T_e to T_f; while at the same time the temperature at the core of the rock slab would fall slightly to T_{c3}, as lower gradients would now exist throughout the slab. The area AA′F′C₃F will be the same as A′E′C₂E, as the 'historical' amount of total heat removed will remain the same.

It is here advisable to consider the concept of *drawdown*, which can be represented by the difference between the original rock temperature and the produced fluid temperature. Drawdown can be of two kinds. The *dynamic* or *flowing drawdown* is illustrated in Fig. 7.5 by $(T_a - T_b)$ in case I and by $(T_a - T_e)$ in case II. The quantity of extracted heat is shown by the cross-hatched areas in Fig. 7.5. On all practical time scales, this depletion is 'permanent' and results in a *stagnant drawdown* as illustrated by $(T_a - T_d)$ in case I and by $(T_a - T_g)$ in case II. If the system is shut down, temperatures will tend to recover toward this stagnant limit (T_d or T_g). With reduced water flow, the flowing drawdown would be diminished from $(T_a - T_e)$ to $(T_a - T_f)$ as in case III.

The actual field definition of drawdown has been simplified in terms of measured fluid temperatures in the production well and can be expressed in terms of the produced fluid temperature at a specific time t ($T(t=t)$ in °C), the initial fluid temperature ($T(t=0)$ in °C) and the temperature of the reinjected fluid (T_{rej} in °C):

$$\frac{T(t=t)-T_{rej}}{T(t=0)-T_{rej}}$$

This graphical description of the way in which heat is transferred from rock to water gives a simplified picture of what is in fact a more complicated process (see Chapter 10). In the first place the direction of water flow through the fissures should really be perpendicular to the plane of the paper, and not upwards, otherwise the temperature could not be the same at all points in the fissure section shown. Secondly, heat will also be extracted from the outer regions of rock to the right and left of the two fissures shown. Here the processes will be similar to those illustrated, but the core of the rock is at an infinite distance, so that temperature attenuation would be more gradual, and assymptotic to T_a rather than attaining a maximum along the C axis. Thirdly, the slab of cooled rock is a concept that implies a finite depth at right angles to the paper, whereas in fact there would be lumps of various sizes within the reservoir – some in the form of 'island' blocks and some connected to an infinite matrix of rock beyond the confines of the reservoir, thus giving a totally three-dimensional aspect to flow in the reservoir. Lastly, though extracted heat has been shown as an area, it would in fact be represented by volumes depending on the slab dimensions in depth. Despite these limitations, the heat transfer process is fairly rationally illustrated in Fig. 7.5.

Thus the temperature at the rock/water interface will be a function both of the flow $\dot{q}$ through the fissure and also of the time that has elapsed since the beginning of exploitation. It is emphasized that the temperatures, shown qualitatively rather than quantitatively in Fig. 7.5, are *in the rock* at the interface: the water temperatures at the same interface would be very slightly lower so as to allow for the heat transfer from the rock surface to the water.

7.8 Ability to match supply and demand

One of the minor constraints in using *hydrothermal* energy for power generation is its relative inability to match energy supply to the load factor of the demand: it is more easily and economically exploited at constant load. This is of little consequence if the hydrothermal system is exploited in parallel with some other energy source capable of absorbing the fluctuations of demand, so that the hydrothermal energy may take care of the base load only. With heat mining the question of long term matching of supply and demand has already been examined in Chapter 5; but for hour-to-hour matching, it might be useful if the supply could be matched exactly with the demand, simply by varying the quantity of primary circulating water through the reservoir. When the demand is reduced, energy could then be conserved and the life of an exploited HDR zone thus lengthened. With any form of capital investment in engineering works a high utilization factor is of course desirable, but in the case of hydrothermal energy there are technical difficulties of accepting a fluctuating demand without wastage: it is necessary to consider whether any such constraints would apply to a heat mining project. In view of what has been said in Section 7.7 above, it must be accepted that at times of light load and reduced circulation rates the underground water temperatures will rise; and at the same time the reduction in frictional heads would have the effect of changing the net hydraulic pressures at all points in the circuit.

Let it be supposed, as a first approximation, that the circulation rate is halved and that the frictional pressure drops are everywhere quartered – since they are approximately proportional to the square of the flow rate. Referring again to Fig. 7.3, if the heat-exchanger still extracts all the heat that is practically available, and rejects at 40°C at A, then the temperatures throughout the circuit would not be greatly affected, as will now be shown.

Owing to the slower passage of the water through the reservoir, it might be expected that the water temperature at D reaches about 235°C instead of 230°C, and that owing to the doubled time of ascent from D to E the temperature loss rises from 8°C (at full flow) to, say 15°C – resulting in a water temperature of 220°C at E. In the injection bore the heat pick-up might rise from 10°C (at full flow) to perhaps 18°C, so that the water temperature at C would be 58°C. Thus the average water temperature in the production bore would rise from 226°C at full flow to 227.5°C, while the average water temperature in the injection bore would rise from 45°C (at full flow) to 49°C. It is improbable that these small temperature changes would cause any serious problems to the bore casings or grout. As to pressures, the hydrostatic pressures would be affected only very slightly: 38.8 MPa (5626 psi) from B to C, 28.5 MPa (4133 psi) from D to E, and 4.4 MPa (638 psi) from C to D. Thermosyphon conditions would prevail, so that the circulating pump could be bypassed, and it would be necessary to throttle the flow before the point E. If the pressure drop across the throttle valve were 3.5 MPa (508 psi), the differential pressures throughout the circuit would be approximately as follows:

at A		0	
at E	0 +0.69/4	=0.17 MPa (25 psi)	
at D	0.17+3.48+28.5+0.97/4=32.4	MPa (4699 psi)	
at C	32.4 +4.4 +6.9/4	=38.5	MPa (5583 psi)
at B	38.5 −38.8+1.1/4	=0	

The vapor pressures at D and E would be 3.06 and 2.32 MPa (444.4 and 336.5 psia) respectively, so that it would be necessary to pressurize the system by no less than 2.32 MPa (336.5 psia) plus an adequate boiling suppression margin at E. As the delivery pressure of the pressurizing pump is 2.62 MPa (380 psia), this would provide a boiling suppression margin of 0.3 MPa (43.5 psi) – some 0.09 MPa (13 psi) more than at full flow conditions. The system should be both safe and economical at half load.

Now although water friction in pipes varies more or less in proportion to the square of the flow rate, experience has shown that in a fracture reservoir flow friction is more nearly directly proportional to the flow rate, as laminar flow conditions are approached. Thus in the above example it would be more realistic to assume 3.45 MPa (500 psi) pressure drop through the reservoir from C to D at half flow. This would enable the throttling to be reduced from 3.48 to 1.79 MPa (504 to 260 psi) in order to ensure half flow under thermo-syphon conditions. With the same pressurizing pump, boiling would still be adequately suppressed at all points in the circuit.

It has thus been demonstrated that neither at no flow nor at half flow would there be any risk of boiling; nor would the bores be subjected to any drastic changes of temperature. Matching the heat supply to the demand should therefore be quite practicable with heat mining, though for economic reasons it would of course be preferable to operate the circuit as nearly as possible to full flow continuously.

There would be no risk of cavitation at the entry to the circulating pump, as the pressure at that point would never be less than 1.94 MPa (281 psia) under any flow conditions.

7.9   Consideration of alternative working fluids

In the development of hydrothermal energy for power generation much attention has been devoted to binary cycles in which use is made of low boiling point organic liquids, such as isobutane, propane and halogenated hydrocarbons (freons), in place of water as the working fluid in a turbine cycle. Such fluids have the advantage of producing high vapor pressures at moderate temperatures so that very compact self-starting turbines can be used, and the occurence of sub-atmospheric pressures at any point in the cycle can be avoided. They can also provide useful pressures at much lower temperatures than in the case of water/steam, so that heat mining operations at relatively shallow depths could be applied to power generation.

Binary fluids could of course, be used quite easily in the closed secondary circuit of a heat mining project. The question arises as to whether binary fluids could perhaps be used in the primary circuit and used as the sole working fluid, instead of water. Could such fluids perhaps be injected directly into the HDR, forced through the reservoir, flashed into vapor at or near the surface, passed successively through a turbine, condenser and feed pump, and reinjected into the HDR? In theory the idea has attractions, but it is improbable that it would ever be put into practice. Binary fluids are expensive, and the combined volume of the two bores and the reservoir labyrinth would involve a high investment in providing a large working stock of the fluid. Then again, unless a labyrinth's boundaries were totally impermeable, or almost totally loss-free, further cost would be involved in making good the fluid losses. Moreover, it is at shallow depths (for which the use of binary fluids would theoretically be the most suitable) that system losses are at their greatest. Finally, since binary fluids are usually both volatile and toxic, even the slightest leakage from a labyrinth could perhaps lead to contamination of farming environments or water supplies.

7.10   Unavoidable temperature drops

The question of unavoidable temperature drops has been raised already in Chapter 4, Section 4 and again in Chapter 6, Sections 2 and 3. For the simplification of the arguments raised in those chapters an arbitrary total drop of 25°C was assumed to occur between the unexploited rock and the point of use of the extracted heat. A glance at Fig. 7.3 will show that this problem is really far more complex than has hitherto been tacitly assumed, for the temperature drop from the rock to the water (ΔT) has been shown in that figure as (275 − 50), or 225°C at C and as (242.5 − 230), or 12.5°C at D. Obviously it is the temperature drop at the terminal point of the labyrinth that is of particular significance – i.e. at the point where the heated water enters the rising production bore. In the example of Fig. 7.3 it was shown how the water in the reservoir would progress from one hot spot C to a less hot spot D, and that the water temperature would rise throughout its path between these points.

In the upper graph of Fig. 7.6 is represented a reservoir and two bores similar to the system shown in Fig. 7.3. The water entering the reservoir at C would splay out in a divergent flow pattern and would later reconverge on point D. The broken lines in the right hand sketches of Fig. 7.6 represent the mean water flow paths. In the upper left hand diagram are shown the initial rock temperatures T_{rC} and T_{rD} at points C and D respectively, and a series of water temperature patterns representing increasing rates of flow from 1 (min.) to 4 (max.) at a particular point in the reservoir's lifetime. For any one curve the difference between the initial rock temperature T_{rP} at point P and the water temperature at that point is shown as ΔT. As ΔT declines from C to D, so the rate of heat transfer (i.e. the slope of the water temperature curve) also declines; hence the curvatures of the four water temperature characteristics. At a high rate of water flow the temperature curve would be only moderately curved, but at decreasing flow rates the curvature becomes more pronounced, until at a very slow flow rate (1) the water may even reach higher temperatures at some points than that attained at D; i.e. the water would actually be giving back some of its heat to the rock near the end of its passage through the reservoir as indicated by the crossing of the water temperature and initial rock temperature profiles. With falling flow rates, not only does the curvature of the water temperature characteristic increase, but also the finally attained temperature at D rises. The temperature profile will generally become more s-shaped or sigmoidal as time proceeds in the reservoir's life. This is characteristic of the depletion of heat from the entrance region of the reservoir and asymptotic approach of fluid temperatures to the conditions in the exit region.

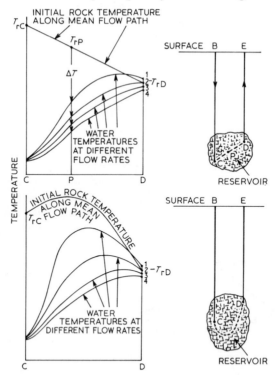

Fig. 7.6 The pattern of water-heating in a reservoir at an intermediate time in the life of the reservoir.

– – – mean flow paths
slowest flow rate 1
highest flow rate 4

The lower graph of Fig. 7.6 shows a similar qualitative analysis for a reservoir that extends to much greater depths below the injection point at C. Here the mean flow path would be somewhat as shown which includes the effects of buoyancy, and the rock temperature curve would usually rise to a higher level than T_{rC} before declining to T_{rD}. In such a reservoir the water temperature curves except at high flow rates, would probably attain maximum values before reaching point D.

Despite the lower temperatures at D, high flow rates would extract heat more rapidly than low flow rates, for the temperature loss would be more than compensated by the greater flow. Moreover, the slight fall in temperature during the water's ascent through the production bore from D to E would be even less at high flow rates.

Although reversed flow, from D to C would conform with the normal heat exchanger practice of counter-flow, the fact that pumping would have to work in opposition to the thermo-syphon or buoyancy effects would render this impractical and could in fact lead to enhanced short-circuiting as the natural tendency would be for the cooler more dense fluid to fall directly toward the lower outlet at C.

Although Fig. 7.6 gives a broad notion of how the water temperature changes during its passage through the rock from the injection bore to the production bore, it fails to take time into consideration; for the pattern of the water temperature from C to D will change with time as heat is extracted.

The initial rock temperatures (T_r) shown in Fig. 7.6 represent the natural conditions before any heat has been extracted by mining (i.e. they correspond to T_a in Fig. 7.5); but as shown in Fig. 7.5, the rock temperatures at the rock/water interface will be appreciably lower – in fact they will be very nearly the same as that of the water. At first, when cold water enters the virgin hot rock, the temperature rise near the base of the injection bore will be very rapid, especially at low flow rates; but as the rock heat is gradually depleted, the temperature pattern will shift, somewhat in the manner shown in Fig. 7.7.

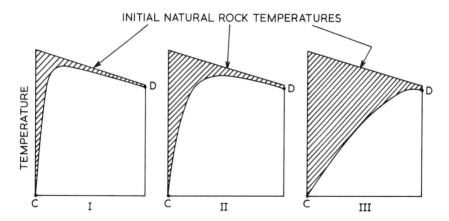

Fig. 7.7 The pattern of rock/water interface temperatures with the passage of time.
  I Early
 II Later
III Much later
C The base of the injection bore (see Fig. 7.6).
D The base of the production bore (see Fig. 7.6).
It is assumed that the interface temperatures shown by the curves CD are the same for the rock and for the water. The shaded areas are a measure of the mined heat.

In Fig. 7.7, it is assumed that the rock and water temperatures at the interface are the same, although there must always be a small temperature drop at the interface 'skin' to ensure heat flow from the rock to the water. A low flow rate has been chosen for Fig. 7.7 because the change of the temperature pattern with time is more marked than for a high flow rate. The passage of time will have the effect of shifting the point of maximum interface temperature to the right (towards D) and also of reducing that maximum temperature. (With high flow rates the maximum interface temperature is attained only at D, so no shift to the right could occur; but the curvature of the temperature profile would flatten with the passage of time). The cross-hatched area in Fig. 7.7 could loosely be regarded as a measure of the heat mined, since its area can be measured in m°C which, as pointed out earlier, is a unit of heat. The progression of the sigmoidal 'cold wave' or 'front' as time increases is easily seen.

One other point needs to be brought out. Fig. 7.6 and 7.7 indicate that the produced fluid temperature is always below the initial rock temperature at the outlet point. In cases where the vertical extent of the reservoir from C to D is very large or where the gradient is extremely steep, it is very likely that the produced fluid temperature will exceed the initial rock values at the outlet as heat will be mined from deeper, hotter regions of the reservoir and is temporarily deposited in shallower regions. Eventually as production continues, this heat too will be mined by the injected fluid. At the opposite extreme, where gradients are low or the reservoir is effectively horizontal from C to D, the produced fluid temperature will asymptotically approach the initial rock temperature.

A point of interest is the differential between T_{rD} and the water temperature at D. In Section 7.5 above, an arbitrary value of 12.5°C was assumed; but this was purely illustrative. Obviously the figure would greatly depend on the attainable flow rate, and as this rate would be very low with existing technology the upper water temperature curves (1) of Fig. 7.6 are probably more realistic than the lower ones. Although it is not possible to be precise, a suggested breakdown of the 25°C differential assumed in Chapters 4 and 6 might be somewhat as follows:

Rock to fissure wall differential = 12.0°C
Rock to water interface = 0.5°C
Temperature loss in upriser bore = 5.5°C
Terminal temperature difference across the surface heat-exchanger = 7.0°C
Total = 25.5°C

It may be noted that the suggested figure of 5.5°C for the temperature loss in the upriser bore is lower than the 7.5°C assumed for full flow in Section 7.5 above. The figure would almost certainly fall with time, as the rock surrounding the bore warms up; so that even if a fairly high drop has to be tolerated at first, a lower figure would be representative of established conditions at high flow rates.

In Chapter 6 various rock temperatures were specified as the minima necessary for certain applications: direct use, low efficiency power generation, and high efficiency power generation. These temperatures would have to apply to the rock at the depth of the fluid entry into the upriser bore; the reinjection bore would have to penetrate deeper (by 500 m in the example of Section 7.5 above).

7.11 Basic problems to be overcome

For heat mining to become commercially practicable, three main obstacles must be mastered:

(1) The art of rock fracturing, whereby artificial permeability may be imparted to HDRs, must be perfected.
(2) The art of creating a reservoir with maximum heat extraction efficiency from large volumes of rock must also be perfected.
(3) The art of very deep penetration at commercially acceptable cost must be greatly improved if high grade heat is to be won in non-thermal areas.

The first of these obstacles is already well on the way towards solution, as will be explained in Chapter 9. The second obstacle is being addressed in several extensive field programs discussed in Chapters 9 and 10. There is no reason to doubt that technological advances will continue to be made that will steadily improve the economics of endowing HDR with optimally designed artificial permeability. As to the third obstacle, this is less urgent; but some progress has already been made and further improvements in drilling will doubtless be achieved in time (see Chapter 11 for more details).

Part II

The present state of the art

Chapter 8

The theory of rock fracturing

8.1 General

In Chapter 7, Section 1, it was shown that the prime requirement for heat mining in HDR is to create artificial permeability where none existed before or, more realistically, to enhance the natural permeability of deep rock masses to sustain fluid circulation at useful levels for heat extraction. This can only be achieved by *stimulating* the rock in such a way that a flow path (or paths) is formed between two or more bores in the mass of hot rock to allow the passage of a working fluid such as water. Low permeability zones in hot rock masses may be in 'non-geothermal' areas or in unproductive regions in a hydrothermal field. In the latter cases stimulation could perhaps be an effective way of converting these regions into producing zones.

In the oil and gas mining industries, induced rock fracturing has been successfully practiced since the late 1940s as a means of boosting productivity; and it seems very probable that hydrothermal fields may also benefit from the judicious application of this technology. Unfortunately, the results of a major experimental program in the US have been disappointing (Hanold and Morris, 1982). The simplified fracturing theories described in this chapter are mainly directed towards the concept of extracting heat from hot dry rock that is initially devoid of any significant permeability and is unjointed. These theories do not apply in detail to rock masses with pronounced joint patterns that are not aligned in principal stress directions. We are defining a joint as a discontinuity (or crack) that can only transmit shear stress under compressive load – it has no tensile strength.

Fracturing can be effected artificially in intact rock by applying tensile stresses at the right place either hydraulically, explosively or by other means which will be described in detail in Chapter 9. In order to understand how this can be done it is first necessary to study the stresses to which rocks are subjected in nature.

Distinction must be made between 'competent' and 'incompetent' rock. Although the difference is easy to conceive it is rather hard to define succinctly. *Incompetent* rocks are those which have little or no resistance to deformation if unsupported – e.g. sand or clay. These materials have low shear strengths and can carry compressive loads only when confined. It would not be possible to fracture incompetent rocks. Hot magma within plutons or volcanoes also should be regarded as incompetent. *Competent* rocks have a natural rigidity that enables them to carry substantial stresses without collapsing – e.g. granite, limestone, ignimbrite, basalt, etc. A vertical, or even an overhanging cliff of competent rock could survive, whereas a heap of unsupported incompetent rock would slump and take up its natural angle of repose. It would be difficult to conceive of heat mining in anything but basically competent rock capable of being subjected to compressive, tensile and shear stresses with enough intrinsic strength to permit the rock fabric to span joint openings and maintain an artificial dilation sufficient to permit fluid circulation. The presence of weakly sealed joints, for example, would not render a rock mass incompetent. In order to study the stimulation process it is useful for illustrative purposes to consider first the very simplified, idealized condition of homogeneous, isotropic rock devoid of natural cracks. It is emphasized, however, that such ideal conditions do not occur in nature, where *all* rock formations feature some degree of natural cracking and jointing. Thus, the practical art of reservoir stimulation involves not only the cracking of virgin matrix rock but also the intelligent exploitation of natural cracks and joints.

8.2 Rocks in repose

The simplest condition in which a part of the earth's crust might exist would be one where no appreciable forces act upon the rock other than gravity. The rock would then be in a state of repose, and all of it would be under compressive stress. The intensity of this stress will depend upon the depth and density of the rock. In Fig. 8.1 the compressive stress in the rock at point P, measured vertically, is represented by the vector S_L. This stress is known as the *lithostatic pressure* at depth d, and is equal to the weight of the overburden all the way from the ground surface to P, per unit of horizontal area. If the mean density of the overburden is assumed to be 2700 kg/m^3 then with $g = 9.8$m/s^2

$$S_L = 26\ d\ \text{(in MPa)} \qquad \text{where } d \text{ has units of km} \qquad (8.1)$$

Under elastic deformation conditions, in homogeneous, isotropic rock in repose, the horizontal compressive stress at P is represented in Fig. 8.1 by the vector S_H, having a value

$$S_H = \frac{S_L v}{1-v} \qquad (8.2)$$

where v is Poisson's ratio – a function of the elasticity of the material. In practice, equation (8.1) is applicable in simple, lightly loaded soils but not in deep rock masses (Hoek and Brown, 1980).

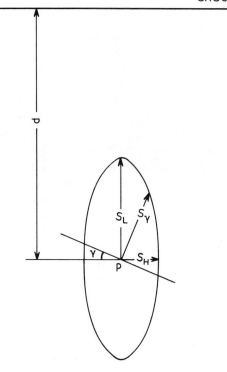

Fig. 8.1 Ellipse of stress for rock in repose.
d=depth in km (feet)
S_L=lithostatic pressure in Pa (lb/in²)
S_H=horizontal compression in Pa (lb/in²)
$S\gamma$=compression in direction γ in Pa (lb/in²)
ν=Poisson's ratio

$$\frac{S_H}{S_L}=\frac{\nu}{1-\nu}$$

As drawn, ν=0.27

The lithostatic stress S_L, though exerted vertically, will of course be borne by the horizontal plane passing through P. Strictly speaking this implies that the horizontal components of strain are zero and that all volumetric strain is exerted in the vertical direction. In homogeneous, isotropic rock in repose, S_H would be the same across all vertical planes passing through P, though it should be remembered that there is no such thing as perfectly homogeneous rock. The results of many field measurements show that the natural inhomogeneity of the rock and the presence of secondary forces in addition to gravity invalidate the concept of 'rock in repose'. Furthermore, this nonhomogeneity affects the interpretation of various parameters such as Poisson's ratio, which for a linearly elastic material is the negative ratio of lateral expansion strain to longitudinal contraction strain where the strains are induced by a uniaxial stress. In addition, the normal defining relations as cited by Jaeger and Cook (1979), pp. 109–111 (1979) for a linear elastic material no longer apply.

8.3   Tectonic stresses

Clearly, conditions will never be quite as simple as suggested by Fig. 8.1. Quite apart from the complication of 'creep' and long term viscoelastic behavior, no measurements of deep stresses have shown the earth's crust to be in a state of repose; for superimposed upon the gravitational forces there will be tectonic forces that will complicate the stress pattern and cause it to depart from the simple concept of an ellipse of stress symmetrically placed about a vertical major axis and symmetrical also in azimuth. The effects of these superimposed tectonic forces would be to distort the scale and the symmetry in azimuth of the ellipse of stress and also to tilt it so that its major axis would be inclined to the vertical at some angle which, in extreme cases, could approach 90°.

In volcanic areas many complex stresses can be induced by severely anisotropic forces at faults, or by thermal influences arising from volcanic activities; but even at inactive places far removed from such tormented areas, horizontal shearing stresses can be caused by the slow, but irresistable, movements of the tectonic plates (Wegener, 1915 as discussed by Armstead, 1978). These plates are being simultaneously subjected to the drag exerted from beneath by the flow of sub-Moho mantle material, and the resistance to that drag offered by a colliding adjacent plate. As a result of all these several tectonic forces, the pattern of stress distribution in deep crustal rocks becomes too complicated to predict with any precision, and can be determined only by measurement.

8.4   Hydrogeochemical influences on crustal stresses

Where underground waters flow or seep through permeable rocks or between adjacent geological strata, soluble materials may be slowly removed from one location and moved to another – usually to streams and rivers. In limestone country, for instance, the formation of caves in this way is a common feature, resulting from the solubility of the rock itself in meteoric waters containing carbon dioxide: such waters may penetrate to considerable depths through fissures or porous flow paths. Elsewhere acid waters may seep through permeable or porous rocks containing alkaline ingredients, or alkaline waters may react with acid ingredients in the rocks. As a result of all such processes, deep voids of one sort or another may be formed in the course of time, and large masses of rocks may become inadequately supported from below. If no surface subsidence occurs, the weight of overlying rocks will be redistributed laterally by an 'arching effect', with the result that some deep rocks may be carrying more than the theoretical lithostatic pressure while others will be carrying less. The mere existence of a temperature gradient may lead to underground convection currents of liquid water and may cause the formation of voids in hot reservoir regions by dissolution and the subsequent reprecipitation of material in other voids in cooler regions.

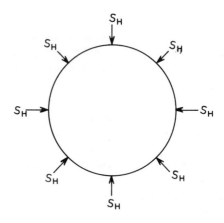

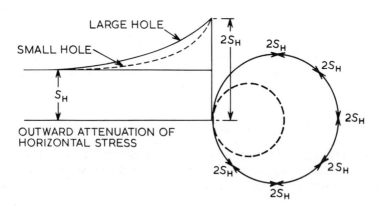

Fig. 8.2 Attenuation of circumferential stress in the vicinity of an empty bore, for isotropic horizontal stress conditions.

8.5 Disturbance to stress pattern caused by drilling

If a well is sunk into competent rock, no longer will every particle of rock be supported equally on all sides by the horizontal compressive stresses. The removal of matter by the drill leaves the borehole walls unsupported on one side; yet the horizontal compressive stresses will continue to act radially, as shown in the upper part of Fig. 8.2. Competent rock, however – even semi-competent rock – will not collapse under such conditions unless the stress magnitudes are high or are very anisotropic. Instead, the walls of the bore become self supporting by the 'arch' action of the bore curvature; and the rock locks itself into position by means of a circumferential stress. Assuming for the moment that the horizontal stress field is uniform, it can be shown that this compressive circumferential stress at the wall of a bore is twice the prevailing horizontal stress at any depth – i.e. $2S_H$ – regardless of the bore diameter. Moving outwards from the bore, the circumferential stress attenuates, approaching asymptotically the normal horizontal stress S_H, somewhat in the manner shown by the curve to the left of the lower part of Fig. 8.2. With a bore of different diameter the rate of attenuation would differ; for example, the stress pattern for a smaller bore would be somewhat as shown by the broken curve in Fig. 8.2. This theory assumes a symmetry in underground conditions which is absent in practically all actual situations. As a result of tectonic stresses, or of slumping tendencies in the vicinity of inclined plane faults, there may be quite wide variations in S_H with azimuth at any one point. But although asymmetry may modify the theory, it does not invalidate its implications. Drilling a vertical well in no way disturbs the local pattern of *vertical* compressive stress which remains at S_V.

This asymmetric or anisotropic horizontal stress field is usually characterized by defining principal stresses S_X and S_Y that act at right angles to each other in the horizontal plane (Fig. 8.3). The circumferential stresses then become a function of the azimuthal angle θ which defines a direction relative to these orthogonal principal stresses. This problem has also been solved theoretically for the case when the borehole axis is vertically aligned to give the circumferential stress at a distance r from borehole center as

$$S(r,\theta) = 1/2(S_X + S_Y)(1 + r_w^2/r^2) - 1/2(S_X - S_Y)(1 + 3r_w^4/r^4)\cos 2\theta \quad (8.3)$$

At the borehole wall $r = r_w$ so

$$S(r = r_w, \theta) = (S_X + S_Y) - 2(S_X - S_Y)\cos 2\theta \quad (8.4)$$

Two features are apparent from equations (8.3) and (8.4). First, the borehole wall stress $S(r = r_w, \theta)$ will vary from $3S_Y - S_X$ at $\theta = 0$ to $3S_X - S_Y$ at $\theta = \pi/2$. Second, the stress attenuates as r increases in an analogous manner to that depicted in Fig. 8.2. For many cases of interest, we will be interested in stresses far removed from the wellbore ($r > 10r_w$) where the circumferential stress attenuates to a value

$$S(r,\theta) \simeq 1/2(S_X + S_Y) - 1/2(S_X - S_Y)\cos 2\theta \quad (8.5)$$
$$r > 10r_w$$

which varies from S_Y at $\theta = 0$ to S_X at $\theta = \pi/2$.

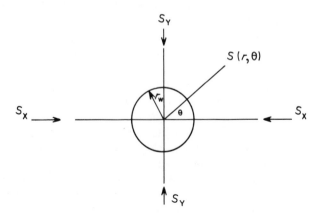

Fig. 8.3 Azimuthal dependence of horizontal stress around a vertical wellbore.

8.6 Fracturing pressures

The application of a sufficiently high hydraulic pressure at some point in a bore where the wall is uncased will cause the rock to fracture. Whether or not the resulting crack will lie in a vertical, a horizontal, or an inclined plane will depend upon certain factors that will now be examined.

8.6.1 Vertical fractures

If a steadily rising hydraulic pressure is applied at an uncased section of a bore sunk into homogeneous isotropic rock in repose, it will be seen from Fig. 8.2 that by the time that pressure has risen to a value of S_H, the bore walls will be supported internally to the same extent as they were before the bore had been excavated. At that point the circumferential compressive stress in the bore wall will have fallen from $2S_H$ to its original value of S_H: the water, in effect, has taken the place and function of the original rock filling of the bore. Now let the pressure be raised further, until it reaches a value of $2S_H$. The bore walls will then have been relieved of all compressive stress and will be in a totally unstressed state. In order to create the fracture, however, it is necessary to raise the pressure still higher so as to reverse the wall stress from compression to tension, to an extent that will overcome the tensile breaking strength of the rock. When this point has been reached the rock must yield, and a vertical crack will be formed, as shown in Fig. 8.4(a). The azimuth of the fracture could be in any direction, since theoretically the 'hoop' stress in the bore wall reaches fracturing intensity at all points of the circumference simultaneously in homogeneous isotropic rock in repose. No rock, however, is perfectly homogeneous, and the crack, of course, will initiate along the weakest plane. Put in the simplest mathematical terms:

Isotropic case

$$P_{fv} \geqslant 2S_H + S_t \qquad (8.6)$$

where P_{fv} = hydraulic pressure needed to initiate a vertical crack

and S_t = effective tensile breaking strength of the rock.

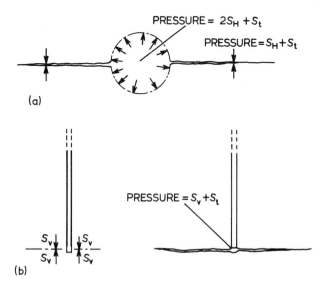

Fig. 8.4 Rock fracturing by hydraulic pressure in competent rocks in repose. (a) Formation of vertical fracture (plan view). (b) Formation of horizontal fracture (sectional elevation).

The stresses arise from a distribution of strains. Here again, as so often happens, practice may differ from theory. If weakly sealed natural joints should traverse the uncased part of the bore, these may determine the location of the fracture and will yield almost as soon as the horizontal stress has been fully relieved by hydraulic pressure. This has the effect of making S_t almost zero.

If we return to the theoretical treatment of the anisotropic stress case, fracturing criteria can also be established. If the axis of the borehole is aligned with the vertical stress S_V direction, equations (8.3) and (8.4) apply and the hydraulic pressure required to initiate a fracture is for the case of $S_X > S_Y$ and $S_V > S_Y$ as follows:

Anisotropic case

$$P_{fv} > (3S_Y - S_X) + S_t \qquad (8.7)$$

The term $(3S_Y - S_X)$ corresponds to the minimum circumferential stress occuring at $\theta = 0$ or parallel to the x-horizontal direction. Now the azimuth of the fracture is oriented perpendicular to the least principal horizontal stress S_Y which is sometimes referred to as S_3 or S_h.

Another point concerning the use of equations (8.6) or (8.7) is given by Haimson (1968,1978) who used the concept of effective stress to account for the presence of fluid in the pores of the rock. For finite permeability systems, hydraulic communication occurs between the fluid in the borehole and the surrounding rock formation. This alters the natural pore pressure and modifies the effective stress on the borehole; thus equation (8.6) or (8.7) is not quite correct for all cases. If the formation is permeable to the borehole fluid and starts with a finite pore pressure P_o, equations (8.6) and (8.7) are modified to

Isotropic case

$$P_{fv} - P_o \geqslant (2S_Y + S_t - 2P_o)/K \tag{8.8}$$

Anisotropic case

$$P_{fv} - P_o \geqslant (3S_Y - S_X + S_t - 2P_o)/K \tag{8.9}$$

where K is a poroelastic parameter relating the rock's pore pressure effects to the ratio of the rock's matrix to bulk compressibility and Poisson's ratio. If the rock is *almost* impermeable to the borehole fluid, K approaches unity and

Isotropic case

$$P_{fv} - P_o \geqslant 2S_H + S_t - 2P_o \tag{8.10}$$

Anisotropic case

$$P_{fv} - P_o \geqslant 3S_Y - S_X + S_t - 2P_o \tag{8.11}$$

If the rock were *totally* impermeable, the original equations (8.6) and (8.7) would hold. For many of the rock types important to HDR development, such as granites or other crystalline rocks at great depths, these permeability and pore pressure effects on the effective stress can be safely neglected. In our subsequent discussions we shall assume this to be true to simplify matters.

8.6.2 Horizontal fractures

The vertical stress S_V in an uncased section of a deep bore may be almost anything (see Sections 8.3 and 8.4 above), but in the simplest case of rocks in repose it will equal the lithostatic pressure S_L appropriate to the depth. A hydraulic pressure of S_L in the bore would just relieve the bore walls of all vertical stress, while the addition of further pressure to the extent of S_t could initiate a horizontal crack, as shown in Fig. 8(b). Again, expressing this in the simplest mathematical terms $P_{fh} = S_V + S_t$ where P_{fh} is the hydraulic pressure needed to form a horizontal crack.

Where rocks are in repose, S_V will be the same as the lithostatic pressure S_L, and

$$P_{fh} = S_L + S_t \qquad (8.12)$$

8.7 Orientation of a fracture

A fracture initiated by hydraulic pressure applied to the uncased lower extremity of a bore will form along the plane of least resistance. If, as a first hypothesis, it is assumed that the fracture lies either in a horizontal or in a vertical plane, and not at an inclined slope, then

	Uniform horizontal stress field $S_H = S_X = S_Y$	Anisotropic horizontal stress field $S_X > S_Y$	
Cracks will be vertical if	$2S_H + S_t < S_V + S_t$	$3S_Y - S_X + S_t < S_V + S_t$	(8.13)
Cracks will be horizontal if	$2S_H + S_t > S_V + S_t$	$3S_Y - S_X + S_t > S_V + S_t$	(8.14)

Since S_t is common to both expressions, it follows that the crack will be vertical or horizontal according to whether $2S_H$ or $(3S_Y - S_X)$ is *less* or *greater* than S_V, respectively.

The presence of weakly-sealed natural joints may well determine the orientation of fractures. As a result of past tectonic upheavals, bedrock will often have an almost 'block-like or macro-granular' structure caused by multiple faulting and the subsequent 'cementing' of adjacent parallel blocks of rock by substances that have been precipitated from chemically supersaturated waters that have percolated through the faults after they were formed. Such formations frequently outcrop at the surface. If adjacent blocks of rock are subjected to the same stress field, and if the jointing is very strong, then an artificial fracture could pass from one block to another with scarcely any change of direction; but if the jointing is weak, the crack could be deflected along the joint in preference to onward propagation. Thus, the fact that a fracture may *start* to form in one particular direction does not necessarily imply that it will continue in the same direction as far as its outermost extremities. Cracks initiated in one stratum of rock could well extend as far as the boundary of another stratum having different characteristics, and a growing crack might perhaps be deflected along the boundary, especially if the boundary slopes at an angle to the horizontal.

Other conditions that can affect the orientation of a fracture are the presence of stresses due to tectonic or hydrogeological influences (see Sections 8.3 and 8.4), random irregularities in rock formations, or the influence of viscoelastic forces. These last factors (see Section 8.2) are unlikely to be significant except at great depths and at high temperatures, where 'hydrostatic' conditions could be approached and S_H and S_V are approximately the same; then the orientation of a crack could be in any direction, as determined by random departures from homogeneity.

An important distinction needs to be made regarding crack initiation at the borehole wall ($r = r_w$) versus crack propagation far away from the borehole ($r >> r_w$). As one can see from equation (8.5), the circumferential stress attenuates to a minimum value independent of r and only dependent on azimuthal angle. If S_Y is truly the minimum principal horizontal stress, then the criteria for propagation of a vertical fracture become

$$P_{fv} > S_Y + S_t^* \text{ if } S_Y < S_X \tag{8.15}$$

where the asterisk refers to the net effective tensile strength of a propagating crack. Any theory applying to hydraulic fracturing will need to incorporate the conditions given by equation (8.15) as the appropriate criteria for fracture propagation. One should keep in mind that it would be possible for a fracture initiated in a horizontal plane to change its direction into a vertical plane within a fairly short distance of the bore, and vice versa.

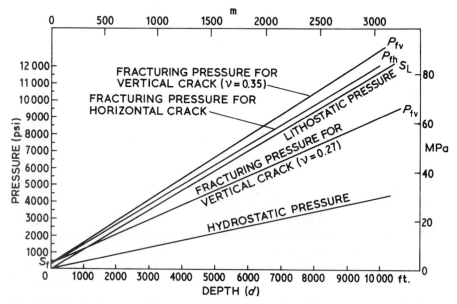

Fig. 8.5 Lithostatic, hydrostatic and fracturing pressures at different depths for rock in repose.

Assumptions: mean rock density $= 2.7\,\mathrm{g\,cm^{-3}}$ (2700 kg/m³);
tensile breaking strength $= 400$ psi (2.76 MPa) $= S_t$

Note: If the uncased borehole should intersect a weakly sealed rock joint, S_t could be very small and the 'fracturing' pressure would be reduced accordingly.

In practice, most hydraulically induced rock fractures are assumed to lie in a vertical plane for formations deeper than 1000m (3280 ft). This suggests that the resultant principal stresses are minimal somewhere in the horizontal plane.

It may be noted in passing that although large pressures are needed to fracture rocks hydraulically, only part of those pressures need be provided by pumping. In Fig. 8.5 a curve of hydrostatic pressure is shown, which makes allowance for the decrease in water density with increasing depth owing to temperature rise (the curvature is very slight). For the earth stresses shown hydrostatic pressure alone could provide something approaching half of the total required pressure.

8.8 Hydraulically induced propagation and dilation of fractures

8.8.1 Fracture initiation concepts

As stated earlier, although a high pressure is necessary to *initiate* a rock fracture at the uncased bottom of a bore, less pressure is needed to *extend* a fracture. Figure 8.6(a) shows (in sectional plan view) a vertical crack at the instant of its formation: it will extend to points T and T' at some distance from the bore. The horizontal stress S_H tends to hold the crack closed, while the far greater pressure P_{fv} [$\geqslant(2S_H + S_t)$ or $(3S_Y - S_X) + S_t$] simultaneously tends to open the crack. Just before the precise moment of the fracture, the hoop stress at the bore wall is S_t; but in the next moment, just as the fracture initiates, this hoop stress is suddenly released and ceases to exist, so that the full pressure within the bore P_{fv} will start to force the two halves of the bore apart and at the same time to open up the crack and admit high pressure water into it.

The term S_t has been so far generically used to represent the tensile strength of the rock that must be overcome to allow fracturing. Many years of effort in the area of theoretical rock mechanics have attempted to unravel the complex aspects of material failure in rocks (see also Howard and Fast, 1970, Chapter 2, and Jaeger and Cook, 1979, Chapters 4 and 12). Out of this effort have arisen widely differing theories and predictions as to how fractures occur and propagate. Early theoreticians assumed that the rock surrounding the wellbore behaved as a homogeneous, elastic solid. They then used a maximum stress or strain criterion to model failure. In some cases, a Mohr failure envelope was drawn using principal stress information. More recently, theories following Griffith's (1921) approach have been developed. Griffith hypothesized that the small flaws inherently present in any solid material can intensify imposed stresses to the point where failure occurs at much lower stresses than would be estimated for a 'perfect', flaw-free specimen. Subsequent investigations have claimed that randomly distributed 'Griffith cracks' of the order of grain size dimensions control fracture initiation in rock. Walsh and Brace (1964) modified this approach by assuming that an additional set of ordered, longer

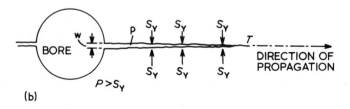

$P_{fv} \geqslant 2S_H + S_t$ (UNIFORM STRESS)
$P_{fv} \geqslant (3S_Y - S_X) + S_t$ (NON-UNIFORM STRESS)

(a)

$P > S_Y$

(b)

Fig. 8.6 Formation and extension of a vertical fracture in a uniform horizontal stress field.
S_Y (uniform horizontal stress field $= S_H$)
S_Y (non-uniform horizontal stress field), $S_Y < S_X$
(a) Pressures at the moment of fracture. (Sectional plan view.) (b) Subsequent extension of fracture after opening. (Sectional plan view.)

cracks existed in the material along with the smaller Griffith cracks. Failure could be caused by either set of cracks, depending on the orientation of the applied load. Cleary (1979) has also cited the importance of fluid permeation and its influence on the effective stress field by altering the pore pressure. This results in a stabilization effect during crack growth that he terms 'back stress'.

Much of the experimental work used to test the theories cited above has dealt with testing small specimens of rock. For example, Howard and Fast (1970) present theoretical and experimental results for internal rupturing pressures of rock cylinders. Even for the cases of rocks that are nearly homogeneous, agreement between theory and experiment is poor: the results differ often by a factor of ten or more. At this point, one might argue that the theories should be abandoned in favor of direct experimental measurements of tensile strength which could then be incorporated into appropriate field equations. However, laboratory-scale measurements can also be misleading because of their idealized conditions. For example, major flaws exist in wellbores that could dominate all initiation processes. For instance, as mentioned earlier, the presence of weakly-sealed joints or cracks could serve as ideal sites for fracture extension with $S_t = 0$. Field experiments at Los Alamos appear to show this behavior (see Chapter 9).

Figure 8.6(b) shows, on a grossly exaggerated scale, the crack just after it has opened up. The water within the crack will be at a pressure P, only slightly less than P_{fv} and this pressure will exceed the initial resisting pressure S_H or S_Y offered by the walls of the crack by an amount approaching S_t. This excess pressure will compress the rock behind the crack walls to an extent permitted by the elasticity of the rock, and the crack will thus widen near its opening. The leverage of the internal pressure P exerted along the whole wall of the crack will, by a sort of 'wedge' action give rise to a highly concentrated tensile stress at the tip of the crack T with the result that the crack will extend.

The effect may be likened to the action of tearing a cloth, which will resist a very forceful pull until a small incision is made with scissors, after which only a slight effort is required to extend the tear because the tension has become concentrated at a single point. Alternatively, it is analogous to the splitting of a wooden log with an axe. The tendency for a crack to extend depends on the geometry of the crack – i.e. upon the ratio of the aperture or width w at the entrance to the length of the crack, or upon the small angle at the apex of the crack, T. The smaller this ratio and angle, the less will be the tendency of the crack to extend. As the movement apart of the two halves of the bore is determined by the hydraulic pressure and the elasticity of the rock, the width of the crack aperture w at the entrance is more or less fixed. As a result, the crack will propagate outwards from T to such a distance that its geometry becomes too elongated to sustain any further splitting action at the tip. If, at the moment of initiating a crack extending to T only, the hydraulic pressure were to collapse, the crack could be persuaded to extend far beyond T by applying a hydraulic pressure much less than the original fracturing pressure P_{fv}. A similar process would occur with a horizontal crack, where S_V would be substituted for S_H or S_Y, and P_{fh} for P_{fv}; but since P_{fh} exceeds S_V by much less than P_{fv} exceeds S_H, the extension of horizontal cracks would be much less than in the case of vertical cracks.

In summary, it may be stated that in a homogeneous, isotropic elastic material the direction in which a hydro-fracture will propagate will be perpendicular to the least principal compressive stress; but that random influences may decree otherwise.

After the removal of the hydraulic pressure the crack would reclose. In theory reclosure would be complete, and the crack walls would mate precisely into their original conformation. However, if shear stresses were present before the moment of fracture there could be a very slight relative lateral movement between the walls, so that surface irregularities of order 0.1 to 1.0 mm in the rock walls would prevent complete closure. The breaking off of particles of rock from the walls at the moment of fracture could also obstruct reclosure to some extent. This phenomenon has been named self-propping and may be very important to heat mining methods because permanent increases in permeability may be possible without continuous pressurization. The reapplication of the original fracturing pressure would again dilate the fracture to its former aperture width w at the edge of the bore, regardless of whether reclosure was complete or incomplete.

Batchelor (1978, p.34) has stated that in a core sample of Cornish granite it requires 119 J/m^2 to generate a fracture, and 140 000 J/m^2 (i.e. 1176 times as much energy) to dilate it to 5 mm width. The relationship between the dilation energy and the crack width is complex and requires a model for failure of the material (see also Section 8.8.2). Cracks do dilate as they grow larger in area, but the aspect ratio of crack length (or radius) to crack aperture is very large, of order 1000 to 10 000 for even idealized cases. The point is that individual fractures will always tend to be flat and pancake shaped.

From Batchelor's observations and many others, by pumping in energy at modest levels of order 120 J/m^2 a crack can be extended with only minimal dilation. At the first Fenton Hill experiments at Los Alamos (see Chapter 9) a pressure of about 90 bar (1300 psi) was found to be sufficient to hold open a vertical fracture to a width of 'a few millimetres'. By raising the pressure above 90 bar with continued pumping some increased dilation results, but the major effect was that the crack was extended. Extension and dilation are interrelated by a complex set of equations which depend on the geometry of the crack and, of course, on the physical properties of rock and the magnitude of the stress field.

8.8.2   Current theories of hydraulic fracturing

The two aspects of the fracturing process that need to be modeled are *crack initiation* and *crack propagation*. Given that the horizontal stress field surrounding the borehole is either uniform $S_Y = S_X = S_H$ or non-uniform with $S_Y < S_X$, fluid pressures required for vertical crack initiation depend directly on the term S_t, corresponding to the effective tensile breaking stress of the rock. As discussed earlier, estimating S_t from theory is difficult because of the complex, superimposed effects of rock property anisotropy and inhomogeneity, pore pressure changes due to fluid permeation, and the presence of pre-existing fractures in the formation. In practice, values of S_t can be obtained crudely from field measurements; but this too is difficult because *in situ* geometry and stresses are not always known. Consequently, virtually all of the existing theories for hydraulic fracturing assume that initiation at a given site has occurred and that the principal stress orientations and magnitudes are known. The theoretical modeling problem now becomes one of *crack propagation*. Propagation is a transient process which requires fluid both for inflating and for extending the fracture as it grows, as well as providing fluid for permeation losses as the surrounding formation is pressurized above its normal level. Thus, the effects of fracture growth under the earth's confining stress and permeation loss are coupled, as both consume a fraction of the injected fluid. Important factors in this propagation process are described by Howard and Fast (1970, Chapter 4), Daneshy (1973), and McLennan and Roegiers (1979). They include:

(1) *Geometric effects* – Aperture, or opening, of the crack, as a function of lateral fracture length or radial position from the point of injection at the borehole is required.

(2) *Inhomogeneities including interaction with lithologic boundaries* – Gross changes in physical properties may limit crack propagation.

(3) *Rock deformation mechanism* – Fracture aperture changes as the internal pressure distribution is altered. Behavior is usually assumed to follow a simple linear elastic law (Sneddon, 1946; Sack, 1946; and England and Green, 1963) or a quasi-equilibrium criterion for propagation (Barenblatt, 1962 and Khristianovitch and Zheltov, 1955).

(4) *Fluid flow inside the fracture* – Some type of 'flow law' is required to describe the internal pressure distribution as a function of fluid velocity and fluid properties. Flow is usually assumed to be laminar or 'Darcy-like' where the local velocity is linearly dependent on the pressure gradient along the fracture length, inversely dependent on fluid viscosity and proportional to the square of the fracture aperture. Newtonian fluids such as water whose viscosity is independent of shear rate or non-Newtonian viscoelastic fluids such as 'gelled frac' fluids whose viscosity varies with shear rate may be used in some applications.

(5) *Fluid loss to the formation surrounding the fracture* – As the fracture propagates, fluid pressure levels, locally around the fractured region, are higher than the ambient pore pressure. This pressure gradient results in fluid permeation into the formation surrounding the fracture. The effect is commonly called fluid loss.

(6) *Proppant transport* – In many fracture treatments, solid particles such as sand are suspended in the fracture fluid and pumped into the fracture to prevent closure when the pressure is reduced. The particles act as proppants to maintain a finite aperture against the surrounding confining earth stresses. The passage within the fracture might resemble a packed bed where fluid meanders through the interstices at a low impedance.

The objective of any fracture model is to incorporate these effects into a mathematical framework to permit prediction of fracture geometry. Field measurement capabilities are usually limited to injection rate, total injected volume, and pressure as functions of time. Consequently with no direct measurements of fracture size (width or aperture and radial or lateral extent) and orientation, we must rely on an appropriate model to estimate these features (Nierode, 1983). In designing a fracture treatment, one might ask the question, 'What pumping rate and injection times are required to produce a fracture of a desired surface area?'

Given the complexities of this model development effort, investigators have employed a number of simplifications, including:

(1) *Linear versus radial two-dimensional shapes* – Fig. 8.7 shows the two overall geometries employed in most two-dimensional models. In the

linear case, the fracture is confined to a bilaterally symmetric lenticular shape by formations above and below. Thus a vertical lithology bounds the system. Oil and gas producing reservoirs are often modeled in this manner because the 'production zone' is usually contained in a well-defined layer. For many HDR systems, fracturing will occur in a more or less homogeneous region of hot rock; therefore, the vertical lithologic bounds germaine to an oil or gas reservoir may not be appropriate for an HDR system. A radially symmetric fracture that propagates outward from the borehole idealizes this case. This type of fracture is commonly called a 'penny-shaped' crack and was theoretically justified by Sneddon and Sack in 1946.

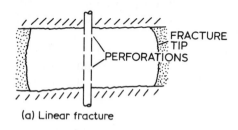

(a) Linear fracture

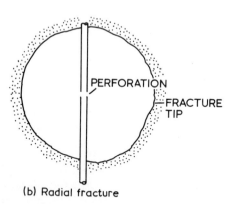

(b) Radial fracture

Fig. 8.7 Comparison of fracture propagation geometries for linear confined and radial unconfined models. (a) Linear fracture propagation bounded by lithology vertically. Elevation view parallel to S_Y direction. (b) Radial fracture propagation unbounded by lithology vertically. Elevation view parallel to S_Y direction.

(2) *Constant versus variable aperture.* The two simplifications most frequently used in theoretical modeling are to assume that the crack aperture is constant or that it tapers elliptically from a maximum width at borehole wall towards the fracture tip.

(3) *Fluid loss versus fracture inflation volume* – In all hydraulic fracturing operations, there are two factors that consume fluid. The first is the actual volume of fluid contained inside the fracture and the second is fluid lost to permeation into the rock surrounding the fracture walls. Thus, fracture volume and fluid leak-off compete for portions of the injected fluid. If our design objective is to maximize fracture surface area, for a given injected volume of fluid, then we would want to minimize permeation losses. Also in order to increase lateral or radial extent in an optimal fashion, we would want to limit the aperture or fracture opening to something reasonably small. For instance, by mixing certain additives with the injected water, viscosity can be increased to reduce fluid losses. However, higher viscosities also result in larger apertures which is an undesirable side effect. In an optimal treatment design, fluid properties, pumping rates, etc., should be selected to minimize permeation losses while maintaining aperture profiles large enough either to permit adequate proppant transport into the fracture or to ensure that self-propping occurs. If apertures are large enough to accommodate the inherent roughness of the asperities of the fracture surfaces, sufficient clearance will exist to allow fracture surfaces to slide over one another. This relative displacement will lead to self-propping. When the fluid pressure is released, the fracture will not completely close leaving an open labyrinth for fluid to pass.

In the remaining discussion of this section we have subdivided the major theoretical approaches into four categories.

(a) PERKINS AND KERN/NORDGREN ELLIPTICAL APERTURE, LINEAR AND RADIAL MODEL (OR PKN MODEL)

The basic geometric assumptions for a linear or radial fracture following the Perkins–Kern–Nordgren or PKN model are depicted in Figs. 8.8 and 8.9. Both vertical and horizontal cross-sections are more or less elliptical. For the bilaterally symmetric linear model the fracture height is fixed while for the radial model the fracture is symmetric about the midpoint where fluid leaves the wellbore at the perforation shown in Fig. 8.7. The PKN model also assumes:

(1) laminar Newtonian flow in the fracture such that for a linear fracture of height H

$$\dot{q}_x = \frac{-\pi w^3 H}{64\mu} \left(\frac{\partial P}{\partial x} \right) \qquad linear \qquad (8.16)$$

where
  $\dot{q}_x$=local volumetric flow rate at position x
  w=aperture of fracture
  μ=fluid viscosity
  P=fluid pressure
  x=dimension along the length of fracture

or for a radial fracture at a radius r

$$\dot{q}_r = \frac{-\pi w^3 r}{6\mu}\left(\frac{\partial P}{\partial r}\right) \qquad radial \qquad (8.17)$$

(2) Fluid pressure at fracture tip equal to minimum earth stress S_Y.
(3) The fracture walls deform in a linear elastic manner.
(4) The effective rock tensile strength is small (S_t=0).

England and Green (1963) showed that the combination of *in situ* earth stress acting against fluid pressure in a linear crack results in an elliptical aperture profile given by

$$w(x)=\frac{2(1-v)L(P-S_Y)}{G}[1-(x/L)^2]^{1/2} \qquad (8.18)$$

where a uniform pressure P and stress S_Y is assumed over the length of the fracture L and

  G=Shear modulus
  v=Poisson's ratio.

One limitation to equation (8.18) is that it does not apply in the immediate vicinity of the crack tip, where the geometry is not elliptical (as seen in Fig. 8.9).
  By using the results of England and Green's (1963) analysis and the assumed flow law, Perkins and Kern (1961) were able to derive a relationship for fracture aperture as a function of position x and time t for a linear fracture:

$$w(x,t)=2.00\left[(1-v)\dot{q}^2\mu/GH\right]^{1/5}t^{1/5}\left[1-(x/L)\right]^{1/4} \qquad (8.19)$$

where $\dot{q}$=the injection flow rate.

Using Nordgren's (1972) correction to the time variation of aperture, the length of the fracture can also be written as

$$L(t)=0.45\left[\frac{G\dot{q}^3}{(1-v)\mu H^4}\right]^{1/5}t^{4/5} \qquad (8.20)$$

  The radial case is treated by Perkins and Kern (1961) somewhat differently. For a penny-shaped crack with neglible pressure drop within the crack and no leak-off, the minimum wellbore pressure $P(t)$ necessary to extend a fracture of radius R is given by (Sack, 1946):

$$P(t)-S_Y=\left[\frac{\pi\alpha_s E}{2(1-v^2)R(t)}\right]^{1/2} \qquad (8.21)$$

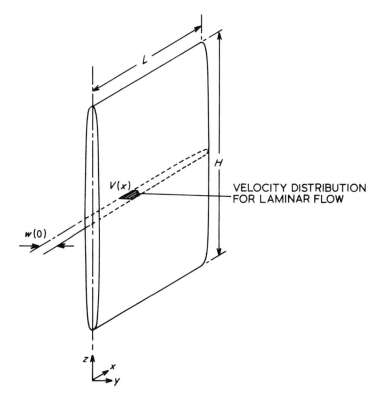

Fig. 8.8 Conceptualization of a linearly propagating fracture of the Perkins and Kern type. (Adapted from Geertsma and Haafkens, 1979).

where

α_s=specific surface energy of the rock
E=Young's modulus
v=Poisson's ratio

Using Sneddon's (1946) stress analysis for the linear elastic radial case, it is found that crack aperture varies elliptically with radius as it did with length for the linear case. Assuming that $E = 5/2G$ (which is equivalent to assuming that v = 0.25) one obtains an expression for fracture radius R:

$$R(t)= \left[\frac{3}{16} \dot{q}t \left(\frac{5G}{\pi \alpha_s(1-v^2)} \right)^{0.5} \right]^{0.4}$$ (8.22)

The term $\dot{q}t$ is the total volume of fluid pumped; and for the case of zero leak-off is equivalent to the fracture volume (V_f). Note that the radial case does not depend on any fluid property parameters (such as viscosity) nor does it depend on any pumping parameters other than the total injected volume ($\dot{q}t$).

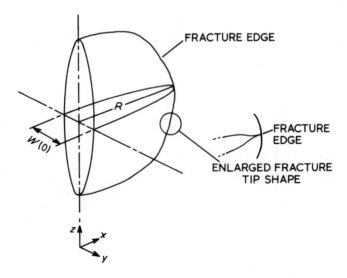

FRACTURE EDGE

FRACTURE EDGE

ENLARGED FRACTURE TIP SHAPE

Fig. 8.9 Conceptualization of a radially propagating vertical fracture of the Perkins–Kern and Nordgren type. Fracture aperture *w* varies elliptically from the centerline to the vicinity of the tip.

If we were to relax the no leak-off assumption and permit finite fluid permeation, the analysis becomes more complex. However, the asymptotic case of a small fracture volume compared with fluid losses to permeation can be treated by assuming a uniform internal pressure in the fracture and integrating the partial differential equation that describes one-dimensional transient diffusion of fluid into the rock surrounding the fracture. For Darcy-like flow into a medium of constant permeability k and compressibility c, the local flow rate into the formation $\dot{q}_y$ is proportional to the pressure gradient in the y-direction, perpendicular to the fracture face and is given by:

$$\dot{q}_y = -2\pi R^2 \frac{k}{\mu} \left(\frac{\partial P}{\partial y} \right)_{y=0} \tag{8.23}$$

The equation for $\dot{q}_y$ can be combined with the continuity equation for incompressible flow to give

$$\left(\frac{k}{\mu c} \right) \frac{\partial^2 P}{\partial y^2} = \frac{\partial P}{\partial t} \tag{8.24}$$

which when solved for the case of a fixed total injection rate $\dot{q}$, gives an expression for fracture radius R

$$R(t) = \left[\frac{\dot{q}^2 \mu t}{\pi^3 k c (\overline{\Delta P})^2} \right]^{0.25} \tag{8.25}$$

where $\overline{\Delta P}$=average difference of pressure between the fracture and the original pore pressure.

Nordgren (1972) analyzed fluid loss effects for the linear fracture model to give for large times (or large loss rates)

$$L(t)=\dot{q}\sqrt{t}/2\pi C_{\rm L}\,H \tag{8.26}$$

$$w(x,t)=3.36\left[\frac{2\mu(1-v)\dot{q}^2}{\pi^3 GHC_{\rm L}}\right]^{1/4}t^{1/8}(1-(x/L)^2)^{1/2} \tag{8.27}$$

$$P(t)=3.36\left[\frac{2G^3\mu\dot{q}^2}{\pi^3(1-v)^3 C_{\rm L}H^5}\right]^{1/4}t^{1/8} \quad \text{(pressure at the wellbore)} \tag{8.28}$$

where $C_{\rm L}$=fluid loss coefficient which is approximated by the following formula:

$$C_{\rm L}=u\sqrt{(t-t_{\rm o})} \tag{8.29}$$

with

u=fluid velocity under permeation conditions
t=time
$t_{\rm o}$=time when permeation started.

Cases other than these extremes can be treated by numerical approaches such as used by Murphy and Aamodt (1979) where the permeation effects (equations (8.23) and (8.24)) are coupled to linear elastic displacement results.

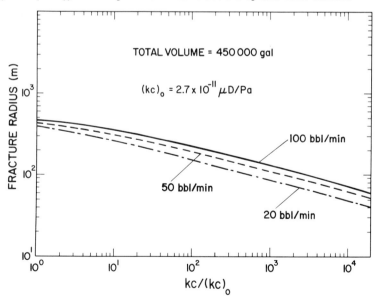

Fig. 8.10 Perkins–Kern–Nordgren model results for fracture radii as a function of the formation's permeability–compressibility product with pure water as the fracture fluid at 200°C where $\mu=1.4\times10^{-4}$ Pa-s (or 0.14 centipoise).
Injection flow rates given in bbl/min where 1 bbl=42 gal=0.159 m³.

For the case of plain water as the fracture fluid and radial fracturing, fracture radius as a function of the ratio of kc to a base value of $(kc)_o = 2.7 \times 10^{-11}$ μ darcy/Pa is given in Fig. 8.10 for several injection rates.

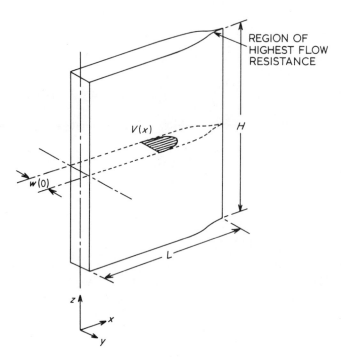

Fig. 8.11 Conceptualization of a linearly propagating, constant aperture fracture of the Geertsma–de Klerk and Khristianovich–Zheltov type. (Adapted from Geertsma and de Klerk, 1969).

(b) GEERTSMA AND DE KLERK/KHRISTIANOVITCH AND ZHELTOV – CONSTANT APERTURE, LINEAR AND RADIAL MODEL (OR GKKZ MODEL)

The second fracture model (GKKZ) was first proposed by Khristianovitch and Zheltov (1955) and later developed by Geertsma and de Klerk (1969). The geometric aspects given in Fig. 8.11 show that the major difference from the PKN model lies in the assumption of a constant rather than elliptical aperture. Furthermore, the GKKZ model attempts to account for the stress-concentration that exists at the tip of the fracture which was completely ignored by Perkins and Kern. They essentially incorporated the concept of equilibrium fracture propagation from Barenblatt (1962) whereby the distribution of the fracture fluid pressure must be such that the fracture faces close smoothly at the tip as shown in Fig. 8.11. The GKKZ model has also been developed to treat

the cases of no leak-off or finite fluid permeation for both linear (fixed height H) and radial fractures. Rather than present the derivations here, we refer to the papers cited earlier. The results of their analysis can be summarized as follows.
For the linear fracture model with no leak-off:

$$L(t)=0.48\left[\frac{G\dot{q}^3}{\mu(1-v)H^3}\right]^{1/6}t^{2/3} \tag{8.30}$$

$$w(x,t)=1.32\left[\frac{(1-v)\mu\dot{q}^3}{GH^3}\right]^{1/6}t^{1/3}[1-(x/L)^2]^{1/2} \tag{8.31}$$

$$P(t)-S_Y=\frac{1.91}{2H}\left[\frac{G^3\dot{q}\mu H^3}{(1-v)^3(L(t))^2}\right]^{1/4} \tag{8.32}$$

For the linear fracture model with finite fluid permeation, Geertsma and de Klerk (1969) present numerical results, as shown in Fig. 8.12, for various injection flow rates to fracture height ratios ($\dot{q}/H$) with a fixed loss coefficient ($C_L = 0.01$ cm/s$^{1/2}$).

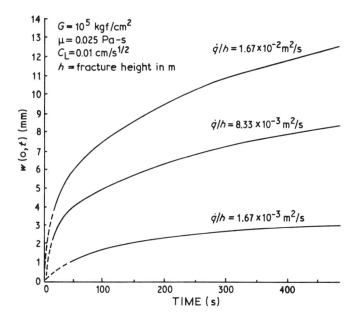

Fig. 8.12 Dependence of fracture aperture at the wellbore ($w(0,t)$) on time for various fluid injection rates $\dot{q}$ in m³/s (after Geertsma and de Klerk (1969)).

The radial case with no leak-off has also been solved by them by combining expressions for radial Darcy-like flow with Sneddon's displacement model. The results are

$$R(t) = \left[\frac{15}{16\pi} \dot{q} t (G/\mu\dot{q})^{0.25} \right]^{0.444} \tag{8.33}$$

$$w(r,t) = 2[\mu\dot{q}(R(t))/G]^{0.25}(1-r/R)^{1/2} \tag{8.34}$$

$$P(t) - S_Y = -\frac{10}{4\pi}[G^3\mu\dot{q}/R^3]^{0.25}\ln(r_w/R) \tag{8.35}$$

where r_w = wellbore radius.

By following an analogous procedure to that employed for a linear fracture, they also developed similar expressions and numerical results to those depicted in Fig. 8.12. Murphy and Aamodt (1979) also treated the radial case numerically and presented their results in the same form as Fig. 8.10. A comparison of Figs. 8.10 and 8.13 shows that the radius prediction for fractures following the GKKZ model are very similar, in fact almost identical with, the PKN model predictions.

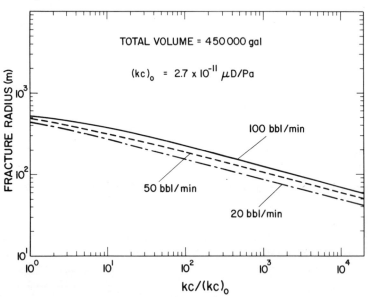

Fig. 8.13 Geertsma–de Klerk–Khristianovitch–Zheltov model results for fracture radii as a function of the formation's permeability–compressibility product with pure water as the fracture fluid at 200°C where $\mu = 1.4 \times 10^{-4}$ Pa-s (=0.14 centipoise). Injection flow rate given in bbl/min where 1 bbl = 42 US gal = 0.159 m³.

One complicating factor that has been observed in several field tests of prototype HDR reservoirs has to do with the pressure-dependence of formation permeability, porosity, and compressibility (see Chapter 9 for details). The importance of this effect is illustrated in Figs. 8.10 and 8.13. As the kc product increases, fracture radii will decrease and effective reservoir heat transfer areas will be lowered. What has been observed so far indicates that as pressures in the formation approach S_Y, the kc product becomes very large thereby reducing the growth potential of the fracture. Thus in order to ensure proper reservoir development, these pressure-dependent effects must be well-characterized.

In some cases, fluid loss additives may be available to lower k to appropriate levels to increase fracture growth. If, however, these additives also increase viscosity as is commonly the case, the negative effects of increasing aperture rather than radius must be considered. An increase in viscosity, with kc held constant will result in substantially lower increase in fracture radius. Practically speaking, a tenfold increase in viscosity results in a 30 to 50% increase in radius.

(c) CLEARY PSEUDO THREE-DIMENSIONAL HYDRAFRAC MODEL (OR P3DH MODEL)

Cleary at the Massachusetts Institute of Technology recognized the limitations of the earlier theories and attempted to modify the existing theoretical approaches to be more realistic in the prediction of fracture length and sizes (Cleary, 1982). His concerns include:

(1) complex phenomena involving crack growth or retardation across interfaces between strata;
(2) importance of non-linear anisotropic and mechanistic effects in determining effective *in situ* confining stresses;
(3) unstable behavior of many laboratory hydraulic fracture measurements versus the stable quasi-static growth of fractures in the field;
(4) characterization of multiple fracture networks both to sustain multiple crack propagation and provide linkage in generating underground fracture networks.

Cleary's general approach utilizes lumped parameter models to illustrate the effects of injection pressure, formation strength moduli, and fluid properties (e.g. viscosity). By proper non-dimensionalization of separate equations for fracture length and height growth, time scaling factors are established. This allows him to model growth in two dimensions to produce different shapes as well as to produce a variety of aperture profiles. In this way, the result is a Pseudo-Three Dimensional Hydrafrac Model, or P3DH as it is generically called.

An example of a specific application would be to express the aperture, height and length in differential equations as:

$$\frac{\partial H}{\partial t}=\bar{\gamma}_2 H/\tau_c \qquad \frac{w}{H}=\gamma_1\sqrt{\bar{\bar{E}}}(P-S_y) \qquad \left(\frac{\partial L}{\partial t}\right)^{m^*}=\bar{\Gamma}_2^{m^*}H^{m^*+1}/L\tau_c^{m^*} \qquad (8.36)$$

where

γ_1, $\bar{\gamma}_2$ and $\bar{\Gamma}_2$ are model dependent parameters and m^* is a power-law exponent.

Also

$$\tau_c = \text{characteristic time constant} = [\bar{\eta}(\overline{E}[(P-S_y) \vee \overline{E}]^{2n^*+2-m^*}H^{2n^*-2m^*})]^{1/m^*}$$

with $\bar{\eta}$ another model parameter and n^* another power-law exponent.

The format of equations (8.36) is such that parameters can be produced tha are consistent with PKN or GKKZ model growth or any other model for tha matter. The P3DH model can include effects caused by fluid loss and variation: in effective confining stress due to fluid migration.

Cleary's claim is that the approach is simple yet more general than previou: investigations. What is important to keep in mind is that the P3DH scheme stil requires certain 'adjustable' parameters to be specified and that numerica integration techniques are required to solve for explicit values of H, L, and w at particular times of interest. Cleary's research group at MIT under industria: sponsorship continues to work on developing more generalized theories and models for hydraulic fracturing, and will no doubt be publishing its findings in the near future.

(d) CUNDALL AND BATCHELOR TWO-DIMENSIONAL MULTIPLE FRACTURES MODEL FOR JOINTED ROCK (OR FRIP MODEL)

The last model to be discussed has been developed by the UK HDR project following the initial formulation by P. Cundall (see Cundall, 1978 for details). The model's backbone is a two-dimensional rock mechanics code called FRIP (which is short for *Fluid Rock Interaction Program*). In its formulation, rock joints are placed on a regular rectangular grid and the code permits interactive coupling of flow fluid with confining stresses and deformations. The basic grid is shown in Fig. 8.14. The code handles fully-transient situations. For example, a pressure excess on a specific block during one computational cycle will result in compression of that block and opening or dilation of the joints adjacent to it. This will in turn change the permeability and pressure distribution. Conse-quently, the FRIP model can compute discrete changes in the porosity, permeability, and fluid flow characteristics within the matrix of joints. The model does require certain specifications in addition to the joint pattern and injection point location(s), including

(1) A suitable law for flow within the joints – typically, laminar or Darcy-like behavior as described by equation (8.16) is assumed.
(2) Rock strength properties and moduli E, G, K, and v are needed.
(3) Initial confining stresses, S_Y and S_X and pore pressure.
(4) Fluid viscosity and injection rate.
(5) Residual or minimum joint aperture.
(6) Boundary conditions – fixed or semi-infinite in extent to simulate a large rock formation.

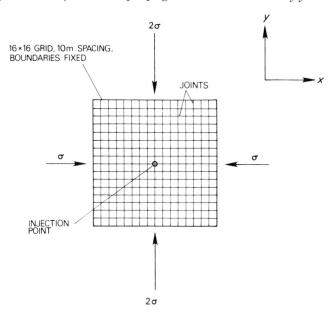

Fig. 8.14 FRIP grid used for classical hydraulic fracturing with joints aligned with principal stress. In this case $\sigma=S_Y$ and $2\sigma=S_X$. $G=5\times10^9$ Pa; $K=10\times10^9$ Pa; $v=0.29$; Minimum fracture aperture $=100\,\mu$m; $\mu=3.5\times10^{-4}$ Pa-s $=0.35$ centipoise; $\dot{q}=0.0265\,$m^3/s per 1 m of fracture width. (Published with permission of A.S. Batchelor, Camborne School of Mines).

Displacements are calculated by a double integration of net accelerations on each block due to net resultant forces. For the classical fracturing case where major joints are aligned with S_Y and S_X (see Fig. 8.14), FRIP produces the expected pressure-time response shown in Fig. 8.15. Also given in the figure are the fracture aperture profiles and pressure distribution surrounding the fracture after five to ten seconds of pumping.

FRIP can also handle shear joint motion during stimulation which will occur when joints are not aligned with the principal stresses. This feature becomes important when considering the field results obtained at Fenton Hill, Rosemanowes, and elsewhere which are discussed further in Chapter 9.

The essential characteristic that gives the FRIP model versatility is its two-dimensional joint/block structure that more closely resembles a real reservoir. However, FRIP, like the previous three models, still requires 'adjustable parameters' to be specified in order to calculate practical results.

Interested readers should consult papers by Batchelor *et al.* (1983) and Murphy (1982) for more details on the theory and application of the FRIP model to hydraulic fracturing in HDR systems.

One word of caution is in order regarding all these models of fracture extension and propagation. While the physics of the hydraulic fracturing

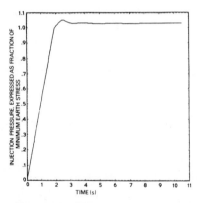

(a) INJECTION PRESSURE HISTORY

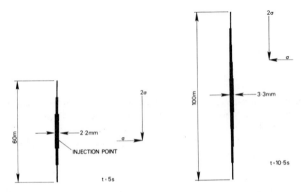

(b) FRACTURING PROFILE AFTER 5s   (c) FRACTURING PROFILE AFTER 10.5 s

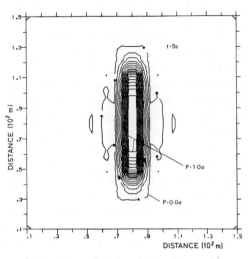

(d) PRESSURE DISTRIBUTION AFTER 5s

process have been reasonably described for tensile failure and growth if the rock behaves as a homogeneous, isotropic medium, the real situation is likely to be very different. Model predictions and fracture designs are obviously less quantitatively useful as heterogeneities, non-linearities, and/or other fracture failure mechanisms become important.

8.9   Thermal effects and secondary thermal stress cracking

A further complication in the matter of crack extension is that of thermal stress. It has been suggested, and there is good circumstantial evidence in the field to support the suggestion, that as soon as relatively cool water is circulated through cracks in hot rock, the resulting thermal contraction of the rock will cause cracks to become self-propagating outwards. This tendency would diminish in the outer extremities of the crack, since for a given hydraulic pressure the crack would become narrower as it extends outwards and would thus offer increased flow resistance to the water, which would then have a longer time in which to become heated to temperatures that would cause insufficient thermal shock to extend the crack further. If flow through the crack is governed by Darcy's Law, then the effective permeability within the crack is proportional to the square of the aperture (or w^2). Thus as the crack tip is approached, w tends to zero, the fracture permeability vanishes and the fluid no longer flows. All cooling to the crack tip area then must be by a conduction rather than by a convection mechanism and is governed by the rate at which heat diffuses through the solid rock itself. Due to these slower heat transfer rates, thermal shock at the crack tip will be virtually eliminated.

Thermal shrinkage could also affect the degree of dilation, for the walls of a crack would shrink very slightly in a direction normal to the surface as they become chilled by circulating water. This shrinkage would be very small at first, when only a thin skin of rock has been chilled; but after a long period of heat extraction it could become significant. Even a miniscule widening of a narrow fissure, however, could have a significant and beneficial effect upon flow resistance as cited above.

Let it be supposed that a vertical fracture of undetermined shape has been formed within a mass of hot competent rock. Figure 8.16(a) shows such a fracture in elevation. Initially the wall of the fracture will be subjected at all points to *compressive* stresses – S_H in a horizontal direction and S_v, or S_L if the formation is in repose, in a vertical direction. Usually S_L will be the greater. Now let it be supposed that circulating water is allowed to flow across the surface of the fracture. The water temperature will initially be much lower than

Fig. 8.15 FRIP results for an example of classical fracturing. (a) Injection pressure versus time of injection. (b) Fracturing profile after 5 seconds. (c) Fracturing profile after 10.5 seconds. (d) Pressure distribution after 5 seconds.

that of the rock; in the course of time it will remain lower, but the temperature difference between the water and the rock will become less. This temperature difference will cool the surface of the rock and, to a decreasing extent, the rock *behind* the surface. This will provoke a tendency for the rock to shrink, especially at and near the surface; but at first the pre-existence of compressive stresses will prevent actual shrinkage. Instead, the compressive stresses in the rock will be reduced at and near the surface; the extent of the reduction being taken up by somewhat increased compressive stresses in the body of the rock further behind the fracture.

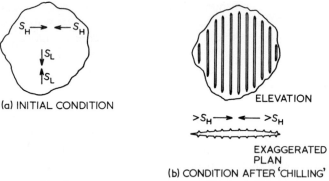

(a) INITIAL CONDITION

ELEVATION

(b) CONDITION AFTER 'CHILLING'

Fig. 8.16 Thermal cracking in a vertical fracture.

Depending upon the intensity of the stresses, the linear coefficient of thermal expansion of the rock, Young's modulus of elasticity for the rock, and the rate at which the water removes heat from the walls of the crack, a time may come when the horizontal compressive stress S_H is fully relieved and the skin of the rock will pass from a state of compression to one of tension. This point will be reached, it should be noted, before the greater initial *vertical* compressive stress has been relieved; so that at some time the skin of the rock in the fracture will be in horizontal tension and vertical compression simultaneously.

If the cooling process continues far enough, the horizontal tension in the skin of the rock will become greater than the tensile breaking stress of the rock S_t, and vertical hair cracks will start forming. Further shrinkage will widen these cracks and admit water more deeply into the hotter body of the rock behind the original crack surface, and the cooling effect will become more pronounced. Ultimately, a series of rib-like secondary cracks will form all over the main fracture surface in the manner shown (to a greatly exaggerated scale) in Fig. 8.16(b). With further temperature reduction at the rock surface it is possible that even S_L may be overcome and the vertical stress reversed from compression into tension, so that ultimately horizontal cracks may appear across the surface of the original fracture, thus forming a sort of 'waffle' grid of grooves.

Thermal secondary cracking would of course not occur suddenly in all parts of the primary crack, for when circulation is first established there will be intense cooling near the point of water entry and far less cooling near the point of water exit. In the course of time the 'cooling wave' will spread towards the exit. Thermal cracking will therefore appear gradually and progressively and should reduce water flow resistance and aid the heat mining process.

Whether, and to what extent, secondary thermal cracking will occur will depend upon the depth, the temperatures of the rock and of the water at their contact interface, the thermal expansion coefficient and Young's modulus for the rock, the rate of water circulation and the total time lapsed for heat extraction.

In some cases in the field, it may be difficult to distinguish between changes in the behavior of the reservoir caused by thermal stress cracking or by other effects. For example, continuous pressurization of a two-borehole system during extended operation could give rise to secondary fracturing away from the primary hydraulic fractures. This could also increase the effective heat transfer surface area as much as thermally-induced fracturing. Some evidence of secondary, pressure-induced fractures has been observed in the seismic data at Fenton Hill and Rosemanowes (see Chapter 9 for more details).

8.10 Survival of original voids in rocks

Permeability tends to decrease with increasing depth. This is simply the result of the rising lithostatic pressure, aided sometimes by high lateral tectonic stresses. A mass of shingle can retain a high volumetric proportion of voids near the surface; but if more and more material were loaded on top of it, some of the individual stones would be crushed by the sheer weight from above, and the smaller fragments would displace some of the void space so that the overall proportion of voids would be reduced. Another apt analogy would be a sponge confined between two horizontal plates and subjected to increasing pressure from the upper plate.

It is of course true that a decrease in void space or porosity must lead to a decrease in permeability. The Kozeny–Carmen equation, although idealized, gives a simple relationship between permeability (k) and porosity (ϕ) as:

$$k = a^* \phi^3 / (1 - \phi) \qquad (8.37)$$

where a^* is an empirical constant dependent on the material under study. One would expect in practice that a more complex relationship might exist between k and ϕ; however, some data for low porosity materials ($\phi < 0.01$) show a surprisingly good correlation with k varying with the cube of ϕ.

At very great depths, the intense lithostatic pressure tends to squeeze all voids out of existence. It is maintained by some that a small degree of vestigial permeability will persist right through the crust as far as the Moho; others believe that no permeability could survive below some unspecified depth.

Observation at Los Alamos confirms that permeability of the competent rock matrix decreases with increasing depth. However, one would expect that an asymptotic value of permeability would be reached corresponding to the practical maximum compaction limit for that particular rock type. If this were an established 'law' it would support the theory that *some* permeability persists even at great depths.

The importance of permeability at depth to heat mining is three-fold. In the first place it will be recalled that heat mining requires the formation of an underground labyrinth at great depth. Because a labyrinth is no more than a reservoir of artificial permeability, a major question is whether it will survive. The answer is probably no, unless the reservoir is permanently dilated by the application of an adequate hydraulic pressure within the voids and fissures by self-propping due to shear displacement or by the use of proppant materials of high strength. Secondly, deep underground artificial fractures, even in so-called impermeable rocks, are not completely water tight – at least to the depths (of up to about 4.5 km (14 000 ft)) at which such fractures have so far been made. Small, but significant, losses of circulating water are experienced. This point has been mentioned in Chapter 7, Section 3. If we ever succeed in penetrating to far greater depths, could we expect to form a water-tight labyrinth? That is at present an unsolved problem. Finally, although it is known that voids of one sort or another certainly exist in most rocks down to depths of at least three or four kilometres, and although such voids are scattered about the crust at random, would it be possible to take advantage of them at some point quite a long distance away? This may sound an odd question to raise, but it could have a practical application. If we were to apply a huge pressure at one point, underground, could we squeeze out of existence all the natural voids within a certain distance from that point so as to make room for new void space at a point of our own choosing?

It is conceivable that the answer could be yes, if the applied pressure were sufficiently high to cause massive displacement of a significant volume of the earth – as, for example, in the case of an underground nuclear explosion. But the complete annihilation of voids may not be necessary. New voids could probably be created simply by *reducing* the size of pre-existing voids in their neighborhood, the degree of reduction diminishing with the distance from the location of the new voids. In other words, voids newly created, either by hydraulic pressure or by the forces of a nuclear explosion, could come into being by a process of displacement; so that pre-existing voids close to the new voids would be greatly reduced in size while those further away would be robbed of less of their original volume.

8.11   Fracture patterns

It has been shown in Section 8.6 that the steady application of a high hydraulic pressure at a point lying deep within a mass of competent rock can form monoplanar cracks. Even though a crack may change direction at a point of discontinuity in the geological structure, cracks created by hydraulic pumping will still effectively be two-dimensional in a normal stress field. Single fractures of this sort by no means conform with the desired qualities of an ideal underground labyrinth as set out in Chapter 7, Section 2, for the 'heat capture zone' of such a fracture would be confined to a fairly thin slab of rock, either flat or bent, as shown in Fig. 8.17.

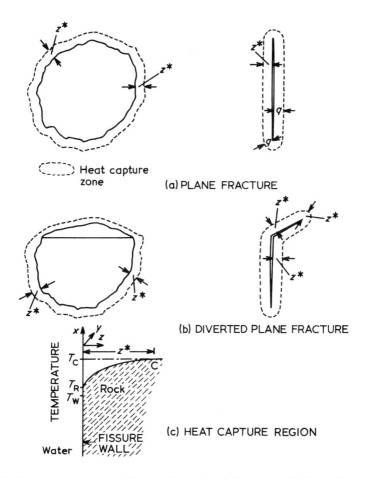

Fig. 8.17 Heat capture zone with two-dimensional fractures. (a) Plane fracture. (b) Diverted plane fracture. (c) Heat capture region.

The heat capture zone, as illustrated in Fig. 8.17, is more of an abstract conception than a precise factual entity. Figure 8.17(c) represents diagrammatically the sort of temperature distribution that could be expected within a zone of rock close to the fissure, in accordance with the arguments advanced in

Chapter 7, Section 7. The water temperature adjacent to the rock wall in the fissure would have some value T_w, while the rock at the same wall would be at T_r: the very small temperature drop $(T_r - T_w)$ being necessary for the transfer of heat from the rock to the water. As described in Chapter 6, Section 2, this differential would be less than 1°C for normal heat transfer conditions with water flowing through fractures. Within the body of rock close to the fracture the temperature would follow some such curve as $T_r - C$, which will be asymptotic to T_c, the temperature of the 'core' of the rock at some distance z^* from the fracture where the effects of heat extraction would be virtually unfelt. The use of the word 'virtually' is unavoidable if z^* is not to be infinite, since the temperature curve is asymptotic to T_c. If z^* be defined as the distance at which the rock temperature is never lowered by the heat mining process by more than, say, 1°C within some acceptable 'life' of the mining zone, then the heat capture zone can be endowed with some pragmatic, though to some extent arbitrary, meaning.

Figure 8.17(a) shows the conceptual heat capture zone for a planar fracture; Fig. 8.17(b) for a *diverted* fracture. The volumes of the heat capture zones for these two cases would not differ greatly, but they are both primarily of a two-dimensional nature in that they have only moderate width but large area. (It is emphasized that Fig. 8.17 is not drawn to scale.)

Figure 8.18 illustrates some conceptual *three dimensional* patterns. Figure 8.18(a) shows a tri-axial concentric cluster of radial planar cracks: Fig. 8.18(b) shows a rubble-filled cavity. In the first case an approximately spherical heat capture zone would result; in the second case the heat capture zone could be somewhat amorphous; but in both cases the zones are essentially three-dimensional in that length, breadth and height are all of comparable magnitude. It is also clear that in both cases the ratio of rock/water interface area to the rock volume of the heat capture zone will be much greater than in the case of the single two-dimensional fracture, which means that a hot rock zone of *given volume* could be mined more rapidly than with a single planar fracture. This is of considerable economic importance, as will be shown in Chapter 10. Yet another approximation to a three-dimensional fracture pattern is shown in Fig. 8.18(c), in which a series of parallel planar fractures is arranged side by side at such a spacing that their heat capture zones can meet, or slightly overlap. Of course, as the fracture density is increased, thermal interactions will occur until the heat transfer characteristics resemble a fully rubblized zone.

In comparing these cases we have examined the extremes of two different heat extraction strategies – areal sweep through individual isolated fractures where the rate of heat conduction through the adjacent rock limits the heat extraction rate, versus volumetric sweep through a rubblized zone or a true labyrinth where only the heat contained in the rock volume limits the total amount of heat that can be mined. There is, of course, still a small residual heat conduction resistance within the rock lumps that limits the rate of heat extraction.

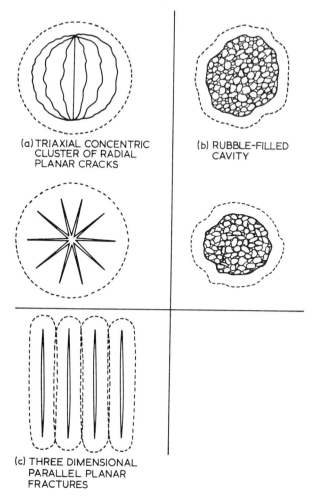

Fig. 8.18 Heat capture zones with three-dimensional fracture patterns. (a) Triaxial concentric cluster of radial planar cracks. (b) Rubble-filled cavity. (c) Three dimensional planar fractures. – – – Boundaries of heat capture zones.

Whether and how the various fracture patterns of Fig. 8.18 can be achieved in practice will be considered in Chapters 9 and 10; for the present, it is only the possibilities and potential advantages of three-dimensional fracturing that is being considered. Whereas the rubble-filled cavity of Fig. 8.18(b) would provide a true labyrinth, the star-shaped assembly of planar fractures shown in Fig. 8.18(a) would not. However, an assembly of several such star-shaped clusters, placed fairly close to one another, would provide opportunities for fractures to intersect with one another, and thus to form the rudiments of a true labyrinth. Alternatively, the fractures might be placed so close to one another that there could be a good chance of lateral thermal cracks (see Section 8.9 above) interconnecting the main fractures with one another: this could perhaps be a possibility.

In summary, three-dimensional fracture patterns would be more conducive to efficient heat extraction than two-dimensional patterns. A fully rubblized three-dimensional cavity would be best, while a system of parallel planar fractures would be the next best option.

Chapter 9

Rock fracturing practice

9.1 Overall objectives of field programs

After regions of hot rock have been located and holes have been drilled to gain access to the resource, the next step is the creation of artificial permeability for heat extraction purposes. In low-permeability formations a suitable network of fractures must be produced for heat mining. Fracture initiation and growth have been treated theoretically in the previous chapter and must now be considered in terms of actual field experience. The discussion of reservoir performance under continuous circulation and heat extraction conditions is deferred to Chapter 10.

Ever since 1970, several countries have been actively engaged in field experiments aimed at demonstrating the technical feasibility of heat mining. The United States and the United Kingdom have been the most active with large field programs and several reservoirs either tested or in the developmental stages. The West Germans, Japanese, and French have programs of their own while the West Germans, Japanese, Swedes and Swiss participated jointly with the American and British in the so-called MAGES project [*M*an *M*ade *G*eothermal *E*nergy *S*ystems] (see Section 9.4.3 for details). The US work on HDR has been almost exclusively conducted at the Fenton Hill site located in Northern New Mexico on the western flank of a large extinct volcanic complex, the Valles Caldera. This site was selected for many practical and logistical reasons and because of its high geothermal gradient (approximately 60°C/km on average) and reasonable accessibility to crystalline basement rock of low permeability at depths of less than 1 km. The Los Alamos National Laboratory directs the US project under sponsorship from the US Department of Energy (US DOE) with financial and technical participation from the Federal Republic of Germany and the Government of Japan. From 1979 on, the Federal Republic of Germany with its agent Kernforschungsanlage Jülich GmbH became a 25% cost sharing participant; and in 1981, the Government of Japan became a second 25% partner. At present about fifty professional scientists and engineers are involved with the project at Fenton Hill.

The British project is similarly sized in terms of technical staff and financial backing. The British site is in a granite quarry at Rosemanowes in the Cornish

161

granite located in the extreme southwestern part of the UK. As with Fenton Hill, Rosemanowes was selected because of practical logistics as well as its higher than normal geothermal gradient of approximately 30 to 40°C/km, with heat flows ranging from 100 to 130 mW/m² (Garnish, 1976). The Camborne School of Mines officially directs the project with financial support provided by the UK Department of Energy and the European Economic Community, (EEC or Common Market) as a part of their research programs.

The West German, French, and Japanese field efforts in HDR are much smaller in scope and financial support levels but active nonetheless. The West Germans have experimented with hydraulic fracturing at two sites, a 3300 m borehole in Urach and a 300 m deep borehole at Falkenburg. The French have focused on shallow 200 m hole fracturing tests at Mayet de Montagne, 25 km southeast of Vichy. Future plans call for a 20 MWt system in 4500 m deep bores at a site north of Limoges. The Japanese have worked at two different sites in shallow holes but plan to develop a full-scale HDR system at 2000 m in 200°C rock at Kurayoshi.

The next two sections in this chapter deal with field techniques for forming an HDR reservoir and with measurement techniques for determining what has been created. The final section of the chapter covers field results. Because the American and British programs are so large in scale by comparison with other activities, results from their tests will dominate our discussion of rock fracturing practice.

As discussed in Chapters 7 and 8, in order to optimize the heat mining process it is important to place the injection and production wells properly so as to maximize the rock area and volume that the fluid sweeps across as it passes through the underground labyrinth. Reasonably low fluid losses must also be maintained. In most cases the host rock will initially be almost impermeable, so that fractures must be created to permit low-impedance flow. If this creation process proceeds from an optimally positioned well, there are basically two steps to be followed: fracture initiation and propagation. To date the field programs have focused primarily on developing suitable techniques to initiate and propagate fractures in low-permeability crystalline rock as well as on the required diagnostic tools to characterize the geometry of the fractures produced.

9.2 Techniques for forming HDR reservoirs

9.2.1 Fracture initiation

Open natural fractures or joints contained in the host rock are likely candidates to be used for heat mining purposes. Unfortunately, many of these natural fractures have become tightly sealed from hydrothermal deposition of secondary minerals at depths of interest. Consequently activation or stimulation is required to reopen them. If the rock is totally competent, cracks must be initiated by some process. Whether the process is one of mere activation or true initiation, open fractures need to be placed at particular points along the borehole so as to maximize the swept area or volume of rock in the fracture network. Four possibilities exist for treating isolated sections of the wellbore so as to exceed the confining stresses and effective tensile strength, thus causing formation breakdown.

(a) *Hydraulic pressurization* by pumping water or other fluids.
(b) *Rapid gas pressurization* by igniting a suitable propellant or gas generating system.
(c) *Explosive fracturing.* The Camborne method of detonating an explosive of modest intensity to induce shock waves that create a radial array of cracks growing outward from the borehole (Batchelor, 1982; Batchelor *et al.,* 1980).
(d) *Thermal stressing* created by injecting cold or hot fluid into the borehole.

Although all four of these techniques have been used either for stimulation work in oil or gas reservoirs or in natural geothermal reservoirs, hydraulic pressurization and explosive fracturing are now receiving the most attention in the US and UK HDR field programs. The main distinction between these four initiation methods centers around the time scales involved. As shown in Table 9.1, the rise times and extents differ by several orders of magnitude. Consequently, the failure mechanism and resultant geometry of the initiated cracks will be quite different in each case.

Table 9.1 Time scales associated with fracture initiation and propagation

	Time scales (s or seconds)	
	Rise	*Extent*
Hydraulic pressurization	10^1—10^3	10^3—10^5
Gas generation pressurization	10^{-3}	10^0
Explosive fracturing (general and shaped charges)	10^{-6}	10^{-3}—10^{-2}
Thermal stressing (Liquid nitrogen pulse or long-term injection of cold water)	10^1—10^5	10^3—10^6

Low-rate initiation methods which have time scales of the order of seconds to hours occur so slowly that only a 'single' fracture may be created or reopened at the weakest point in the open borehole. Then, no matter how complex the reservoir or labyrinth of fractures connecting the wells becomes, the near-wellbore region in either the injection or production well will always have a locally high flow impedance because of the relatively high velocities and limited open area in those regions. High-rate initiation techniques should produce multiple fractures within the borehole and thus reduce the near-wellbore impedance problem.

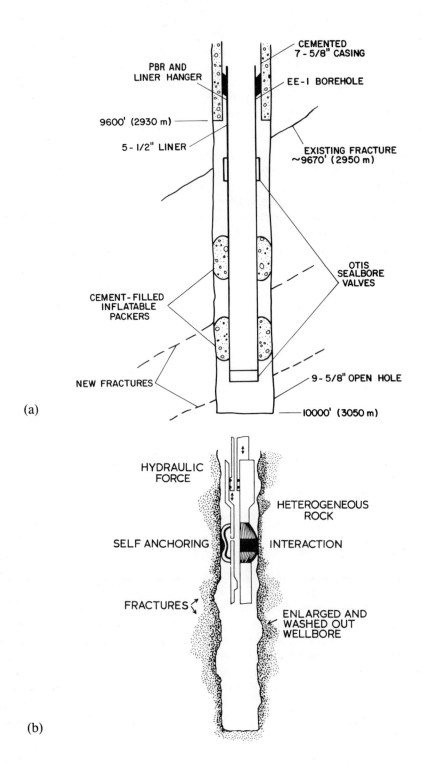

(a)

(b)

(a) IDEALIZED HYDRAULIC PRESSURIZATION

For the idealized hydraulic pressurization, one would expect behavior more or less to follow what was described in Chapter 8. On theoretical grounds, a fracture should initiate with its orientation perpendicular to the least principal horizontal stress. As will be shown in Section 9.4, some experimental justification exists for this but an equally likely mechanism is for the fracture or fractures to initiate at weak points (such as weakly-sealed joints) in the formation section that is being pressurized. In practice, a limited length of the borehole is hydraulically isolated using piping and elastomeric packers, cement plugs, or other sealant devices as depicted in Fig. 9.1. Assuming that the tubing string and plug/packer arrangement can be properly placed without undesirable leaks either in the tubing at joints or around the packers or plugs, then fluid may be injected into the well by using suitable positive displacement pumps. The pressure increases rapidly and linearly, controlled by the compression of the wellbore fluid, until a breakdown occurs, at which point the pressure begins to level off. In most hydraulic treatments, the second stage of propagation would commence immediately, pumping rates being increased and special fracture fluids added if desired. This is in contrast to the other three methods of initiating which would be conducted as separate operations before fracture extension.

(b) RAPID GAS PRESSURIZATION

The rapid gas pressurization method looks quite promising on theoretical grounds because it potentially eliminates the need for packers or plugs as well as any downhole tubing to deliver fluids (Schmidt, *et al.* 1981). A wireline or cable could be used to locate the gas generator at the desired depth horizon. In practice, a rapid chemical reaction which produces gaseous reaction products is used to generate a 'gas bubble'. The gas products are released so quickly via a deflagration-type mechanism that a high-pressure wave is produced over a limited section of the well, effectively confined by the borehole walls and fluid. The rise time for the pulse may be of the order of one milli-second while the total duration would be approximately one second. One commercial firm, Kinetech, markets a proprietary method they call KINEFRAC which supposedly ignites a rocket propellant fuel to generate the pressure pulse. They claim great success in pre-treating oil and gas wells to improve productivity. Unfortunately, experiments at Fenton Hill by the Los Alamos team were not able to produce satisfactory ignition of the KINEFRAC system. This failure was attributed to complications caused by the high temperature environment (200°C) of the Los Alamos wells.

Fig. 9.1 (a) Schematic showing tubing string and packer or cement plug arrangement to achieve hydraulic isolation of certain zones of boreholes. PBR is a polished bore receptacle that is designed to form a seal between the surface-connected fracture string and the liner. (b) Present packer concept for open-hole applications (after Dreesen and Miller, 1985).

(c) EXPLOSIVE FRACTURING

The 'ultimate' in the use of explosives for creating HDR reservoirs would be to detonate a nuclear explosion underground. This would create a vertical rubble-filled cavern that would fulfil all the ideal qualities of an underground permeable labyrinth, as listed in Chapter 7, Section 2. The properties of such a cavern – or 'chimney', as it is usually called – have been extensively studied by several investigators (Kennedy, 1964, Burnham and Stewart, 1975 and others) and have been summarized by Armstead (1978), pp. 292–297. Obviously this method would have to be confined to remote sites and the subsequent extension of chimneys would seldom be practicable, but the notion has obvious attractions as well as potential problems. Remote sites are needed because ground shock may be excessive. Although it has been claimed that most of the radioactive fission products of long half life would be trapped within a vitreous layer of fused rock, there could be a problem of contamination – not only of circulating working fluid but also of neighboring aquifers through migration. On balance, the political and social objections are at present considered to be insurmountable.

In the present context we are concerned with reservoir pretreatment by means of underground *chemical* explosions. Such pretreatment is designed to produce self-propped fractures radiating outward from the wellbore wall. This should provide a low impedance path at desired locations for fluid entering or leaving an HDR labyrinthine fracture network. It also permits wireline operation, thus avoiding the hardware complications of packers or plugs and pipes. Apart from the obvious safety aspects of handling explosives in the field, there are other considerations as Batchelor, *et al.* (1980, 1982) point out. Batchelor's detailed treatment of how properly to design an explosive pretreatment is summarized below.

Theoretically, the rapid explosive process on a micro-second scale creates shock waves of sufficient intensity to cause almost geometrically isotropic failure overcoming any anisotropic wellbore stress concentration and local heterogeneities of natural, sealed and unsealed joints intersecting the well. However, if the charge is too strong, it may result in plastic deformation and actually seal off the region of rock surrounding the well. Batchelor has illustrated this effect in the laboratory in transparent plastic (see Fig. 9.2). Furthermore, the explosively-induced stress must be not too far above the elastic range of the rock to avoid massive shear failure that may result in hole collapse.

The major design requirements center around: (1) proper specification of the effective compressive strength of the rock under dynamic load and (2) employment of the proper failure criteria with an anisotropic stress field ($S_X > S_Y$).

The Hoek and Brown (1980) empirical relationship is used to describe shear failure. Using measured values of dynamic-compressive strength on cores as well as field measured *in situ* horizontal stresses, the limiting stresses are given by:

$$\Delta\sigma = S_\theta - S_r < \sigma_c[(S^* + M^*S_\theta/\sigma_c)]^{1/2} \qquad (9.1)$$

where: $-\Delta\sigma=S_r-S_\theta=$the so-called 'deviator stress'
$\sigma_c=$dynamic uniaxial compressive strength
$S_\theta=$maximum tangential stress on borehole wall
$S_r=$maximum radial stress on borehole wall
$S^*,M^*=$material constants

Both S_θ and S_r are functions of the principal horizontal stresses S_X and S_Y and other rock strength properties.

The design problem now consists of controlling the deviator and S_θ stress magnitudes generated by the shock wave. Having selected an insensitive explosive type on safety grounds, Batchelor (1982) estimates the amount to be used and its configuration in the wellbore by iterative calculation using a 'state of the art' two-dimensional Lagrangian (2DL) computer model developed at Los Alamos National Laboratory.

For practical reasons, Batchelor's team has selected a well-characterized explosive known to be stable at temperatures up to 300°C. The charge used in field experiments at Camborne is a mixture of triaminotrinitrobenzene (TATB) and an inert substance. They also use a commercial hexanitrostilbene detonator ignited by an exploding foil which is electrically activated from the surface.

Systems are typically designed to fracture the well over an interval of 20 to 50 m. To treat a 30 m section of wellbore, a downhole tubular device containing

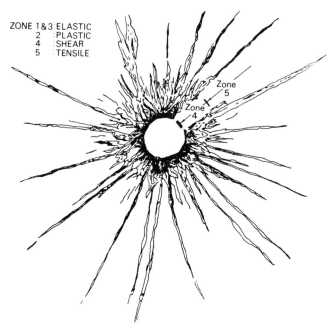

ZONE 1 & 3 : ELASTIC
2 : PLASTIC
4 : SHEAR
5 : TENSILE

Zone 5
Zone 4

Fig. 9.2 Explosive fracturing around a model wellbore in PERSPECS (after Batchelor *et al.*, 1980). Zones 4 and 5 shown represented in terms of stresses in Fig. 9.3. Zones 1–3 are not shown here but can be seen in Fig. 9.3 as a part of the early period of the stress build up.

a 5 m length of explosive (18 kg of TATB mixture) and 3 m of firing electronics would be used. These materials are housed in a steel or aluminium container with a 7.5 mm wall thickness to limit the explosive stress on the rock wall by taking advantage of the attenuation caused by the water annulus between the explosive device and borehole wall. The entire package is lowered in the hole and centered vertically over the treatment interval and radially to a symmetric standoff distance of 3.7 cm in a 15 cm-diameter bore. The 2DL code's simulation of this particular formulation is shown in Fig. 9.3. The predicted radial fracturing pattern zones are also located on the figure.

Fracturing over the 30 m length results both from the actual explosion and the shock wave produced that travels along the well in both directions. Because of the attenuation of the shock wave along the borehole, most of the energy is dissipated in a relatively small length, thus eliminating the need for any physical confinement of the package in the hole.

After firing the charge, the wellbore is cleared of debris and logged to character-ize the fracturing produced. Ideally, the explosive initiation procedure will generate a large number of radial fractures of length of order 1 m. Hydraulic stimulation to extend these fractures to develop the reservoir usually follows.

Actual field results of explosive pretreatments at Rosemanowes have been quite good. Aside from the obvious visual indication of detonation with a massive water plume blasting out of the borehole (as shown in Fig. 9.4 for an

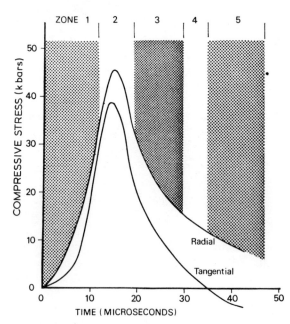

Fig. 9.3 The 2DL, two-dimensional Lagrangian code prediction of the rock wall stress–time behavior for a TATB charge contained in a 7.5 cm diameter device 8 m long in a 15 cm-diameter granite bore. (Batchelor, 1982).

early test at Rosemanowes) downhole logging showed that a 40 m section of borehole was extensively fractured. This can be translated into an effective dynamic tensile strength five times the static value if the two dimensional model is correct. Subsequent hydraulic stimulation and testing of this and other wellbores will be discussed in Section 9.4.

(d) THERMAL STRESSES

The final method of fracture initiation involves the inducement of thermal stresses in a wellbore so as eventually to cause local failure at weak points. This could be done in an active mode where a very hot or very cold fluid is introduced into a section of the wellbore. The resultant thermal stresses will rise with increasing temperature difference, ΔT_r, between the local and ambient or undisturbed rock temperature at that point. The effective maximum thermal stress ($\Delta\sigma_{th}$) can be estimated if we assume the rock behaves isotropically and elastically:

$$\Delta\sigma_{th}=\beta_T E\Delta T_r \tag{9.2}$$

where

β_T = thermal coefficient of expansion
E = Young's modulus
ΔT_r = rock temperature difference = $T(t)-T(0)$

By using a hot combustion flame ($T \cong 2000°C$) or boiling liquid N_2 ($T \cong -200°C$) sufficiently large stresses can be introduced rapidly to cause fracturing. In fact, these methods may also have application for drilling as will be seen in Chapter 11. However, aside from their potential economic disadvantages, active methods have another drawback in that the induced stress levels are so large and so close to the inside of the borehole wall that penetration of the fractures away from the well is limited. Perhaps it would be better to use a more passive treatment with lower induced stresses and deeper penetration. Since the penetration depth of the induced thermal wave front depends on the thermal diffusivity of the rock, which is inherently low, it might be better to use a smaller ΔT 'driving force' at the wellbore and wait longer to reach the critical value of $\Delta\sigma_{th}$ to cause failure. This is exactly the mechanism involved with reservoir thermal stress cracking as discussed in Chapters 7 and 8. However, in this instance although the secondary cracking from thermal stresses may be useful in prolonging reservoir life, it cannot be regarded as a method for creating a fractured reservoir of useful size at the outset. Furthermore, when stimulating one well we cannot take advantage of the connected labyrinth between bores to allow for once-through cooling because no such connection is presumed to exist before the reservoir has been 'created'. So a coaxial, cooling system designed to work in a single well would be required which would not be as efficient as a once-through system.

Given these alternatives and the present state of technology, hydraulic activation of existing weakly sealed joints and explosive pretreatment would be our first choices for fracture initiation at selected borehole locations.

Fig. 9.4 Exit vent plume from a 17 kg explosive charge at 290 m at Rosemanowes, Cornwall for pretreatment of granite (courtesy of Camborne School of Mines, 1980)

9.2.2   Fracture propagation

All techniques of growing fractures involve fluid injection at elevated pressures. As discussed in Section 8.8.2 'Current theories of hydraulic fracturing,' certain design selections must be made to optimize the treatment for producing the desired fracture geometry. These choices can be subdivided into several categories:

(1) fluid choice;

(2) proppant or no proppant;

(3) downhole hardware selection and plumbing set-up;

(4) pumping schedule (pressures, flow rates, etc.)

Selection may seem straightforward but one must recognize that an entire service-based industry has evolved to provide an enormous variety of possible schemes for hydraulic fracturing with very little hard data to back up claims. Choices range from water to exotic polymer-based fluids costing $1 US per pound or more. These special fluids are designed to reduce friction in the tubular casing and piping in the borehole and to increase viscosity up to several orders of magnitude above pure water, which would be useful in increasing fracture aperture as well as in suspending and transporting solid proppants within the fracture and to reduce fluid losses by filtering out on the surfaces of fractures as they are formed.

The next set of choices involves whether to use proppants, and if so what type and size range will give the best results. In all of the HDR stimulation work done so far except at Urach and a single experiment at Fenton Hill, solid proppants have not been used. The reasons for this are partly due to the higher costs and lack of performance data for proppants in high-temperature (200° to 300°C) hydrothermal environments. For example, many proppants are silica-sand-(primarily SiO_2)-based and are apt to dissolve in very hot water. Possibly using aluminum oxide-based materials (Al_2O_3) or epoxy-resin coated sand could minimize this problem, but they are considerably more expensive than sized sand particles. The other, perhaps more significant, reason for their lack of use is because self-propping by asperities on the rock fracture faces is apparently sufficient to reduce flow impedances to acceptable levels and avoid any need for proppants. However, this does not preclude their use in future HDR developments. In fact, in the limited HDR field tests, large reductions in flow impedance at the fracture-wellbore connection were observed. However, other testing with proppants at Urach and in tests to stimulate poorly producing hydrothermal wells with hydraulic fracturing have been unimpressive or shown only limited success (Sinclair *et al.*, 1984).

The third category of choices involves the set-up of hardware and tubular goods in the wells. Fluid is pumped either inside the well casing itself or in a smaller diameter fracture tubing string that is frequently run inside the casing to carry packers or to allow for ultra-high pressure levels above that possible in the larger diameter casing. It is usually desirable to minimize frictional pressure losses in the injection string to permit the highest possible downhole pressure at the fracture initiation point. This also helps to keep pumping power requirements within bounds. Once the borehole piping arrangement has been selected, several other pieces of hardware need to be specified including a downhole-activated valve mechanism for pressurizing packers, centralization devices, and flow control valves for multiple zone injection.

Finally, surface pumping equipment and an appropriate pumping schedule and control strategy need to be selected. The usual options pursued are either to set a flow rate or sequence of rates for the duration of the treatment or to adjust

Fig. 9.5 Gas turbine driven positive displacement pump trucks on site during a hydraulic fracture treatment at the Fenton Hill, New Mexico Hot Dry Rock site (photo courtesy of Los Alamos National Laboratory).

the rate to maintain a fixed injection pressure or a specified sequence of pressures. Because of the high injection rates employed, skid- or truck-mounted positive displacement pumps powered by internal combustion or gas turbine engines are arranged in tandem with suitable manifolding and centralized controls. Figure 9.5 shows a set-up for a large fracture job at the Los Alamos Fenton Hill site. The largest units of this design can inject fluid in excess of 100 bbl/min (560 kg/s) at wellhead pressures of 5000 psi (34.5 MPa) or more. If chemicals or proppants are to be used, then mixing tanks or blenders must also be included. For straight water fracture jobs, water would be pumped from a storage area, filtered, and fed to the positive displacement pumps using lowhead centrifugal pumps or 'superchargers' as they are commonly called. Friction reducer additives might also be added directly to the water as it is pumped.

9.3 Reservoir diagnostics

The actual pumping operation can be remotely monitored for seismic activity in the far field away from the borehole by using a suitable array of sensitive 'listening devices' placed in nearby observation holes. In addition, other borehole diagnostic techniques are frequently used before and after the stimulation to identify and characterize any changes to the formation. These latter borehole tests can be subdivided into two broad categories: (1) physical structure and property measurements and (2) hydraulic property or flow characterization tests. In the first category are included cores taken during drilling to identify regions with natural fractures as well as a host of geophysical logs such as sonic televiewer, resistivity, induced or self potential (IP/SP) and multi-arm caliper. The second group would include pressurization tests to determine *in situ* permeability along the wellbore surface and temperature, radioactive tracer and turbine flow (spinner) logs to locate production or injection regions. These are discussed in more detail below.

9.3.1 Far-field monitoring (seismic and other methods)

The basic physical principle employed in the seismic method is to locate micro-seismic events within the reservoir by monitoring acoustic transmissions from the system, before, during and after hydraulic stimulation takes place. Usually arrays of geophone seismometers, hydrophones or accelerometers are located at fixed, known positions in observation wells surrounding the injection bore (see Fig. 9.6). Arrival times of a particular event are identified from the observed seismic wave pattern. By knowing the sonic velocity in competent rock, and the specific location of the monitoring devices; triangulation methods can be used to locate the event epicenter. Alternatively, the structure of the seismic wave pattern as schematically represented in Fig. 9.7 can be dissected to relate the time difference between the compressional P-wave's and shear S-wave's first arrival to a distance away from the seismometer. The seismic package shown in Fig. 9.6 has three orthogonal sets of geophones with spatial orientation in the borehole determined by the inclinometer array. This permits estimations of the directional angle of the incoming seismic wave by the sense of first motion using P-wave amplitudes on each of the three orthogonal geophone sets and a hodographic analysis of the combined waveforms (see Fehler, House and Kaieda, 1986 and 1987, Keppler, 1983 and House, 1986). Although this method is inherently not as accurate as a travel time triangulation method from a multiple array or seismic net, it can be used with only *one* observation well. Both of these methods have been successfully deployed in field tests at Los Alamos and Cornwall.

One key feature of seismic methods is their ability to collect information far from the wellbore. If induced microseismicity can be directly related to the initiation and growth of the hydraulic fracture itself, then they provide a powerful mapping tool useful to both the proper design and operation of the HDR system with respect to heat transfer and hydraulic performance (see Section 9.5). Although this is a hotly debated point among seismologists, the majority opinion is that microseismic emissions occur where significant pressure increases exist above the 'ambient' initial pore fluid pressure levels

(Pearson, 1981 and Albright and Pearson, 1982). Two primary mechanisms of displacement can possibly generate the observed signals: a shear mechanism whereby planar joints or fractures slide past each other and a tensile mechanism associated with fluid-driven tension cracks induced in competent rock (Batchelor, *et al.*, 1983). Most data collected to date strongly suggest that the shear mechanism is dominant.

The early Los Alamos work relied almost entirely on hodographic analysis of microseismic signals detected with single downhole geophone tools. Normally, a three-axis array of geophones was oriented with respect to a known wellbore position by accurate depth and angle measurement from cable length, and the output of a two-axis inclinometer. After the package was lowered to a prescribed depth in the borehole, it was acoustically coupled to the borehole wall by a retractable locking arm (see Fig. 9.6). Absolute orientation was then determined by comparison of the instrument package orientation relative to a known borehole survey. The geophones themselves have a natural resonating frequency of 10 to 20 Hz, but above 100 Hz the response of the instrument package is essentially linear. Microseismic signals from each direction are carried up a multiconductor cable and recorded as separate continuous signals on magnetic tape. For the tool shown in Fig. 9.6, amplification gains of 40 000 were possible. This permitted measurement of extremely small events with Richter magnitudes between -6 and -2 or about four orders of magnitude below the threshold that can be felt by humans (see Fig. 9.7 and Pearson and Albright, 1984).

Refinements in instrumentation and data processing were later introduced at Los Alamos and Camborne to provide for rapid signal processing on-line during fracturing operations. For example, at Rosemanowes, 8 kbytes per channel of signal information from their surface-mounted vertical axis accelerometers and their downhole hydrophone array were digitized after automatic triggering by the first arrival of a minimum-sized event. The digitized data were then transmitted directly to a VAX 11/780 computer for disc storage and preliminary analysis (Batchelor *et al.*, 1983). Figure 9.8 is a block diagram showing the major processing steps for their seismic data.

The acoustic methods described so far are *passive* in nature in that they 'listen' to events caused by the stimulation process and hopefully map the fracturing process itself. A second class of *active* acoustic measurements has been proposed for fracture mapping. Single hole and hole-to-hole techniques have been proposed to study such things as the attenuation characteristics of the formation under pressurized and unpressurized conditions. If, for example, extensive water-filled open fractured regions exist between boreholes or near a borehole, one would expect some attenuation especially of shear waves. Early tests at Los Alamos involved using an acoustic piezoelectric transceiver device as well as explosive detonator/geophone package arrays for generating and monitoring signals. Time of flight, waveform, and energy spectral analyses were then used to quantify changes due to fracturing. Although the results from these

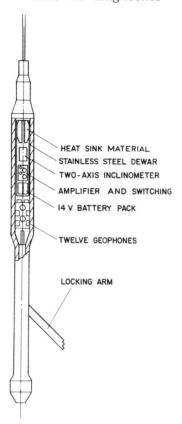

HEAT SINK MATERIAL
STAINLESS STEEL DEWAR
TWO-AXIS INCLINOMETER
AMPLIFIER AND SWITCHING
14 V BATTERY PACK

TWELVE GEOPHONES

LOCKING ARM

Fig. 9.6 Downhole siesmic package developed by Los Alamos National Laboratory for *in situ* microseismic measurements. (From Albright and Pearson, 1982).

active tests showed significant changes to S-wave (shear) and P-wave (compressive) velocities and power attenuation to both shear and compressive wave components that could be correlated with structural changes, no definitive mechanism could be proposed (Brown *et al.* 1979, Pearson *et al.* (1983), Aki *et al.* (1982), and Fehler, 1981). Subsequently, passive acoustic emission methods have been mostly used in field testing because they seem to provide less ambiguous results in regard to fracture location. The exception to this is the use of active methods for calibration purposes – whereby sound speeds and distances can be measured with great accuracy (Pearson and Albright, 1984).

Other far-field measurement methods have been proposed. Accurate measurement of the displacement field surrounding the stimulated region using strain or tilt meter displacement devices would theoretically give a direct measurement of the hydraulic disturbance as well as its orientation. However, the difficulty of doing extensive downhole measurements has limited studies to

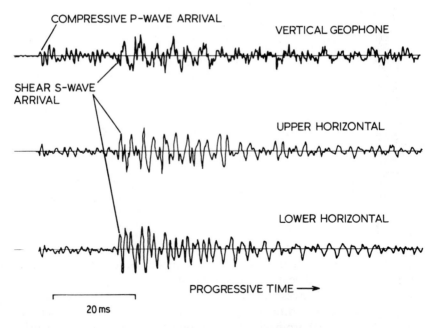

Fig. 9.7 Typical microseismic waveform measured on a three-axis Los Alamos geophone array. (From Albright and Pearson, 1982).

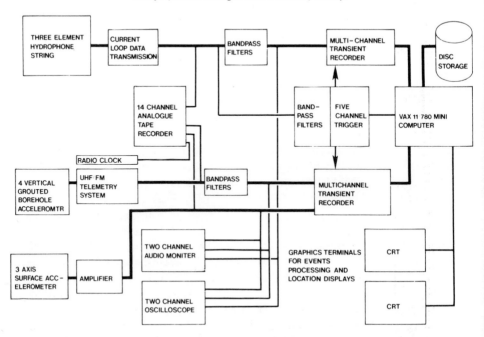

Fig. 9.8 Camborne Hot Dry Rock project microseismic monitoring data acquisition system schematic. (From Batchelor *et al.*, 1983).

the surface where attenuation is large and much ambiguity has been introduced by any heterogenous layers contained between the surface and the active reservoir region. Measurements of changes in the gravitational field or formation resistivity introduced by fluid injection have been tried with only marginal success and many interpretation problems.

Radar imaging theoretically has promise as a far-field electromagnetic monitoring technique. Initial research focused on characterizing the microwave transmission properties of granite and on conceptual downhole tool design (Cremer, *et al.*, 1980). No field tests in HDR reservoirs have been conducted as yet.

Another novel proposal for fracture mapping consists of injecting a magnetic ferrofluid into a hydraulically fractured crack and measuring local changes in the earth's magnetic field. Ground level experiments involving simulated fractures using an ultrasensitive superconducting quantum interference magnetic gradiometer or SQUID have been attempted at Los Alamos with some success. (Cremer *et al.*, 1980). Again downhole implementation has not been attempted as many technical and economic questions still remain.

In a very unusual investigation by Sandia National Laboratories, fracture mapping was achieved directly by excavating a portion of the reservoir for visual observation (Warpinski, 1985). In order to minimize any disturbance caused by the excavation process, fracturing fluids contained cement grouts and colored proppants which hardened in place. These so-called 'mine-back' experiments provided unequivocal proof of the planar, stress-direction dependent orientation of hydraulic fractures near the borehole as well as the extent and distribution of proppant transport. It is important to recognize that the Sandia tests were conducted in reasonably uniform, unjointed rock and that only the near-wellbore region was studied.

Because of the inaccessibility of HDR reservoirs at depths over 2 km, a mine-back approach would not be economically or even technically feasible. However, redrilling through a pressurized fracture region as was done at Fenton Hill provided direct evidence for fracture intersection. If carried to the extreme of many redrilling paths, although economically unreasonable, drilling itself does provide a fracture mapping tool.

9.3.2 Physical structure and property measurements

Cores of the reservoir rock provide information on the fabric and strength of the material that is to be stimulated by hydraulic fracturing. For example, they can identify weak zones, sealed joints, and changes in mineralogy or permeability that may be important to the design of the stimulation treatments. However, coring operations are expensive; they require separate bit runs, with costly core bit and core barrel assemblies and considerable rig time. Consequently, cores are only taken periodically in an HDR drilling campaign (see Chapter 11). Another factor is that removal of the core from the reservoir can produce irreversible changes that are not representative of actual *in situ* conditions. As the reservoir gets deeper and hotter these changes become more severe.

Geophysical logging of the boreholes is a viable alternative to coring that can provide a continuous record of actual *in situ* conditions along the length of the borehole. Although many logging techniques have been developed for oil and gas formation characterization, relatively few of them are being used in geothermal boreholes. This is partly because of the very hostile, high temperature environment that requires 'hardening' or thermal protection of downhole equipment using 'heat-sink' or heat-absorbing materials or dewar insulation. Even if the instrument itself is protected as was necessary for the Los Alamos geophone tool shown in Fig. 9.6, the multiconductor cable itself becomes a real problem at high temperatures. Above 250°C, sealant and electric insulation properties break down rapidly even for the best high temperature polymers such as irradiated PTFE (polytetrafluoroethylene) or EPDM. Figure 9.9 shows a cross-sectional view of a 7-conductor logging cable that was developed for the Fenton Hill bores for operation at temperatures above 200°C. The Italians, because of their long term research and development programs in geothermal energy exploitations, have successfully logged hydrothermal wells at temperatures up to 250°C (Ceppatelli and Ferrara, 1980).

Given these difficulties, logging efforts in geothermal wells have been limited to hole surveying using gyroscopic and magnetic compass/inclination measurements and relatively simple geophysical measurements including temperature, pressure, hole caliper, flow, gamma emission, and electrical properties such as resistivity and self and induced potential (SP/IP). Notable exceptions involving more sophisticated instruments are microseismic (mentioned earlier), sonic and optical televiewer, and radioisotope tracer measurements.

Although borehole surveying techniques have been well developed for oil and gas drilling, temperature hardening of the tools was required for surveying the

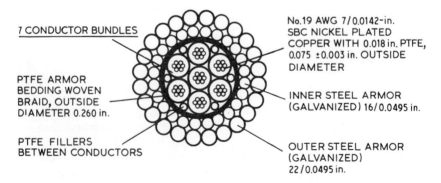

Fig. 9.9 Cross-sectional schematic of a seven-conductor logging cable developed for downhole service above 200°C PTFE=polytetrafluoroethylene polymer (Brown, *et al.*, 1979).

Los Alamos holes at Fenton Hill. Eastman Whipstock improved their heat sink protection and used high temperature film for their magnetic compass/plumb bob device which sequentially photographs a compass face and plumb bob location within the instrument as it is lowered in the hole. In this way, the hole's direction and trajectory can be mapped out. Sperry Sun modified their gyroscopic survey tool for use at Fenton Hill by adding more thermal insulation of their dewar canister that houses the downhole equipment. The gyro tool's orientation is fixed at the surface for calibration. As it is lowered into the borehole, a direct reading of hole direction versus depth is obtained. Even with certain local magnetic oddities inherent in the rock formation which would introduce error into the magnetic/plumb bob tool, agreement between the Eastman and Sperry Sun measurements was good.

The tracer, temperature, pressure and flow measurements will be discussed in the following section (9.3.3) as they relate more to the hydraulic character of the reservoir. Early experiments with resistivity and SP/IP measurements at Los Alamos revealed that they are of limited value in HDR reservoirs in crystalline rock – so that leaves the hole caliper, gamma emission, and sonic televiewer.

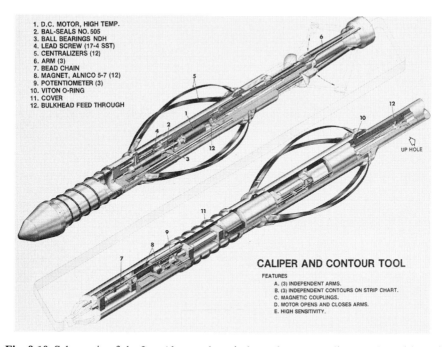

1. D.C. MOTOR, HIGH TEMP.
2. BAL-SEALS NO. 505
3. BALL BEARINGS NDH
4. LEAD SCREW (17-4 SST)
5. CENTRALIZERS (12)
6. ARM (3)
7. BEAD CHAIN
8. MAGNET, ALNICO 5-7 (12)
9. POTENTIOMETER (3)
10. VITON O-RING
11. COVER
12. BULKHEAD FEED THROUGH

UP HOLE

CALIPER AND CONTOUR TOOL

FEATURES
A. (3) INDEPENDENT ARMS.
B. (3) INDEPENDENT CONTOURS ON STRIP CHART.
C. MAGNETIC COUPLINGS.
D. MOTOR OPENS AND CLOSES ARMS.
E. HIGH SENSITIVITY.

Fig. 9.10 Schematic of the Los Alamos three-independent arm caliper tool. Position of each arm is monitored by a separate voltage signal generated by potentiometers independently linked to each arm's movement against the borehole wall. (Courtesy of Los Alamos National Laboratory.)

Rock fracturing practice

Multiple independent arm caliper measurements provide the most detail of the borehole geometry and are the best suited for fracture mapping applications. Los Alamos developed a three-independent arm tool with downhole orientation that has successfully been used above 200°C. Figure 9.10 illustrates the design of the Los Alamos caliper tool while Fig. 9.11 shows a typical set of data. The Italians have also developed tools for measuring temperature, pressure, fluid velocity, and hole caliper, and for fluid sampling at borehole temperatures up to 250°C (Ceppatelli and Ferrara, 1980).

Another feature of crystalline reservoir rocks that can be used to characterize morphology changes is the gamma emissions from radioisotopes contained in the rock, particularly uranium, thorium, and potassium. Changes in rock type that may correspond to boundaries or natural joints can be seen by spectral gamma logging where concentration changes in uranium, thorium, and potassium occur. In one analysis of the Fenton Hill reservoir, a lithologic contact between a gneiss section and an intruded biotite granodiorite body was located by observed potassium concentration changes in spectral gamma logs in three separate wellbores that penetrated the reservoir (Potter, 1977).

The sonic televiewer was originally developed by the Mobil Oil Corporation and later modified by Scott Keyes of the USGS and Sandia National Laboratory for geothermal service. The basic principle is to record sonic waves reflected from the inside surface of the borehole as a way of 'seeing' differences in structures, particularly the presence of fractures. The survey itself provides a three-dimensional picture of the borehole surface by collecting sonic signals on

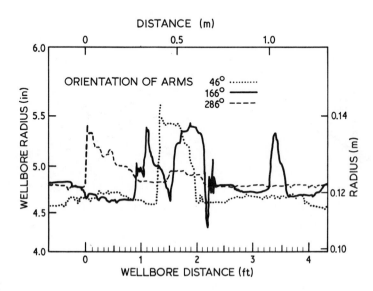

Fig. 9.11 Three-independent arm oriented caliper log of the GT-2B borehole at 2719 m (8920 ft) at the Fenton Hill, NM site.

an oriented rotating transmitter–receiver that moves along the borehole. Typical televiewer scans are given in Fig. 9.12(b) for the same section of Fenton Hill well GT-2B whose caliper log is shown in Fig. 9.11 and in 9.12(a) for a section of the RH-11 well at Rosemanowes. The televiewer scans are reproducible and clearly show the existence of planar fractures with the borehole. By using the techniques outlined in Section 9.3.3, the fractures accepting or producing fluid can be identified and related to the televiewer picture.

In lower temperature reservoirs, it has also been possible to obtain a continuous picture of the inside borehole surface with a small television camera for borehole logging operations. The Camborne project has used downhole TV observation extensively in the well characterization studies because camera exposure temperatures were below 100°C (see Fig. 9.12c). In order to use direct TV imaging in hotter environments, instrument protection and hardening would be required. Los Alamos was interested in such a development for a number of years but now seems to have postponed further work.

The results of all structural and property measurement geophysical logs and core analyses are eventually combined with other data to produce a model of the reservoir. Potter's (1977) model of one of the early Los Alamos reservoirs at Fenton Hill is given in Fig. 9.13 as an example. One can see that other information, particularly data related to flow inlet and outlet regions, is also part of his model. These data come from the second set of borehole tests that deal with hydraulic properties and flow characterizations as discussed in more detail below.

9.3.3 Hydraulic property and flow characterization tests

In situ permeability in low-permeability HDR formations is frequently measured by a transient injection test where fluid is pumped at a constant rate in an isolated section of the well. By assuming that the flow permeates homogeneously through a known surface area (A) in the borehole wall, Darcy's law can be used to estimate the formation's hydraulic diffusivity α_H and its permeability k. This involves using a known analytic solution to the transient diffusion equation where the measured surface pressure response at a fixed injection rate $\dot{q}$ can be used to estimate α_H and k:

$$\alpha_H = \frac{k}{\mu c \phi} = f[\dot{q}, A, P(t)] \tag{9.3}$$

where
    μ = fluid viscosity
    c = effective compressibility of the formation
    t = time
    ϕ = porosity
  $P(t)$ = pressure at a particular time t

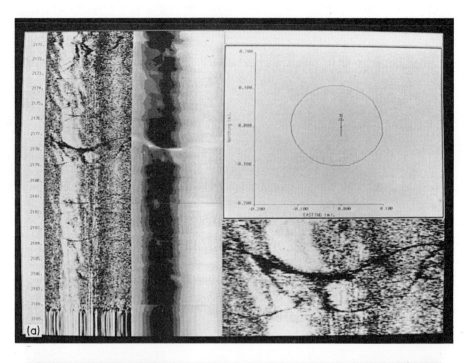

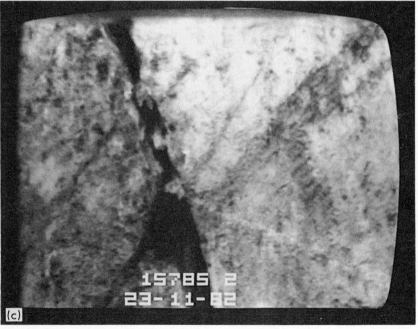

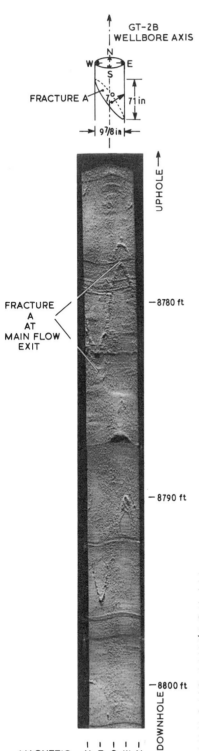

GT-2B
WELLBORE AXIS

N
W E
S

FRACTURE A ─ 7°
o
71 in

9⁷/₈ in

UPHOLE ↑

FRACTURE
A
AT
MAIN FLOW
EXIT

─ 8780 ft

─ 8790 ft

─ 8800 ft

DOWNHOLE ↓

MAGNETIC N E S W N
(b)   GT-2B OPENHOLE

Fig. 9.12 Sonic televiewer and television surveys of borehole sections. (a) Televiewer survey from RH-11 at Rosemanowes. (Provided by the Camborne School of Mines Geothermal Project.) (b) Televiewer survey from GT-2B at Fenton Hill. Fractures that intersect the borehole appear as 'S-shaped' lines because of the unfolding of the 360° sonic survey into the two-dimensional form shown. (Provided by Sandia National Laboratory.) (c) Television (TV) scan of RH-11 borehole at Rosemanowes showing an induced hydraulic fracture penetrating through natural joints at a depth of 1578.5 m. The fracture is propped open with grains approximately 5 mm in diameter.

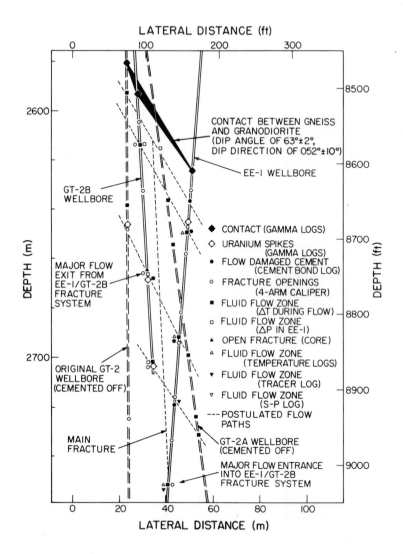

Fig. 9.13 Wellbore-fracture model of the early (Phase I) Fenton Hill system based on geologic, geophysical and hydraulic measurements. Elevation view perpendicular to the direction of the least principal stress. (After Potter, 1977.)

and the functional dependence of equation (9.3) of α_H to $\dot{q}$, A, and $P(t)$ is given by separate equations depending on geometric scaling factors.

For extremely low α_H or k, a one-dimensional geometry is appropriate and

$$P(t)-P_o=\frac{\dot{q}\mu}{Ak}\left[\frac{\alpha_H t}{\pi}\right]^{1/2} \tag{9.4}$$

for somewhat higher α_H or k, radially symmetric flow is appropriate and

$$P(t)-P_o=\frac{\dot{q}\mu}{4\pi kh}\,\mathrm{Ei}\left[\frac{-r^2}{4\alpha_H t}\right] \tag{9.5}$$

where P_o is the initial pressure, h the height of the permeating zone, and t the elapsed time. By fitting the pressure–time reponse to either equation (9.4) or (9.5) values for α_H can be obtained. If c and ϕ are reasonably well established then k can be estimated from known correlations of fluid viscosity to temperature. Comparisons of $P(t)$ behavior in the well before and after stimulation will show the relative increase in k and α_H as a result of the treatment.

Logging of temperature, flow, and radioactive tracer movement in either injection or production wells before and after hydraulic fracturing or after continuous recirculation can locate regions in the wells that accept or produce fluid. If discrete fractures are the primary fluid conduits, rather than diffusive volumetric flow through the rock matrix contained between the injection and production bores, then local peaks in temperature should be measurable. Figure 9.14 shows a series of typical temperature logs in the injection well of the original Fenton Hill reservoir. The progressive localized cooling of the wellbore as well as variable levels of injected flow into discrete fractures is evident by the complex structure of the logs.

The measurement of actual fluid velocities in the borehole, coupled to knowledge of the well's cross-sectional area obtained from caliper logs, can be used to determine the flowrate into or from a specific region. In the field, a so-called 'spinner' logging tool consisting of a fluid turbine with a magnetic pickup to measure rotational speed is used. The tool is calibrated either by lowering it at different measured velocities inside cased portions of the well or by holding it in a fixed position with a measured flow going past it. Figure 9.15 shows a spinner probe developed at Los Alamos and Fig. 9.16 gives the resulting log for a production zone in a Fenton Hill, HDR reservoir. The discrete nature of the flow is clearly evident from the step changes over the depth interval.

Tracer methods have also been used effectively to characterize wellbore flow (Tester *et al.*, 1982). By modifying a device and procedures originally designed for oil and gas reservoir analysis, the Los Alamos team was able to successfully develop a radioisotope injector-gamma ray detector logging tool (see Fig. 9.17). Although some of the early tests used I^{131}, most of the Los Alamos wellbore tracer tests have used Br^{82}, an isotope easily generated by neutron activation in

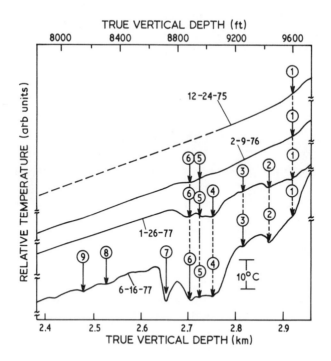

Fig. 9.14 Typical temperature surveys under injection conditions in the early (Phase I) reservoir from the US HDR program at Fenton Hill. Numbers indicate zones where measured flow was observed.

a nuclear reactor. Ammonium bromide, encapsulated in a quartz vial, was irradiated in the Los Alamos reactor to produce Br^{82} by neutron capture. Br^{82} has a short radioactive half life of 36 hr and is a strong gamma emitter and does not adsorb on rock surfaces; so it is potentially an ideal inert tracer that will follow the flow into and out of the borehole. In a typical application, the irradiated capsule containing Br^{82} would be suitably shielded and loaded into the tool, lowered into the well to the desired depth, the capsule broken mechanically by a ram activated by a signal from the surface, and flushed out into the borehole. Fluid would then be pumped down the well to force the tracer into the reservoir. Gamma logging would continue simultaneously during this injection period to record where and how much fluid is left. Also another log would be run in the production well to locate production zones as fluid containing tracers arrives there. A typical result for successive logs in one of Los Alamos's production wells is shown in Fig. 9.18. The location of flow from tracers not only provides an internal check on the spinner survey results

FLUID VELOCITY SPINNER PROBE

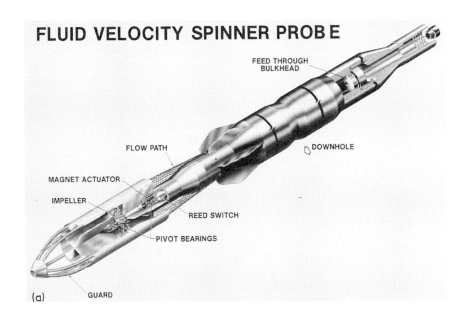

FEED THROUGH
BULKHEAD

FLOW PATH

DOWNHOLE

MAGNET ACTUATOR

IMPELLER

REED SWITCH

PIVOT BEARINGS

(a)

GUARD

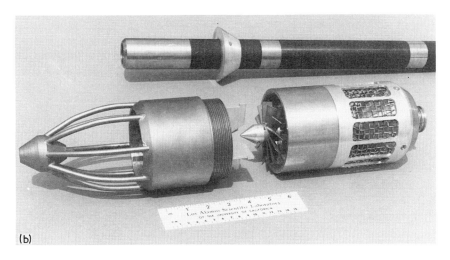

(b)

Fig. 9.15 Fluid velocity spinner probe developed at Los Alamos circa 1978. (a) Assembly. (b) Photograph.

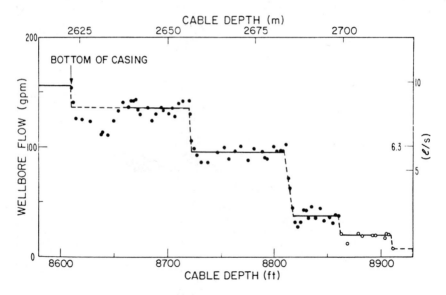

Fig. 9.16 Production well (GT-2B) openhole fluid flow distribution determined by the spinner probe shown in Fig. 9.15.

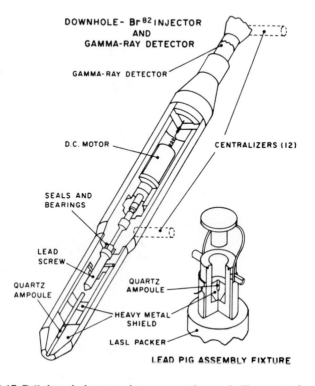

Fig. 9.17 Br⁸² downhole tracer instrumentation tool. (Tester *et al.*, 1982).

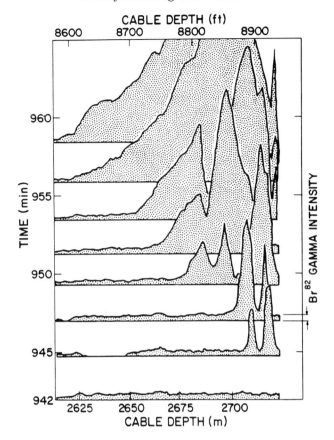

Fig. 9.18 Typical Br^{82} tracer results for a logging sequence in production well GT-2B at Fenton Hill showing successive appearance of tracer at discrete zones of fracture intersection. (Tester *et al.*, 1982.)

but it also gives an indication of the degree of mixing within a specific fracture region which will be important to the actual heat mining process. This point is brought out later in Chapter 10.

One other feature of the radioactive tracer technique is that measurements are possible behind the well's steel casing. This is very useful in testing for fracture flows that do not directly communicate with the wellbore as well as to identify certain 'pathological' conditions such as poorly cemented casing.

9.4 Field fracturing test results

Each project will here be discussed separately. First, the American project at Fenton Hill conducted by Los Alamos National Laboratory; then the British project at Rosemanowes; then the other field programs of Germany, France and Japan will be briefly covered. Finally a synopsis of the entire field fracturing programs will be presented. For each project we shall discuss the geologic setting of the field sites, the fracture design and injection data (pressure–volume–time) and the fracture mapping results including microseismic and geophysical logs. Emphasis will be placed on reservoir formation and initial diagnosis, while actual testing of the created reservoir will be treated in Chapter 10.

9.4.1 American project

(a) GEOLOGIC SETTING

The Fenton Hill site is located in the Jemez Mountains of northern New Mexico on the western flank of a large dormant volcanic complex, the Valles Caldera (see Fig. 9.19). The caldera lies at the intersection of the tectonically active Rio Grande rift and the volcanically active Jemez lineament. The site itself is approximately 27 km west of Los Alamos, 80 km north of Albuquerque and about 120 km south of the Colorado–New Mexico border. Within the caldera itself, about 8 km east of Fenton Hill, is the so-called Baca site, a natural hydrothermal system that was under development by Union Oil Company. Aside from the obvious volcanic character of the region, hot springs and other manifestations of geothermal activity abound.

Recalling that the main development objective of the Los Alamos project was and still is to demonstrate the feasibility of heat mining from hot, low permeability formations, the region outside the ring fault of the collapsed caldera provides an excellent setting because of the higher than normal geothermal gradients and the presence of crystalline basement, granitic rocks at depths less than 1 km.

The first deep borehole near Fenton Hill, named GT-2 for 'geothermal test 2,' was drilled in 1974 to 2.93 km into Precambrian basement rock where the bottomhole temperature was 197°C. The lithology and variation of temperature with depth is shown in Fig. 9.20. Later holes were drilled to a maximum depth of 4.398 km and bottom hole temperature of approximately 327°C. The relatively recent (<0.1 to 1.3 million yr) volcanic activity coupled with a higher than normal heat flow associated with the Rio Grande rift and a small radiogenic contribution from thorium and uranium isotope decay results in an average gradient of 89°C/km in the 720 m (2400 ft) upper section of younger volcanic and sedimentary rocks and from 53 to 80°C/km in the Precambrian basement section averaging about 63°C/km. The general increase in gradient with depth in the Precambrian correlates with the decrease in thermal conductivity of crystalline rocks as temperature increases while the somewhat sharper increase in gradient at depths below 4.2 km may indicate proximity to a magma body at greater depths. Based on a reasonable average conductivity of 2.8 W/mK and the observed range of gradients in the Precambrian corresponds

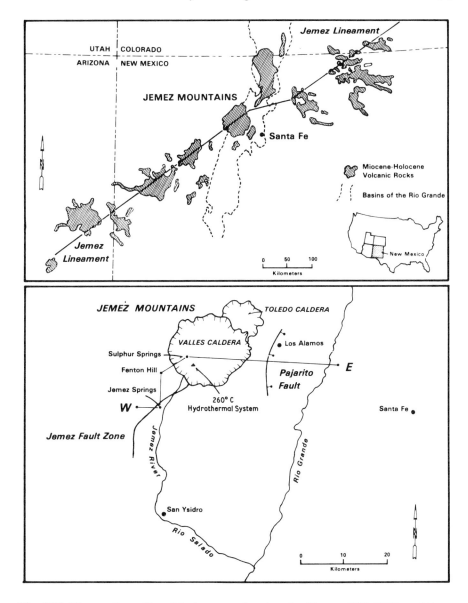

Fig. 9.19 The northern New Mexico area showing the Fenton Hill HDR and Baca location hydrothermal sites. (Murphy *et al.*, 1981.)

to a heat flow of 150 to 180 mW/m², about 2.5 times the worldwide average. Clearly the Fenton Hill site is an example of a high-grade hyperthermal HDR resource where rock temperatures above 200°C can be reached at reasonably shallow depths.

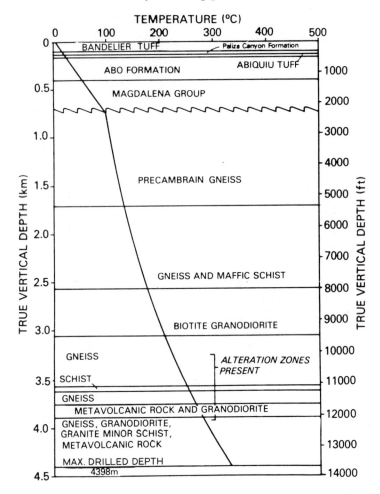

Fig. 9.20 Variation of temperature and lithology with depth at the Fenton Hill site.

Two major reservoir systems at two different depth horizons are under study at Fenton Hill. The first or so-called Phase I system involved two wells (GT-2 and EE-1 or Energy Extraction 1) with an active fractured reservoir in the biotite granodiorite section from about 2.6 to 2.9 km. The Phase I system was developed and tested from 1974 to 1981. The second Los Alamos system, or Phase II, consists of a pair of deeper wells (EE-2 and EE-3) with reservoir region that is presently being developed from about 3.4 to 4.4 km mainly in the metavolcanic section. Work started on the Phase II system in 1980. The Phase I and II systems are schematically depicted in Fig. 9.21 in relation to the Fenton Hill lithology.

Table 9.2 gives average mineralogical composition for the Phase I biotite granodiorite, and Phase II metavolcanic reservoir rocks. Four primary minerals

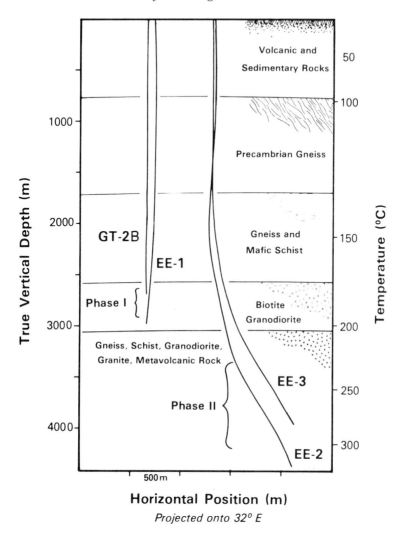

Fig. 9.21 Elevation view of Fenton Hill site showing Phase I (GT-2B/EE-1) and Phase II (EE-2/EE-3) wellbore systems in relation to the lithology.

are involved, quartz (SiO_2) about 20–25%, potassium feldspar about 20–30%, plagioclase feldspar about 35 to 55%, and biotite 1 to 12%. Both rock types contain many joints or natural fractures, at 1 cm to 10 cm intervals, easily identified on cores or geophysical logs. Almost without exception, all these joints are well sealed with a variety of minerals including calcite ($CaCO_3$), potassium feldspar ($KAlSi_3O_8$) and quartz (SiO_2) to give a low natural permeability of 1.0 to 0.01 microdarcies (μD) to the matrix rock. The presence of fluids related to magmatic events over the millenia probably causes the sealing process by dissolution and redeposition under the influence of a temperature gradient. In almost all cases at Fenton Hill, it is believed that hydraulic fracturing initiates at one or more of the weaker sealed joints in a pressurized section of open hole.

Table 9.2. Average mineral and chemical compositions of Fenton Hill reservoir rocks. (Adapted from Laughlin and Eddy, 1977; Laughlin *et al.*, 1983)

	Phase I *Biotite granodiorite* *modal %*	*Phase II* *Metavolcanic* *modal %*
Quartz (SiO_2)	26 ± 2	35 ± 10
Potassium feldspar ($KAlSi_3O_8$)	19 ± 2	25 ± 20
Plagioclase feldspar ($CaAl_2Si_2O_8 + NaAlSi_3O_8$)	36 ± 5	55 ± 30
Biotite ($K_2(Mg,Fe,Al)_6(Si,Al)_8O_{20}(OH)_4$)	12 ± 2	2 ± 2
Sphene ($CaTiSiO_5$)	2	—
Other trace minerals	5	5

(b)   FRACTURE DESIGN AND INJECTION DATA – PHASE I

As cited by Murphy *et al.* (1981) a series of trial hydraulic fracturing attempts were made in GT-2 to evaluate packer performance, characterize *in situ* stresses, and determine whether seismic signals could be detected. Following these, the second well, EE-1 was drilled towards the largest of the GT-2 fractures thus identified in an attempt to complete a heat extraction system. Due to wellbore survey errors during directional drilling, the EE-1 borehole missed the GT-2 fracture by 6 to 20 m, as determined by subsequent gyrosurveys as well as magnetic and acoustic ranging. Despite such close proximity, the intervening rock was so nearly impermeable that the impedance to water flow was too high for a viable heat extraction experiment. Various attempts such as fracturing the EE-1 well to improve flow communication between the boreholes proved unsuccessful, apparently because the fractures in both wells were vertical and parallel and did not intersect (Murphy *et al.* (1977). The EE-1 well was then cased to 2.93 km. Subsequent temperature and cement bond logs showed that the casing cement from 2.74 to 2.93 km deteriorated with time because of the high downhole temperatures as well as the thermal cycling caused by numerous flow and pressurization experiments conducted after setting the casing and cement. This deterioration resulted in the formation of a water flow path in the annulus between the rock and casing. Eventually, a large vertical hydraulic fracture was formed behind the casing in EE-1 at a depth interval centered about 2.75 km.

The first reservoir was completed by deviating GT-2 at 2.5 km and redrilling it toward the top of this large EE-1 fracture in a direction perpendicular to its strike. Two deviated portions were drilled, GT-2A and GT-2B (see Fig. 9.13 for details). The EE-1 borehole was kept pressurized to 7 MPa (1000 psi) during the redrilling operation to allow detection of any intersected fractures. Several pressure changes were noted during drilling. The intent was to produce as large a vertical spacing between the inlet and outlet locations as possible in order to maximize the effective heat transfer area while still achieving the reasonably low flow impedance which was required for high rates of heat extraction. The working pressure at the injection point is ultimately limited by the confining stress field, as excessive pressures during heat extraction could result in continued fracture growth and, consequently, greater downhole water losses. As will be shown in Chapter 10, once the wells are drilled and the initial reservoir is created, further fracture growth does not necessarily result in additional heat production.

Two redrilling attempts were required: the first redrilling path (GT-2A) improved the hydraulic connection to EE-1 (see Fig. 9.13 for detailed geometry). Although a significantly lower impedance was observed than that which had existed in the original system, it was still not low enough for a major heat extraction test. Therefore, a second redrilling was started. Path GT-2B was directed below GT-2A with a trajectory selected to reduce its vertical separation distance to the EE-1 injection zone at 2760 m (9055 ft). At a depth of 2665 m (8744 ft) in GT-2B a major low impedance connection was achieved. Apparently the second path also did not intersect the major vertical fracture directly but did at least penetrate several natural fractures or major joints which communicated with the hydraulic fracture, so that the impedance to flow circulation was low enough to proceed with a heat extraction test. After a further short drilling effort that resulted in a second connection at 2705 m (8870 ft), the new hole was logged and cased to 2580 m (8465 ft).

Characteristics of this first reservoir (GT-2B/EE-1) are summarized in Table 9.3. Performance of the first reservoir was initially evaluated with 75 days of closed-loop operation, termed run segment 2, which was conducted from January 28 to April 13, 1978. Hot water from the production well at flow rates ranging from 7 to 16 kg/s, GT-2B, was directed to a water-to-air heat exchanger where the water was cooled to 25°C before reinjection. Makeup water, required to replace downhole losses to the rock surrounding the fracture, was added to the cooled water and pumped down the injection well, EE-1, and then through the fracture system. Heat was transferred to the circulating water by thermal conduction through the nearly impervious rock contiguous to the fracture surfaces. Heated water was withdrawn through the production well. During this 75-day test, the wellhead injection pressure ranged from 6.3 to 9.4 MPa (910 to 1360 psi), while the wellhead production pressure was maintained at about 1 MPa (145 psi) to avoid flashing.

Table 9.3. Characteristics of reservoir systems studied with the first pair of wellbores, EE-1 and GT-2B at Fenton Hill

Characteristics	First Reservoir (May 1977–Jan. 1979)	Second Reservoir (Jan. 1979–1981)
EE-1 injection hole condition	before recementing	after recementing
Main injection zone location in EE-1	2.75 km (9020 ft)	2.93 km (9620 ft)
Main production zone locations GT-2B	2.6–2.7 km (8600–8850 ft)	2.6–2.7 km (8600–8850 ft)
Average vertical wellbore separation between injection and production zones–the production interval	100 m (300 ft)	300 m (1000 ft)
Production fluid flow rates	7–16 kg/s	6–8 kg/s

During a subsequent 28-day test, termed run segment 3, or the 'high back pressure' experiment, flow impedance behavior under conditions of high mean reservoir pressure was examined. However, during the test the EE-1 casing cement deteriorated further, so that approximately 20% of the water injected into EE-1 flowed through the annulus to the surface.

To correct this condition, and also to investigate the feasibility of creating larger fractures with the same wellbores, the EE-1 well was recemented near the casing bottom at 2.93 km, thus eliminating direct flow communication with the first EE-1 fracture situated at 2.75 km. A new and deeper fractured reservoir was activated in the Phase I system by extending an existing fracture from 2.93 km in EE-1 upward to about 2.6 km. This second reservoir is compared with the first reservoir in Table 9.3. The enlarged fracture system with an inlet-to-outlet spacing of 300 m, three times that of the first fracture, suggests that the effective heat transfer area might be significantly greater. Evaluation of the new reservoir began during a 23-day circulation test, run segment 4, starting on 23 October 1979. A longer term test, run segment 5, began shortly afterward on 3 March 1980 and continued for 286 days until December 1980. During that period, the first electric power was generated from an HDR reservoir. Using an R-113 binary cycle in a prototype design, about 60 kWe were generated from a diverted portion of the produced flow at 150°C. A detailed discussion of the thermal performance and flow characteristics of both reservoirs observed during the heat extraction and circulation tests described above follows in Chapter 10.

Following Potter's early model of the Phase I fracture-borehole system shown in Fig. 9.13, several other representations of the system evolved. Figure 9.22 (a) and (b) shows two later versions of the inferred geometry of both Phase I fracture systems. Model (a) represented the state of knowledge until about 1981 when other data were examined and Model (b) evolved. In essence, the Phase I system contains the original reservoir as an active subsystem of the enlarged reservoir. The first major fracture, whose origin was at 2.75 km in EE-1, is shown as the small vertical fracture with the new, larger fracture to its left. Both major fractures are represented as nearly vertical because the planes of hydraulic fractures are orthogonal to the minimum (least compressive) component of the tectonic earth stress. In most settings this stress is expected to be horizontal at depths greater than about 1 km. The minimum compressive horizontal tectonic stress at the reservoir depth is 37 MPa (5400 psi), about one-half the vertical overburden stress, as estimated by pressurization tests (Murphy *et al.* 1977) and results in a preferred fracture direction in the NW-SE plane. The granitic rock in which these fractures were created is fairly homogeneous and unstratified with respect to its mechanical properties; thus all the fractures discussed here are assumed to be approximately elliptical or circular in shape, rather than rectangular, as is usually assumed for oil and gas reservoirs in sedimentary formation with upper and lower confining regions (see Section 8.8.2 for details.)

All fracturing operations were performed with water alone; no viscosity-increasing or loss-of-fluid agents, or proppants were added. Subsequent pumping tests suggested that upon depressurization the induced fractures remained partially open due to the 'self-propping' of the misaligned rough surfaces produced during fracturing. The new fracture system was activated by injecting a total of 1360 m^3 (360 000 gallons) of water at rates ranging from 25 to 28 l/s (400–600 gpm) raising the downhole pressure to about 50 MPa (7000 psi).

The manner in which the major vertical fractures are connected with the production well, GT-2B, is complex, as illustrated by a number of measurements including downhole temperature and flow rate (spinner) surveys, radioactive tracer logs, caliper logs, and televiewer surveys described earlier in Section 9.2. Major connections apparently consist of a parallel set of natural fractures or joints which intersect both the main vertical hydraulic fractures and the GT-2B wellbore as shown in Fig. 9.22(b). These joints appear to be extensions of the same ones that formed the connections between the old fracture system and GT-2B. The downhole temperatures and velocities of the water in the connecting joints were measured at the joint/well intersections with temperature and flow rate surveys. In order to establish a basis for constructing the models of Figs. 9.13 and 9.22 (a), (b), numerous wellbore logs were examined along with extensive gyroscopic and magnetic/plumb bob surveys of the four wellbore paths (GT-2, GT-2A, GT-2B, and EE-1). The televiewer record given in Fig. 9.12, cores, and spectral gamma logs provide direct evidence that natural, sealed fractures are sufficiently densely located to ensure (on a statistical basis) that there are enough weakly sealed sites capable of being opened by hydraulic pressurization to serve as primary or secondary flow paths. Fracturing experience at Fenton Hill suggests that this is indeed the

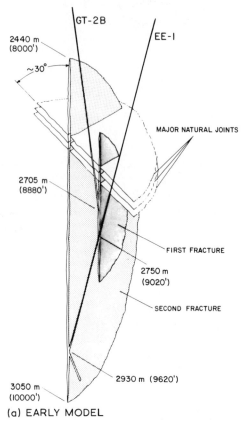

Fig. 9.22 Conceptual models of the Phase I wellbore-fracture systems at Fenton Hill. (a) Early model. (b) Later model. ((a) adapted from Murphy *et al.* (1981) and (b) from Tester *et al.* (1982)).

case. Repeatedly, high injectivity primary fractures have been activated at pressures only slightly above the least principal horizontal stress (S_3 or S_Y). The secondary set of inclined fractures detected by this log are roughly vertical and perpendicular to the major vertical fractures and thus open against the larger of the horizontal stresses, S_2 or S_X. Fracture A in Fig. 9.12(b) is one of a parallel set intersecting the wellbore as shown by the S-shaped features on the televiewer photograph. These fractures are oriented in approximately the north-south direction. The dip to the east is 7° from the axis of the wellbore. Since the wellbore also dips to the east approximately 3 to 4 °, the dip of the fracture set is about 10° from the vertical. There are five or six of these features located at depths in which temperature anomalies were observed, and all had the same orientation. The 10° dip could introduce an additional 21 bar (300 psi) of closure stress, which is added to another 1.7 MPa (250 psi) to account for the 30° azimuth change from the S_Y direction since the difference of S_X and S_Y is

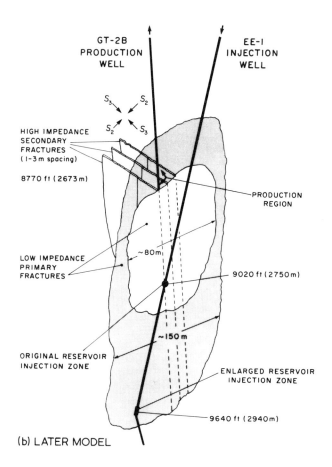

GT-2B
PRODUCTION
WELL

EE-I
INJECTION
WELL

S_3 S_2

S_2 S_3

HIGH IMPEDANCE
SECONDARY
FRACTURES
(1-3 m spacing)

8770 ft (2673 m)

PRODUCTION
REGION

LOW IMPEDANCE
PRIMARY
FRACTURES

~80 m

9020 ft (2750 m)

~150 m

ORIGINAL RESERVOIR
INJECTION ZONE

ENLARGED RESERVOIR
INJECTION ZONE

9640 ft (2940 m)

(b) LATER MODEL

approximately 7 MPa (1000 psi). The entrance points for fluid in the injection and production wellbores were well-defined by tracer measurements and by temperature and velocity changes in the fluid. For example, the thermal drawdown depicted during injection by the sequential temperature surveys of well EE-1 (Fig. 9.14) clearly shows a discrete flow pattern indicative of a series of high impedance vertical fractures crossing the injection wellbore. The nine labeled zones accept various rates of fluid at the same pressure level. As shown by Br[82] tracer logs taken in both wells GT-2B and EE-1, similar to those shown in Fig. 9.18, many fracture connections exist. Substantial agreement was noted between the locations of these zones and anomalies in caliper, temperature, cement bond, and spectral gamma logs. Their grouping at flow points suggests that they mark the intersection of a series of parallel vertical fractures with the wellbores. These secondary fractures then provide a degree of lateral communication between the main vertical fractures that propagate perpendicular to S_Y.

Examination of the geometry of the system shows that this parallel set of fractures cannot represent the main flow path or heat transfer area between the wellbores. In the original reservoir a nearly vertical fracture starting from the major flow exit in EE-1 at about 2760 m (9 055 ft) that intercepts the parallel set of secondary fractures provides the simplest most consistent model (see Fig. 9.22(b)). For the enlarged reservoir, the injection point is lowered to 2940 m (9645 ft), and a similar geometry is assumed. In this case the main fracture is somewhat larger than for the original system to be consistent with the larger inlet to outlet spacing and larger heat transfer area expected. These latter points are covered in detail in Chapter 10.

(c)   FRACTURE DESIGN AND INJECTION DATA – PHASE II

The first field objective for Phase II was to characterize fracture initiation and propagation in the open hole sections of EE-2 and EE-3 as measured along the trajectories of the inclined wellbores, in EE-2 from 3529 m (11 578 ft) to 4660 m (15 289 ft) and EE-3 from 3162 m (10 374 ft) to 4247 m (13 933 ft) where the rock temperature exceeds 325°C. The basic geometry of the bores shown in Fig. 9.23 illustrates the strategy of heat mining in the Phase II system. Multiple hydraulic fractures were intended to be propagated vertically from EE-2 perpendicular to S_Y to intersect the EE-3 borehole in the inclined parallel portion of the wellbores.

As in the Phase I operations, water was used as the frac fluid with no proppants added. Hydraulic pressure from pumping was used to initiate and propagate fractures. A group of positive displacement, truck-mounted pumps was manifolded together to increase flow and injection pressure capability as shown in Fig. 9.24. Even by oil and gas industry standards, many of the Phase II tests correspond to *massive hydraulic fracturing (MHF)* conditions where injection rates will exceed 132 l/s (50 bbl/min) and surface pressures 35 MPa (5000 psi) with up to a million gallons of fluid injected. Injection rates and volumes of this magnitude are required to propagate fractures across the 370 m vertical separation that exists between wells in the active region.

We can now refer back to hydraulic fracturing theory and the mathematical models discussed earlier (see Section 8.8.2) As seen in Fig. 8.9 injection rates of 50 bbl/min (132 l/s) could produce idealized fracture radii of more than 300 m providing that the injection volume is greater than 1700 m³ (450 000 gallons) and that the pressurized value of the permeability-compressiblity product (*kc*) doesn't exceed its low pressure value by more than a factor of 30 or so. However, as we know, any of the theoretical model predictions have to be viewed with some scepticism. Recall also that earlier experience in Phase I showed that 1360 m³ of fluid injected at 10–15 bbl/min (26.5–39.7 l/s) was sufficient to connect EE-1 to GT-2B over a 300 m vertical separation distance. At this point, the Los Alamos team decided on a conservative course of action – namely, to be prepared for the higher rates and larger injected volumes cited earlier.

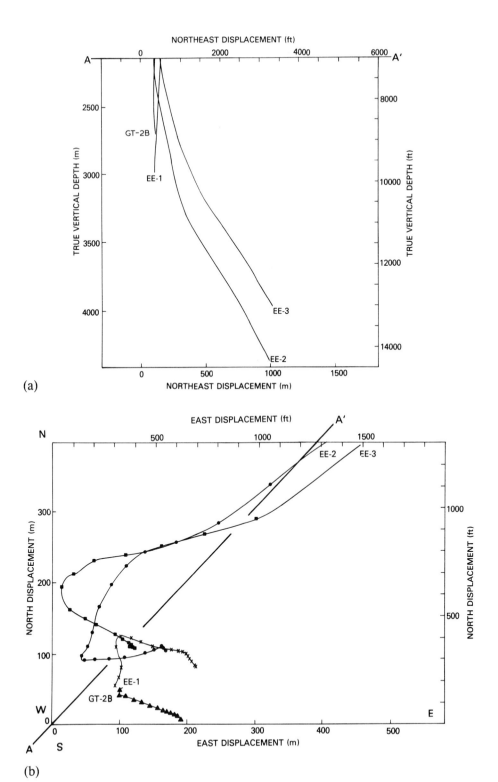

Fig. 9.23 (a) Elevation view of the wellbores at Fenton Hill. (b) Plan view of the wellbores at Fenton Hill.

Fig. 9.24 Diesel and gas turbine driven pump trucks and manifolding for a Phase II massive hydraulic fracturing (MHF) operation at Fenton Hill circa 1982. (Courtesy of R. Pettitt, Los Alamos National Laboratory).

After the EE-2 and EE-3 wellbores were cleaned, fracturing tests first started in EE-2 with the hope that fractures would tend to propagate upward toward EE-3. Unlike Phase I where fracturing was done through a casing string to a relatively small openhole section at the bottom of EE-1, wellbore sections must be isolated for the Phase II tests to restrict initiation to desired locations. Openhole elastomer packers and cemented liners were developed and tested to provide this isolation. Figure 9.25 shows several configurations used for the early tests.

According to Murphy (1982), zone isolation has been the most difficult operational problem encountered in the Phase II tests. He goes on to say that, to date, three methods have been attempted. They involved open hole packers, a cemented liner, and a casing packer. These are discussed below in chronological order.

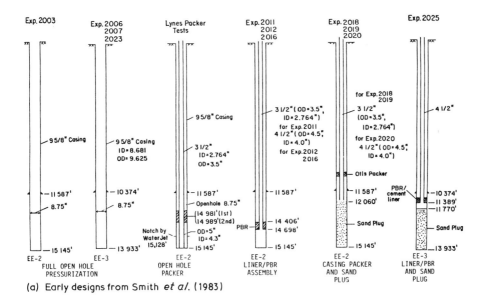

(a) Early designs from Smith *et al.* (1983)

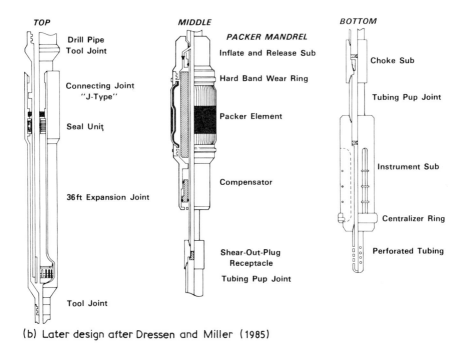

(b) Later design after Dressen and Miller (1985)

Fig. 9.25 (a) Casing and wellbore hardware configurations used during the early phases of Phase II hydraulic fracturing at Fenton Hill. (From Smith *et al.*, 1983). (b) Later Los Alamos design of an open-hole packer system. (After Dreesen and Miller, 1985).

(i) Open hole inflatable packer test

Despite the low probability of operating a packer at high temperature and obtaining an adequate seal against the rough walls of the well, two attempts were made because packers are the most inexpensive and fastest means of achieving zone isolation. Single element Lynes packers were used because of the anticipated difficulties. Unfortunately in this mode of operation the single packer must resist, via frictional shear between the inflatable bladder and the rock wall, the entire upward thrust – about 10^6 Newtons (224 000 lb)–generated by the fracturing pressure. In the two-packer straddle configuration, this thrust is instead reacted by structural members connecting the upper and lower packers. In two attempts the packers were successfully set and were initially able to sustain the fracturing pressure, but eventually began to slip along the well and failed. This may have been a consequence of the anomalously high fracturing pressures required, about 80 MPa, which is 75% of the overburden pressure. Despite their ultimate failures, both packers performed remarkably well considering the harsh environment. This suggests that if packers were rebuilt with better, more temperature and abrasion resistent materials and placed in the borehole with improved operational procedures they could provide an excellent means of fracturing specific zones. In fact, Fig. 9.25(b) shows a later design of a Los Alamos open-hole packer that has worked well in the Fenton Hill boreholes at temperatures in excess of 250°C (Dreesen and Miller, 1985).

(ii) Cemented liner tests

The second major attempt consisted of placing a short length of steel casing, or 'liner', about 130 m long, near the bottom of the well and cementing it in place with specially developed, high silica cement for high temperature applications. The liner was fitted with a polished bore receptacle (PBR) with seals at the upper end, so that drillpipe or a separate fracturing string could be mated to the liner, exposing only the open hole below the liner to fracturing pressure. Accomplishing this seal at 4 km depth is a difficult task. The injection of the cool fracturing fluid from the surface results in a thermal contraction of the fracturing string, so that the seal must accommodate as much as 3 m of relative motion while resisting a differential pressure of 35 MPa without leaking. After several preliminary attempts which were beset by seal problems, a fracturing operation was launched which resulted in the injection of 4900 m³ (1 300 000 gallons) of water at a flow rate of 0.1 m³/s. The course of fracture propagation was monitored by detecting microseismic events with three-axis geophones and accelerometers located 2.8 km down-hole. Distances to the micro-earthquake hypocenters were estimated from compressional and shear wave arrival time differences, and the directions inferred from the sense of first motion and hodographic methods as described earlier in Section 9.3.1. In later experiments, a seismic array placed at various depths into the Precambrian basement was available so that event location could be done more accurately using triangulation techniques (see Fig. 9.29). The microseismic results from those fracturing tests will be described in the next section.

(iii)   Casing packer and sand plug tests

Because the preliminary analysis of the microseismic data indicated that fractures were inclined, the liner at the bottom of EE-2 was temporarily abandoned and the next fracturing attempt took place at the top of the open hole section, where inclined fractures should have an excellent probability of intersecting EE-3. A short interval just below the bottom of the EE-2 casing was isolated by temporarily filling the bottom of EE-2 with sand to within 130 m of the casing bottom. Unfortunately the inner casing in EE-2 had been weakened at 2 km by wear and grooving of the inner wall as a consequence of drilling operations after the casing was set, and to avoid casing rupture a casing packer was set below the weakened section. Thus, the pressurized string consisted of drillpipe to the packer, the full-strength casing below that, and the open hole between the casing and the top of the sand. This fracture attempt, experiment 2018, conducted in August 1982, resulted in the injection of only 1000 m³ (265 000 gallons) of water before the casing packer failed. Fracture propagation was again monitored with downhole seismometers. The pressure and flow history is given in Fig. 9.26. As can be seen, a 'classic breakdown' of the formation was not observed indicating that weaker joints are probably opening at bottomhole pressures near the minimum stress S_Y. A second attempt (experiment 2019) to pump the open hole section through the casing packer failed after only 38 m³ (10 000 gallons) had been injected.

Later experiments (2033, 2032, and 2042) have involved injections of large amounts of fluid at high rates (from 13 to 106 liter/s) but still did not result in a good connection between EE-2 and EE-3.

Results for both phases of testing at Fenton Hill are summarized in Table 9.4 and will be reviewed in the last section of this chapter.

(d)   MICROSEISMIC RESULTS – FRACTURE MAPPING IN PHASES I AND II

Based on field experience at Fenton Hill, microseismic events are associated with massive hydraulic fracturing in low permeability crystalline rock. The source location of these signals have been mapped for both Los Alamos reservoirs and appear to outline planar-like features. The central problem is how to relate these mapped features to hydraulic fracture dimensions and, of course, ultimately to effective heat transfer area.

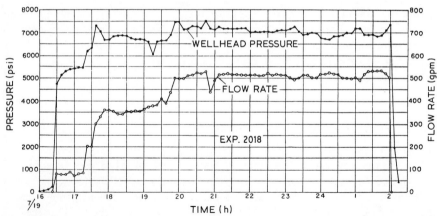

Fig. 9.26 EE-2 wellhead pressure and injection rate, experiment 2018, first MHF attempt in the Fenton Hill Phase II reservoir with casing packer and sand plug used for zone isolation. (Some of the flow and pressure fluctuations are due to equipment malfunction.)

The Phase I microseismic situation has been reviewed in detail by Albright and Pearson (1982) and Pearson and Albright (1984), so we only summarize the salient points here. Microseismic activity (events/min) increases sharply when a critical injection volume is reached. For example, as a fracture system is created by fluid injection, depressurized, and reinflated, this critical volume correlates well with the initial fluid volume injected. Thus, the initial aseismic period observed probably corresponds to a reinflation process. As the fracture is propagated outward from the injection region, joints become exposed to high internal pressures and may slip irreversibly resulting in a *one-time*

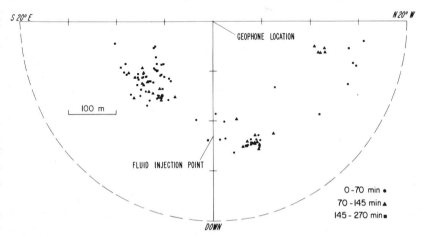

Fig. 9.27 Microseismic event locations during the 14 March 1979 Phase I MHF test at Fenton Hill. Events projected parallel to the fracture plane showing time period of occurrence after injection started into EE-1. (From Albright and Pearson, 1982).

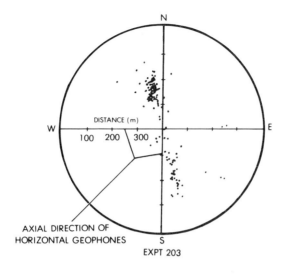

EXPT 203

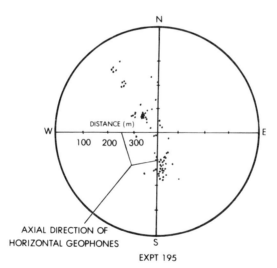

EXPT 195

Fig. 9.28 Microseismic event locations in plan view during two separate MHF tests in the Phase I Fenton Hill reservoir. Events projected to a horizontal plane show a north-by-northwest trending feature.

microseimic event. Upon reinflation, this irreversible process of shear and/or tensile failure does not occur, on average, until the outer extremities of the fracture are fully pressurized.

Figures 9.27 and 9.28 trace the progress of hodographically mapped microseismic events outward from the injection point in EE-1 and upward toward the GT-2B openhole section (see Figs. 9.21 and 9.22 for wellbore location). During all Phase I MHF experiments events fell on a north by northwest trending vertical near planar region as shown by the elevation projection of Fig. 9.28 viewing in that direction and by the plan view. Furthermore, according to Albright and Pearson (1982), there is a clear progression of event density away from the injection point as time proceeds. This result is again consistent with the ideas of fracture propagation introduced in Chapter 8.

Although it is necessary to hypothesize that the microseismic events correlate with actual fracture positions, there is no doubt that the hydraulic pressurization process is responsible for the microseismic activity. This brings us to Albright and Pearson's next concern about pore fluid pressurization and permeation. If the time scale of the fracture job is short compared with the hydraulic diffusion time, that is

$$\text{time} \ll (\Delta Y)^2 / \alpha_H \tag{9.6}$$

where ΔY = distance away from pressurized region
$\alpha_H = k/\mu c\phi$ = hydraulic diffusivity (m²/s),

then the penetration depth for pore pressure to increase above S_y will be small and one should expect microseismic activity to be localized to the near fracture region. Since the observed thickness of the seismic zone is about 30 m (98 ft) for the MHF tests (in comparison with an event location resolution of ± 10m), a permeability k of *at least* 0.01 millidarcy is required to explain the microseismic event distribution. This assumes that the compressibility-porosity product $C\phi$ is 25×10^{-9} Pa⁻¹ $(3.6 \times 10^{-6}$ psi⁻¹) and that the viscosity μ of water at 200°C is 1.4×10^{-4} Pa-s (0.14 centipoise). An effective permeability of 0.01 millidarcy implies that microcracks or very small fractures may be involved rather than the rock matrix itself that is known to have an intrinsically lower permeability of 10^{-3} to 10^{-6} millidarcy. Further data support their model. Microseismic activity observed during long term flow testing over periods of weeks to months are spread more uniformly over a spherical region surrounding the inlet and outlet regions in EE-1 and GT-2B, rather than being concentrated to a two dimensional zone surrounding the plane of the fracture.

One shortcoming of the early microseismic observations is that only one geophone array located near the active reservoir is involved with the measurements. Recent analysis of microseismic data from these earlier tests has indicated that the vertical component of the geophone array is excessively attenuated with respect to the horizontal components. Inadequate coupling of the geophone package to the borehole wall was probably responsible for this attenuation. Although the absolute uncertainty associated with actual event location using hodographic methods as illustrated in Figs. 9.27 and 9.28 increases because of this asymetric behavior, the qualitative features remain intact. Using improved location techniques with multiple receivers and a travel time analysis method, the Phase II microseismic event mapping has yielded results to explain the failure to connect EE-2 to EE-3 (Keppler, 1983, Fehler *et al.* 1986 and 1987, House, 1986). The microseismic array, the so-called surface

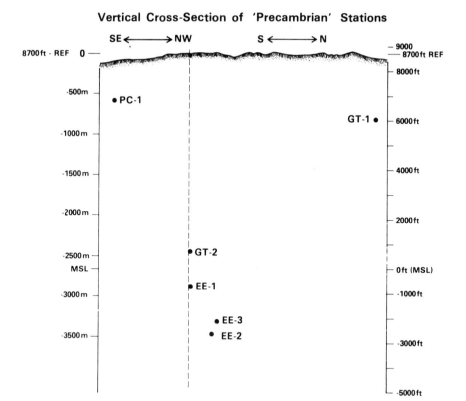

Network of 'Precambrian' Borehole Stations

Experiment 2032, December 1983

Line of Section

GT-1

PC-1

Contour Interval 500 ft

N

Injection, EE-2
3500 m
(11 600 ft DAW)

EE-3
3301 m
(11 200 ft DAW)

EE-1
2856 m
(9400 ft DAW)

GT-2
2437 m
(8000 ft DAW)

0   500 m

0   5 km

Vertical Cross-Section of 'Precambrian' Stations

SE ← → NW   S ← → N

8700 ft - REF   0 —   — 9000
                                                        — 8700 ft REF

                                                        — 8000 ft

-500 m   ● PC-1

                                              GT-1 ●   — 6000 ft

-1000 m

-1500 m   — 4000 ft

-2000 m   — 2000 ft

-2500 m   ● GT-2

MSL   — 0 ft (MSL)

            ● EE-1

-3000 m   — -1000 ft

              ● EE-3
-3500 m   ● EE-2   — -3000 ft

                                                        — -5000 ft

Fig. 9.29 Microseismic net for downhole measurements at Fenton Hill in EE-1, GT-2B, EE-3 and shallower locations in GT-1 and PC-1 that are also in the Precambrian section.

and Precambrian net, employed for a typical Phase II MHF test is shown in Fig. 9.29. An event map superimposed for two particular tests 2032 and 2042 (see Table 9.4 for conditions) is given in Fig. 9.30. The events resulting from injecting into EE-2 seems to shroud the active EE-3 wellbore. A closer examination of the Precambrian and surface net results shows that the seismic features are planar, best oriented in such a direction as to minimize their probability of intersecting the EE-3 wellbore. In contrast, when the hodographic and Precambrian results are combined, the shape of the pattern is less directionally oriented (Keppler, 1983, Fehler *et al.* 1987, House, 1986). In addition, the analysis of seismic signals during pressurization points to shear motion rather than tensile jacking of fractures as the dominant mechanism.

This situation was unfortunate as it suggested that further pumping would not provide a connection unless the natural tendency for fracture orientation was altered. As the Los Alamos team had made a number of unsuccessful attempts and as the casing situation in EE-2 was marginal, a decision was made in late-1984 to sidetrack out of EE-3 and target it toward the seismic cloud that has been mapped (for more details see Section 15.5).

9.4.2 British project

Interest in heat mining in the UK received a tremendous boost from the pioneering efforts of the US scientists at Los Alamos. If the early experiments at Fenton Hill had not been successful, the British program might not have even started. Furthermore, the British benefited from the failures and difficulties of America's learning experience at Fenton Hill. Specifically, the decision to work at intermediate depths was taken because of the temperature problems encountered in the Phase I system at Fenton Hill. Special procedures for inspection of drilling equipment and tubular goods were used because of drilling problems at Los Alamos; and the development of explosive initiation to reduce the wellbore impedances became a prime objective because of the high impedances observed in the Fenton Hill boreholes. The field program at Rosemanowes differs from the American effort at Fenton Hill in two important respects (Batchelor, 1984):

(1) The resource itself is of lower grade and is not associated with any volcanic, hyperthermal activity. Gradients at Rosemanowes are above normal for the UK and Western Europe at about 30 to 40°C/km but are well below the 50 to 100°C/km gradients found at Fenton Hill.

(2) The British have placed much more emphasis than the Americans on the development of fracture initiation techniques at specified locations in a borehole. Their primary focus has been on explosive stimulation in the near wellbore region followed by hydraulically induced fracture growth.

There are, however, many similarities including a two phase approach with Phase I involving several shallow holes (300 m) in granite to study explosive and hydraulic stimulation and to develop diagnostic techniques and Phase II

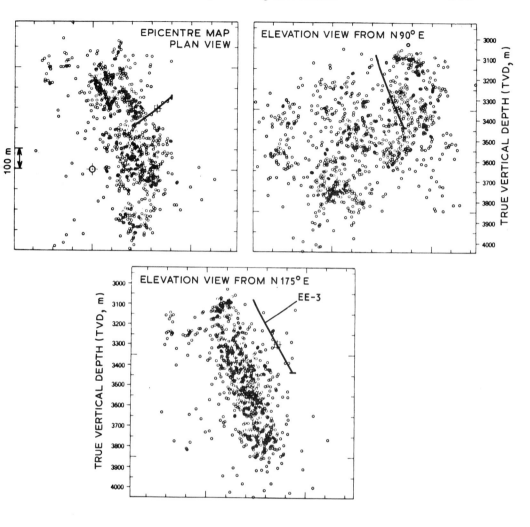

Fig. 9.30 Microseismic event map for an MHF test (experiment 2032 and 2042) in the Phase II Fenton Hill system from single hole hodographic and multiple Precambrian net measurements (Fehler, House and Kaieda, 1987 and House, 1986).

involving stimulation between an inclined wellbore pair at depths of 2000 m and the fact that both projects deal with reservoir development in low permeability crystalline granitic type rock.

From the outset, the British concept for creating an HDR system has been *to activate an existing joint network* to produce the required swept area and volume for producing a commercial sized reservoir. In effect, they assume that all heat transfer surfaces are formed by natural discontinuities and that the elastic and hydraulic properties of the reservoir result from bilateral movement along joints, causing flow primarily through a labyrinth of these joints with

Table 9.4. Summary (no proppants and water as frac fluid except where noted) of HDR fracturing field experiments worldwide

Source: Murphy, Keppler and Dash, Los Alamos National Laboratory

Reservoir	Date	Field experiment number	Injection well	Depth measured along well (km)	Injection volume (m³)	Average injection rate (l/s)	Fracturing duration (h)
Fenton Hill, I[a]	Oct 10, '75	—	GT-2	2.9	44	24	0.5
Fenton Hill, I	June 29, '76	129B	EE-1	2.8	95–144	13	3
Fenton Hill, I	Mar 14, '79	203	EE-1	2.9	600	26	6
Fenton Hill, I	Mar 21, '79	195	EE-1	2.9	760	41	4
Fenton Hill, I	Oct 24–30, '79	215	EE-1	2.9	4 900	20	144
Fenton Hill, II[b]	June 14, '82	2012	EE-2	4.3	3 200	50	16
Fenton Hill, II	June 19, '82	2016	EE-2	4.3	4 900	76	24
Fenton Hill, II	July 19, '82	2018	EE-2	3.4	900	28	10
Fenton Hill, II	Oct 6, '82	2020	EE-2	3.4	3 200	76	15
Fenton Hill, II	Nov 8, '82	2023	EE-3	3.2	148	14	3
Fenton Hill, II	Dec 14, '82	2025[h]	EE-3	3.4	580	50	7
Fenton Hill, II[g]	Nov '83	2033	EE-3	3.2	580	13	11
Fenton Hill, II	Dec 6–9, '83	2032[f]	EE-2	3.5	21 198	106	61
Fenton Hill, II	May '84	2042[f]	EE-3	3.5	7 570	26	81
Rosemanowes, II[c]	Oct 13, '82	2044	RH-12	1.6–2.0	1 000	50	7
Rosemanowes, II	Nov 4–7, '82	2046	RH-12	1.6–2.0	18 500	20–100	72
Rosemanowes, II	Nov 9–11, '82	2047, 8	RH-11, 12	1.6–2.0	40 000	0–100	57
Rosemanowes, II	Nov 11–13, '82	2049	RH-12	1.6–2.0	24 400	20	37
Rosemanowes, II	Nov 13–23, '82	2051	RH-12	1.6–2.0	39 340	15	239
Rosemanowes, II[e]	May 15, '83	2062	RH-11	1.6–2.0	400	50–195	1–2
Falkenberg	Oct 17, '79	Frac	HB-4	0.3	6.8	3	0.6
Le Mayet de Mont.	Nov 5, '81	—	IMAG 3.4	0.2	5.2	9	0.2
Yakedake	Oct 25, '82	4	HY 2	0.3	5.4	3	3

a The Fenton Hill Phase I reservoir was created at depths of 2.6 to 2.9 km, with wells GT-2 and EE-1.
b The Fenton Hill Phase II reservoir was created at depths of 3.2 to 4.4 km with directionally drilled and slanted wells EE-2 and EE-3.
c The Rosemanowes Phase II reservoir was created at depths of 1.4 to 3.0 km with directionally drilled and slanted wells RH-11 and RH-12.
d Le Mayet permeability computed with assumptions that viscosity of gelled water and sand mixture was 1 000 times that of water.
e No seismic events observed, gelled frac job.
f Via triangulation with Precambrian net.
g The zone pumped below EE-3 casing at 3170 m (10 400 ft).
h Frac string into casing packer inside cemented liner, 122 m of open hole, sanded in below open hole zone.

minimal flow in the pore space of the rock fabric. In contrast, the Los Alamos team, early in their efforts, maintained that fractures can be induced as nearly idealized, vertical, planar structures by treating the rock formation as a homogeneous, isotropic continuum in an anisotropic stress field. Figure 9.31 illustrates this contrast conceptually for an HDR reservoir between a pair of inclined wells. Neither of these extreme positions is completely accurate – but this difference in conceptualization of the reservoir has resulted in a productive collaboration between the British and the Americans.

(a) GEOLOGIC SETTING

The British HDR site is located at the Rosemanowes quarry in Southwest England in Cornwall (see Figs. 9.32(a) and (b)). At Rosemanowes, the Carnmenellis granite appears as a surface outcrop of a major batholith underlying much of Cornwall. The Carnmenellis granite is aged at approxi-

Wellhead injection pressure (MPa)	Dimension (m) of MicroSeismic zone in direction of:			Approximate dip angle from horizontal	Approximate strike bearing	Seismic area (thousands of m²)	Seismic volume (millions of m³)	Formation permeability $1\mu D=10^{-18}\ m^2$
	Dip	Strike	Width					
14	—	130	10	90°	N30–60°W	8	0.05	4
12	—	90	10	90°	N60°W	4	0.03	280
19	300	500	200	90°	N20–30°W	120	16	0.1
21	300	500	200	90°	N20–30°W	120	16	1
16	—	400	400	90°	N50°W	380	102	1
44	1 300	700	500	45°	N20°W	710	240	~0
47	1 400	700	600	45°	N10°W	770	310	0.1
48	250	350	120	90°	N20°E	68	5.5	1
45	500	500	170	90°	N10°W	200	22	2
13	120	230	80	90°	N10°W	22	1.2	5
45	350	550	200	90°	N20°W	150	20	~0
12	~200	~200	~100	20°	N60°E	40	4	~0
48	800	900	150	75°	N10°W	720	40–100	~0
36	700	800	100	40°	N	560	56	~0
11	400	—	—	80–90°	N50°W	60	—	260
14	700	430	140	80–90°	N50°W	240	22	260
11	1 400	570	200	80–90°	N50°W	630	84	320
10	1 400	570	200	80–90°	N50°W	630	84	460
11	1 400	570	200	80–90°	N50°W	630	84	460
25	0	0	0	80–90°	—	0	0	320
4	20	40	10	90°	N80°W	0.6	0.005	3 000
12	—	35	20	—	N70°W	0.9	0.013	6 599[d]
12	60	60	40	90°	N30°E	2.8	0.075	0–10

mately 280 million years and seems to have a relatively constant mineralogy over a wide area and to depths of at least 2 km as summarized by Garnish (1976), Edmonds *et al.* (1984) and Pine and Batchelor (1984). It is made up of megacrysts of alkali feldspar set in a coarse fabric of plagioclase, other alkali feldspar, micas or biotites, and trace minerals. The modal composition is approximately:

	modal %
quartz (SiO_2)	30
alkali or potassium feldspar ($KAlSi_3O_8$)	30
plagioclase feldspar ($CaAl_2Si_2O_8 + NaAlSi_3O_8$)	20
muscovite ($KAl_3Si_3O_{10}(OH)_2$)	10
biotite ($K_2(Mg,Fe,Al)_6(Si,Al)_8O_{20}(OH)_4$)	6
tourmaline ($(Na,Ca)(Li,Mg,Fe,Al)(Al,Fe)_6B_3Si_6O_{27}(O,OH,F)_4$)	2
andalusite	1
other trace minerals	1

Grain sizes generally range from 2 to 5 mm with some large feldspar grains (megacrysts) up to 20 mm. Natural joints are normally well-sealed with pervasive hydrothermal mineralization prevalent as indicated by geophysical logs and the large amount of muscovite observed in cores. Major jointing consists of two subvertical sets striking at approximately 155–335° and 75–255° as shown in Fig. 9.33(a). At the surface level, joint spacings are 1–5 m,

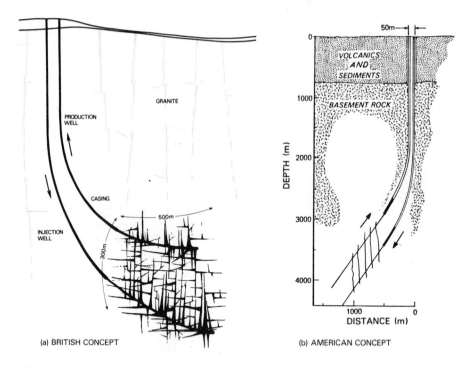

Fig. 9.31 British and American Concepts for HDR reservoir structure between an inclined part of boreholes. (a) British concept. (b) American concept.

Fig. 9.32 (a) Site location and wellbore schematic for the Camborne School of Mines Project at Rosemanowes. (b) Heat flow and geologic map of South West England.

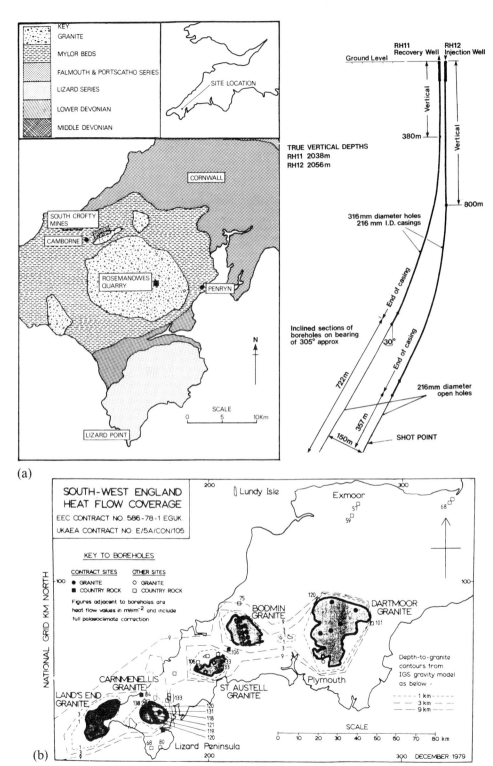

while in the local mines to depths of 800 m they are 3–10 m. These spacings are consistent with logs of the two deep boreholes at Rosemanowes to 2000 m. The mechanical properties of cores taken at Rosemanowes have also been characterized by Pearson (1980) and others. The results of these studies can be summarized as follows (Pine and Batchelor, 1984):

Uniaxial compressive strength, σ_c	100–170 MPa
Brazilian tensile strength, σ_t	9–15 MPa
Young's modulus, E	50–70 GPa
Poisson's ratio, v	0.18–0.22
Hydrofracture tensile strength	
in 10 cm diameter boreholes	10–17 MPa
Hoek–Brown (1980) material constant	
(equation 9.1),M^*	21–35

In situ stress measurements were made in boreholes using hydrofracture techniques and by overcoring methods in mines to 800 m. As reported by Pine *et al.* (1983), both horizontal and vertical components could be correlated by linear functions of depth as shown in Fig. 9.33(b). The appropriate equations are:

$$S_3 = \sigma_h - P = S_Y = \text{minimum principal effective horizontal stress} = 6 + 2.2d$$

$$S_2 = \sigma_H - P = S_X = \text{maximum principal effective horizontal stress} = 15 + 18.2d$$

$$S_1 = \sigma_V - P = S_Z = \text{effective vertical stress} = 16.2d$$

where d = depth in km and all stresses are in MPa.

The orientation of the least principal horizontal stress was relatively constant for all measurements at 130–310° as shown in Fig. 9.33(a). The misalignment between stresses and the joint set turns out to have an important effect on fracturing.

(b)   EXPLOSIVE PRETREATMENT, FRACTURE DESIGN AND INJECTION DATA – PHASE I

The Phase I system consisted of four 15 cm diameter bores that were drilled to depths of about 300 m from the quarry floor using air percussive drilling. Unfortunately the air drilling technique did not provide good directional control so the bores tended to 'wander' somewhat as shown in Fig. 9.34. The well RH9D is at a 24° angle and approximately 42 m away from the other three at its total depth. At 300 m, as shown in Fig. 9.33, both horizontal stresses are greater than the vertical stress – so fractures should be preferentially horizontal as described in Chapter 8. After an extensive series of pressure-flow testing and geophysical logging, the explosive charge was lowered to the shot point position in RH9D and fired. Although post shot logging indicated that the 5 cm standoff

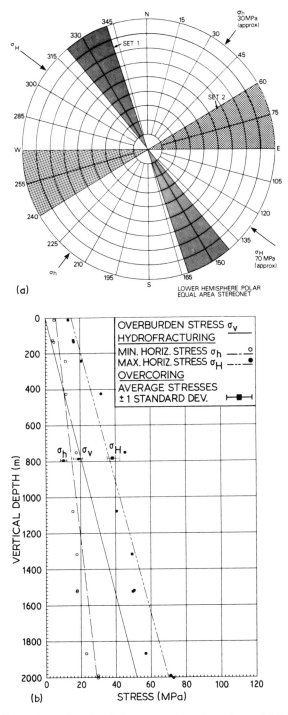

Fig. 9.33 (a) Direction of major joint sets and orientation of horizontal stresses $S_h = S_3 = S_Y$; $S_H = S_2 = S_X$. (b) *In situ* stresses at Rosemanowes to a depth of 2000 m as measured by overcoring and hydrofracture techniques. Values are not corrected for full hydrostatic head. To correct to effective stresses subtract $9.8d$ (source Pine, Ledingham, and Merrifield, 1983).

distance used was too small because the hole diameter at the shot point was over 36 cm, the hole was stable and contained many open, propped fractures easily visible on the borehole television survey over a length of 40 m (Batchelor, 1982).

Hydraulic stimulation of the four well system followed the explosive pretreatment. After two low-flow-rate tests proved that the shot had connected directly into an open joint network, the main hydraulic stimulation got underway. Five subsequent injections into RH9D at flow rates ranging from 1.5 to 21 liter/s (24 to 330 gpm) for periods up to 12 hours were very successful in connecting the four wellbore system with a low flow impedance of about 0.5 to 0.6 GPas/m^3. A total injected volume of about 800 m^3 apparently produced a self-propped contact area in excess of 50 000 m^2 with only 42 m of separation between boreholes. This was indeed an encouraging result and no doubt had a very positive influence on the decision to proceed to Phase II.

(c)   EXPLOSIVE PRETREATMENT, FRACTURE DESIGN AND
       INJECTION DATA – PHASE II

After completing the two inclined wellbores RH11 and RH12 to a total depth of about 2200 m (7200 ft) an explosive shot point was selected near the bottom of the deeper well RH12 at a depth of 2125 m. Figure 9.34(b) shows a plan view and elevation section of the two holes. An explosive configuration similar to that used for the Phase I test as described earlier was used. In this case, however, higher fluid injection rates are required during the fracture propagation stage. Thus the explosive pretreatment will not only locate the fracture initiation point but will also reduce the wellbore-to-fracture impedance so that high flow rates can be accepted with minimal pressure losses. This increases the hydraulic efficiency of the treatment. As pointed out by Batchelor *et al.* (1983), the explosive shot in RH12 was effective. After 10 hours of pumping, about one half the flow leaves the well at the shot point, while the remaining half is divided between three other exit points, presumably at joint intersections.

Following the explosive pretreatment, ten separate hydraulic stimulations were performed during October and November of 1982 with the objective of connecting wells RH11 and RH12 to maximize the interconnected volume at an acceptably low flow impedance. The results of six of the major massive hydraulic fracturing treatments are summarized in Table 9.4 under the headings Rosemanowes, II20ii where the last two digits refer to the experiment number. A more detailed summary of the objectives of each experimental test and the conditions observed is given in Table 9.5. As can be seen, the early tests RT2042–RT2045 focused on system characterization and inlet development while RT2046–RT2049 concentrated on reservoir growth and impedance reduction with RT2051 serving as a preliminary circulation test. It should be noted that the injection volumes are very large − 18 000 to 40 000 m^3 (4 700 000

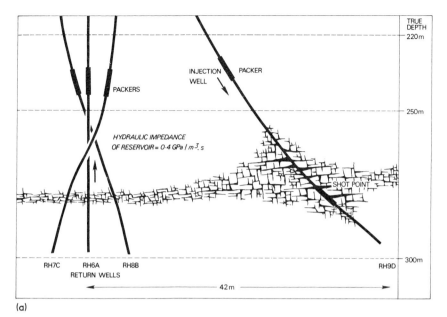

(a)

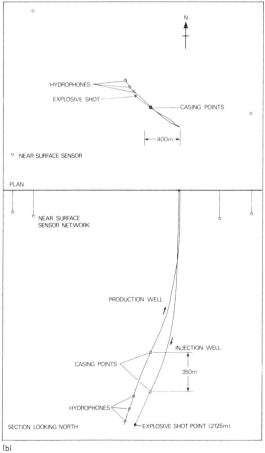

(b)

Fig. 9.34 (a) Conceptual representation of the reservoir created in the Phase I Rosemanowes system (from Batchelor, 1982). (b) Plan view and elevation section of the Rosemanowes Phase II system (from Batchelor *et al.*, 1983).

Table 9.5. Summary of fracturing tests in the Phase II system at Rosemanowes

(a) Tests RT2046–2051

Test date/time	Duration (hr)	Flow rate (liter/s) RH12	RH11	Cumulative volumes (m³) RH12 (inject)	RH11 (vent)	Net stored volume (m³)	Pressures (MPa) RH12	RH11	Temperatures (surface pipework) (°C) RH12	RH11
RT2046										
4 Nov. '82 10:00		100–20	0–5 v	0	0	0	10–14	0–3	13	13–31
9 Nov. '82 09:40	120	av 43		18500	0	18500	10–14	0–3	13	13–31
RT2047										
9 Nov. '82 09:40		0	98 i	18500	−500*	19008	10–09	0–9	13	13 i
9 Nov. '82 14:30	5	0	78–0 v				10–09	0–9	13	<13 v
RT2048										
9 Nov. '82 14:30		100–0	0–58 v	21900	400	21500	12–09	5–0	13–22	<47 v
11 Nov. '82 18:30	52	av 18	av 5 v				12–09	5–0	13–22	<47 v
RT2049										
11 Nov. '82 18:30		20	0–4 v	24400	470	23930	09–10	0–3	8–13	40–10 v
13 Nov. '82 08:00	37	20	0–4 v				09–10	0–3	8–13	40–10 v
RT2050										
13 Nov. '82 08:00		0	0	24400	−2730*	27130	09–08	3–11	11	11
13 Nov. '82 18:30	11	0	95 i				09–08	3–11	11	11
RT2051										
13 Nov. '82 to		0–30	0–8 v	38400	70	38300	8–12	3–11	10–25	10–50
22 Nov. '82 13:00	211	av 18	av 4							
23 Nov. '82 17:00	239	10–30 av 15	7–8 v	40150	720	39430	10–11	3–2	10–30	49
24 Nov. '82 16:30	262	15–25	7 v	41670	1240	40430	11–10	2	10–30	49
25 Nov. '82 16:30	286	15–20	6 v	42900	1791	41109	10	1–1.5	10–30	49
29 Nov. '82 13:30	379	20–05	5–10 v	44775	3870	40905	10–7.5	1.5–0	10–30	49

* indicates a net injected volume in RH11
i indicates injection
v indicates vent (or returns)

(b) Tests RT2042–2045

Test date/time	Duration (hr)	Flow rate RH12 liter/s	Volume RH12 (m³)	Net stored volume (m³)	Pressure RH12 (MPa)
RT2042					
11 Oct. '82 12:00 12:09	0.15	25	11	11	0–7.8 7.8–0.7
RT2043					
12 Oct. '82 09:08 15:42	6.57	5–24	400	411	1.2–10.7 10.7–5.0
RT2044					
13 Oct. '82 14:01 20:43	6.70	20–65	1000	1411	5.0–13.1 11.5–2.2
RT2045					
19 Oct. '82 09:17 16:57	7.67	5–25	530	1941	2.2–11.0

(c) Objectives of tests RT2042–2051

Test	Objective
RT2042	Preliminary systems check prior to inlet development
RT2043	Inlet development
RT2044	Inlet development
RT2045	Investigation of pressure response in RH11 to injections into RH12, and of flowing joints in RH12
RT2046	Major stimulation to grow reservoir
RT2047 RT2050	Injections into RH11, followed by venting RH11, to attempt to reduce system impedance
RT2048 RT2049	Injections into RH12, venting RH11 with back pressure to attempt to reduce impedance
RT2051	Circulation test with variety of back pressures on RH11

to 10 500 000 gallons) and injection rates are high – 20 to 100 liter/s (320 to 1580 gpm or 7.7 to 37 bbl/min). The injected volumes are greater than the Fenton Hill Phase II stimulations by a factor of 3 to 10 but the injection rates are comparable. In all tests the fracture fluid was water with no proppants as was the case at Fenton Hill.

The last test (2062) cited in Table 9.4, conducted on May 15, 1983 was a special case involving a viscous gelled fluid and high injection rates. The results from this experiment were very intriguing. A proprietary cross-linked guar-gum or polysaccharide with no proppants was selected for the test to provide desired rheological properties (Batchelor, 1984). It is a highly non-Newtonian fluid with thixotropic or shear-thinning behavior. In addition to providing high pressures within the stimulated joint, the gel will tend to reduce fluid loss to the surrounding formation while opening the aperture of the joint itself. A ramped increase in injection rate was used to place the fluid properly in the fracture zone – final rates exceeded 195 liter/s (70 bbl/min) with surface pressures at 25 MPa (3625 psi). After the test, a television camera scan of the borehole revealed a series of fresh, angular fractures with apertures of 1 to 2 mm that cut directly through existing natural joints (see Fig. 9.12(c)). However, during the stimulation *no microseismic events were recorded* nor was any reduction in flow impedance observed. The reason for this aseismicity is obviously contained in the microseismic source mechanism and may indicate that 'quiet' tensile-like jacking is occurring with the gelled fluid while water injections stimulate massive shear failure. The failure to reduce impedance may be associated with the lack of intersection between the large axial fracture created and the labyrinth of joints that provides the present communication between RH11 and RH12 or possibly that self-propping did not occur away from the borehole. Apparently, the fracture closed completely as pressure was released.

(d) MICROSEISMIC RESULTS – FRACTURE MAPPING PHASE II

One of the most striking bits of evidence to support fracture growth by hydraulic methods comes from the microseismic measurements during the Rosemanowes Phase II stimulations. Batchelor *et al.* (1983) and Pine and Batchelor (1984) have done an excellent job of describing the results of this first class set of measurements. Using the instruments and analysis methods described earlier in this chapter, Batchelor and his associates carefully monitored the seismic activity during the tests shown in Tables 9.4 and 9.5. In general, intense microseismic activity (12 to 20 events/min) was observed as soon as the system was pressurized above 110 bar at the wellhead. Figure 9.35 gives the event frequency and pressure history for the first large fracture treatment of 18 500 m^3 during test RT2046.

Microseismic event locations followed a stepout trajectory away from the injection point, with subsequent activity in earlier quiet zones to create a general cloud of activity. This effort can be seen by the sequence of A,B,C,D given in Fig. 9.36 for the first 3 hr, 4–11 hr, 12–27 hr, and 0–27 hr of injection during test RT2046. As was the case at Fenton Hill, microseismic event locations are presumed to result from fluid pressure disturbances.

The composite of events from all experiments conducted during October and November, 1982, has also been plotted by Pine and Batchelor (1984) and is presented in Fig. 9.37. The total injection volume was about 100 000 m³ achieved at injection rates up to 100 liter/s and pressures above 10 MPa. Clearly, a confinement of events by the horizontal stress field is seen from the normal and in plane views (a) and (b). The *in plane* view shows that the majority of events are contained in a zone only 200 m wide creating an ellipsoidal shape.

The uncertainty of individual event locations increased with increasing depth corresponding to larger distances between the receivers and the event. However, in all cases the locations in the horizontal plane were probably correct to within ± 20 m. At a 2 km depth, the depth error was about ± 50 m while at 3.5 km it increases to about ± 200 m (Pine and Batchelor, 1984).

Another strong feature of these microseismic maps is the downward growth of the region as injection continues. This behavior is in contrast to what many theories predict about upward hydraulic fracture growth. Batchelor and associates believe this to be associated with a shearing source mechanism interacting with an anisotropic stress field and misaligned joint system (see Fig. 9.33). This downward growth has a number of implications which should be brought out:

(1) It arises because of the different relationships between the gradients of normal stress, shear stress, and hydrostatic stress.
(2) It can lead to uncontrolled fracture growth and consequent high fluid losses because shear failure can occur at much lower overpressures.
(3) The option of using high pressures to jack open tensile fractures is restricted because of the onset of shear failure.

Above all, every field site investigated so far has shown a similar misalignment between the natural joint system and the *in situ* shear stresses. As Pine and Batchelor (1984) point out, this observed phenomenon implies that fracturing will proceed to a natural minimum energy state. If this downward growth can be made eventually to provide an accessible heat transfer area – this may result in a marked improvement in thermal performance and could reduce development costs.

On the whole, the results from the Rosemanowes Phase II tests have been very positive. Again a low impedance connection was established between the wells. Based on the injection volumes of over 40 000 m³, the potential reservoir

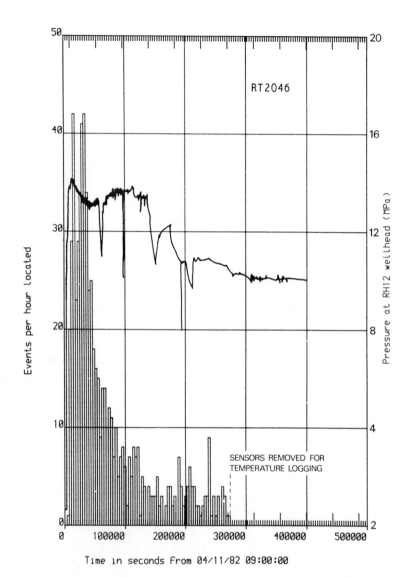

Fig. 9.35 Microseismic event frequency and pressure during stimulation of the Rosemanowes Phase II system (field experiment RT2046, from Batchelor *et al.*, 1983).

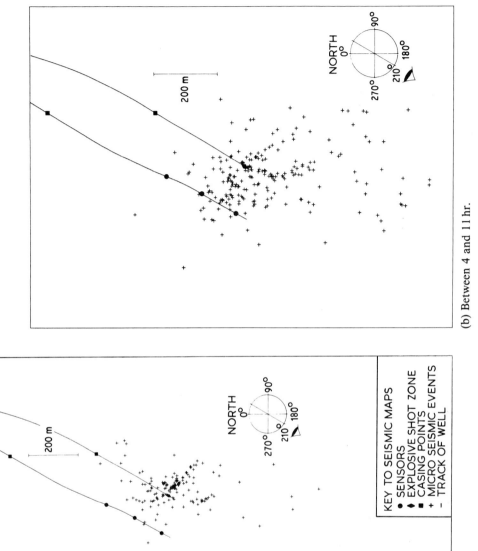

(a) After 3 hours.

(b) Between 4 and 11 hr.

Fig. 9.36 (a, b) Microseismic event locations mapped during a hydraulic stimulation of the Rosemanowes Phase II reservoir (field experiment RT2046). All signals plotted with viewing normal to the plane of the wells.

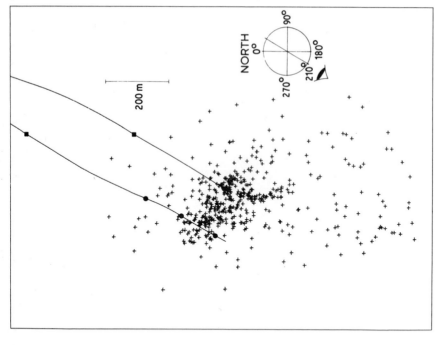

(c) Between 11 and 27 hr.

(d) Composite for 0 through 27 hr.

Fig. 9.36 (c, d) Microseismic event locations mapped during a hydraulic stimulation of the Rosemanowes Phase II reservoir (field experiment RT2046). All signals plotted with viewing normal to the plane of the wells.

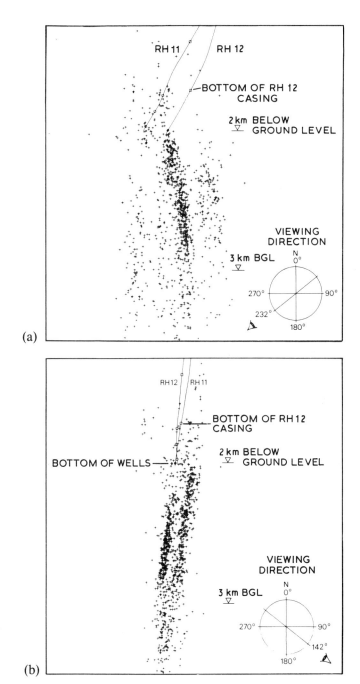

Fig. 9.37 All microseismic event locations mapped during stimulation tests carried out during October–November 1982 on the Rosemanowes Phase II Reservoir. (Batchelor *et al.*, 1984). (a) Viewing normal to the plane of the wells. (b) Viewing in the plane of the wells.

volume as defined by an envelope of microseismic signals was enormous at 400×10^6 m³ of rock. A major problem, however, is that water loss rates are extremely high and may indicate that a substantial portion of the stimulated region is communicating with a low impedance joint system that is at present inaccessible for heat mining. These points will be discussed again in Chapter 10. However, it should now be clear that, similar to the situation at Fenton Hill in its Phase II reservoir, redrilling to connect into this large inaccessible region is likely to be the best strategy to improve the system (for more details see Section 15.5).

9.4.3　Other HDR field tests

Although it is very encouraging to see separate HDR development efforts starting throughout the world, almost all of them are somehow connected or dependent on the outcome of the American effort at Fenton Hill and the British effort at Rosemanowes.

(a)　WEST GERMAN PROJECT

In addition to their participation in the American and British projects involving International Energy Agency and European Commission Agreements, the West Germans have been working at two field sites of their own: a deep single well system at Urach and a shallow fracturing experiment at Falkenburg.

At Urach an ambitious program was undertaken to develop a coaxial single hole heat extraction loop in a well drilled to 3334 m in syenite (see Haenel, 1982, for details). Fracturing was attempted in 1978 below a packer in an open hole section using proppants. The results were promising but bottomhole temperatures were disappointingly low and further work is uncertain.

At Falkenberg, Rummel (1978) and Kappelmeyer (1980, 1984) and their colleagues were involved in fracturing experiments at 200 to 300 m which were attempted with microseismic monitoring. The results from such experiments are given in Table 9.4. Fracture growth was correlated with microseismic activity and event maps were used to locate target holes to intersect the fracture (Leydecker, 1980 and Jung, 1980). One hole did connect with the system and a recovered core showed what could have been a fracture intersection. Subsequent circulation tests with seismic monitoring suggest that a natural joint system has been activated (Kappelmeyer and Rummel, 1984).

(b)　JAPANESE PROJECT

Japan has also been an active participant in the Los Alamos project with resident scientists and engineers on site at Fenton Hill. Their participation is a part of a joint agreement between the US Department of Energy, West Germany, and Japan through the International Energy Agency (IEA). The Japanese, of course, have extensive hydrothermal resources that have been under development for some time. Their interest in HDR is twofold. Firstly, they would like to use the technology as a way to improve productivity at their hydrothermal sites; and secondly, they are interested in developing HDR as extensively as may be feasible to provide an alternative energy source. The latter effort is a part of the so-called Sunshine Project (1982).

Field work in Japan has been at two sites, Yakedake and Kurayoshi. At Yakedake in 1980, hydraulic fracturing tests were performed in slate and very competent sandstone (1 to 2% porosity) at a depth of 300 m in 15 cm diameter holes. As reported by Matsunaga and Kuriyagawa (1983), some 985 microseismic events were detected with hydrophones in shallow bores with 127 being located. These results suggested that a natural joint network was activated by the fracturing. Other relevant details are given in Table 9.4. At Kurayoshi, an HDR system is to be developed in a granitic body at a depth of 2000 m with an initial rock temperature of 220°C. Presumably several holes will be drilled and hydraulic fracturing tests performed in the near future.

(c) FRENCH PROJECT

Research in France has been scattered among a number of groups at different scientific institutions and universities. Field fracturing efforts have been conducted at a site called Mayet de Montagne, 25 km SE of Vichy. Several wells 200 m deep and 16.5 cm in diameter have been drilled into granite. Funds for this work have been provided by the EEC under the auspices of l'Institut National d'Astronomie et de Géophysique (INAG) with supervision provided by Cornet (1980a, 1980b). The French have also attempted seismic measurements using hydrophones similar to those used by the Camborne team and the Japanese at Yakedake. The early tests described by Cornet (1980b) did not detect signals. According to Batchelor (1983), this aseismicity may be expected if joints are stimulated that are in perfect alignment with the principal stresses because they are not experiencing shear displacements. The injection rates have been low (< 6 liters/s) and a high viscosity gel has been used. This would tend to result in tensile failure or 'jacking' rather than shearing. As mentioned earlier these pure tensile fractures along joints would tend to be very 'quiet' seismically.

The French expect to expand their field research into a large program they have named ENERGEROC that would involve stimulation of deep bores at 4500 m to produce 20 MW of thermal power with only 30% drawdown in ten years of operation. A site north of Limoges has been selected for this project but no further word is yet available on its projected completion date. Unfortunately, this project has been delayed indefinitely.

As is evident from the data given in Table 9.4, the scope of the tests at Falkenberg, Le Mayet de Montagne, and Yakedake is much smaller than those performed at Fenton Hill or Rosemanowes. Nonetheless, the results have provided further verification of the ease of producing fractures in low permeability rock and the feasibility of mapping fracture growth with microseismic measurements.

(d) OTHER STUDIES

The authors are aware of other activities on HDR development but these have been paper studies such as the IEA sponsored MAGES (Man Made Geothermal Energy Systems) project involving Sweden, Switzerland, West Germany, Japan, England, and the United States. The MAGES study was designed as a 'brain-storming' effort to complement the strong field programs at Fenton Hill and Rosemanowes. Because all field experiments were concentrating on hydraulic fracturing, people were concerned that we would become too committed to that path without considering other possible techniques some of which might warrant attention in the field. With this rationale, the MAGES group examined a large variety of practical and impractical ideas and concluded at least for the near future that hydraulic stimulation was the optimal technique to be pursued. Outside of the MAGES studies other work has been insufficiently documented in the open literature to discuss here. The Soviet research on fracturing at Kamchatka is a good example of this.

9.5 Current thinking on engineering reservoirs

This chapter documents the extensive efforts at Fenton Hill and Rosemanowes both to create HDR reservoirs and to diagnose what has been created, particularly the geometry of the reservoir. Obviously, these tasks are far from complete as they still remain the central thrust of the US and UK research programs. Basically, we are learning how to optimally use *in situ* conditions of stress and rock lithology and other properties to properly stimulate the formation by increasing effective permeability. This will permit interlinking of the injection and production wells to create a viable reservoir or labyrinth. The overall objective is to *maximize* heat transfer, swept area and volume while keeping flow impedances and water losses within acceptable bounds. The Phase I systems at Fenton Hill and Rosemanowes have demonstrated that interlinking can be achieved and that impedance requirements can be met but these systems were too small from a heat production standpoint to be commercially useful. The Phase II systems are designed to have much larger heat extraction capabilities and to operate with hydraulic characteristics in line with commercial requirements.

Clearly progress has been made in improving our understanding of *in situ* behavior during stimulation. Several generalizations can, in fact, be made at this point based on available field data. First, fracturing is achieved at reasonable fluid injection conditions. Over a wide range of depths from 200 to 4250 m and corresponding range of *in situ* stresses, fracturing pressures have ranged from 4 to 48 MPa (580 to 7000 psia), only modestly above minimum stress levels after pressure losses in piping have been accounted for. Although water without special rheological additives and without proppants has worked well as a fracture fluid in the Phase I systems at Fenton Hill and Rosemanowes, its performance in the Phase II systems at both sites has not been altogether consistent with model prediction or designs. Water can be injected at high rates (30 to 100 liters/s) and reasonable pressures (less than 15 MPa) for prolonged periods during the extension stages of the stimulation. But no system following stimulation has yet been able to sustain production flows of these same magnitudes which are required for commercial operation. Although the Phase II wells are interlinked, overall impedances between wells are still too large to permit these economic production rates. Excessive water losses result from a combined effect of unstable fracture growth and fluid permeation. The research and development effort in this area must be maintained to understand the mechanics of interlinking wells in HDR so that commercial reservoirs can be created.

Some of the field results of hydraulic fracturing experiments support the simple theory that fractures in intact rock propagate in a direction normal to the direction of least principal stress. In addition, the widths of such fractures have been found to be of the order of one to a few millimeters. However, another mechanism of stimulation has been identified in the field work from Fenton Hill, Rosemanowes, LeMayet de Montagne and Falkenberg, the four HDR experimental sites. At each of these sites there is significant anisotropy in the horizontal stresses and the rock fabric is very strong. Each site is located in a jointed crystalline rock. When the pressure within a joint is raised by hydraulic injections it lowers the normal force across the joint. If this reduction is sufficient the joint will move in a shearing motion. This, in turn, causes a small degree of normal dilation because of the ramp angles on the surface irregularities of the joints. The motion takes place in a stick-slip fashion and shear movements of only a few tenths of a millimetre can induce dilations of a few tens of microns over many square metres in a matter of milliseconds. This emits significant microseismic noise and 'consumes' part of the hydraulic energy from the fracturing operations. One way of looking at it is to treat this activity as a highly sensitive periodic or episodic leak off to the secondary joint system that is both pressure and time dependent.

A mechanism by Batchelor *et al.* is described in detail in Appendix A found at the end of this chapter. The quantitative result of this analysis is that the critical pressure for shearing occurs where the maximum shear strength τ_{max} is reduced to below the induced effective shear stress τ:

$$\tau = [(\sigma_H - \sigma_h)/2]\sin 2\theta - P \qquad\qquad (9.6)$$

$$\tau_{max} = [(\sigma_H - \sigma_h)/2 - P - ((\sigma_H - \sigma_h)/2)\cos 2\theta]\tan(\phi + \Omega) \qquad (9.7)$$

$$\theta = 90 - (\phi + \Omega)/2 \qquad\qquad (9.8)$$

where σ_H and σ_h refer to the maximum and minimum principal horizontal stresses and the angles σ and Ω are defined in the figures of Appendix A. Typically $(\phi + \Omega)$ values are 35–40° and, hence, any joint at 25–30° to the maximum stress direction will be at an optimum angle to shear. Pine and Batchelor (1984) show these pressures can be very low and that shearing will occur under anisotropic conditions at pressures less than those needed for tensile fracturing.

If this mechanism is operative, it will defeat attempts to interlink the wells unless it is recognized and controlled by increasing fluid viscosity and restricting injection flow rates during stimulation. This shearing failure mechanism may cause excessive fluid losses in circulation if too high a pressure is used. All of this suggests that the stimulation procedures have to be tailored to what nature has provided.

Current field experiments at Rosemanowes and Fenton Hill are studying rock failure mechanisms in detail, for proper understanding offers the possibility of greatly enhanced heat transfer regions if they can be engineered with a low flow resistance.

The rather limited successes of the Phase II programs in the US and UK so far do not cast a negative shadow on heat mining in general. They only suggest that the original strategy of drilling both injection and production holes and fracturing to interlink them was not appropriate particularly before the underground situation was defined or characterized. The apparent downward growth of fractures in the RH11/RH12 system at Rosemanowes is an excellent example of this. In hindsight, extrapolation of behavior at shallower depths in the Phase I reservoirs was not directly transferable to the deeper Phase II reservoir. Recent developments at Rosemanowes and Fenton Hill have been very positive, details are presented in Section 15.5.

In order to design and engineer such systems for commercialization, one needs a clear picture of how the rock will behave from a fracture mechanics standpoint. This is an absolute requirement for developing optimal reservoir stimulation strategies. The experience gained at Fenton Hill and Rosemanowes has been invaluable in emphasizing this point. Obviously, we are still on a positive learning curve in understanding the *in situ* behavior of large-scale complex and anisotropic rock systems.

As R. B. Duffield so aptly summarized the present situation, "Our plan of action *has* to be in harmony with the way God made the rocks... ". Furthermore, in reviewing our manuscript, he suggested that we redefine the terms 'high grade resource' to be one that not only has a high geothermal gradient but also has rock of predictable properties in regard to creating the reservoir.

The second generalization that can be made deals with developments in the reservoir diagnostics area. Many geophysical techniques have been improved or modified from existing oil and gas technology to provide diagnostic capabilities to trace the effects of fracturing the reservoir. Most of these, such as temperature, televiewer, spinner and radioactive tracer logs provide near wellbore data only. Microseismic measurements are the only working geophysical method for mapping HDR fractures in the far-field; but they too have their limitations.

Clearly, accurate estimates of fracture geometry, particularly aperture distribution and radial or lateral extent, are essential for providing a basis for predicting heat extraction performance as well as for designing and creating the reservoir system in the first place. Basically, the predictive capabilities of the state of the art in reservoir engineering are limited by these geometric uncertainties and the degree of non-uniqueness that accompanies them. For example, in order to fully test the hydraulic fracturing models discussed in Section 8.8.2, field-measured geometries must be available. Without this the fracture models will have limited usefulness aside from providing qualitative trends.

Although microseismic event mapping techniques have progressed tremendously in the last decade with this incentive, much remains to be done before *in situ* fracture geometries can be unambiguously specified. One key uncertainty involves the relationship between the microseismic event map and the location of fracture permeability that participates directly in the actual heat extraction process. At best, the relationship between seismic area or volume and heat transfer area or volume is rather tenuous since the operative seismic source mechanism is unclear.

Regarding the implications of the microseismic data, Murphy, Keppler, and Dash (1983) have made some inroads into interpreting the entire set of data. They have attempted to treat essentially all the data given in Table 9.4. Their major contribution has been to correlate seismic areas and volumes with injected fluid volumes where two-dimensional features have been used to specify areas. All fracture dimensions were taken from microseismic maps of hypocenters located for each test. They have used a subjective analysis to define the areal and volumetric seismic cloud dimensions by excluding some outlying events. They suggest that uncertainties of ± 25% are appropriate.

The results of their analysis are given by Figs. 9.38 and 9.39. Both graphs of seismic area and volume against injection volume suggest linear correlations on log–log coordinates. A linear correlation for seismic area A_s versus injection volume V_{inj} suggests that a predominately planar rather than volumetric fracture model is appropriate as given by Fig. 9.40(b) in comparison with 9.40(a). The position of the unit slope line on Fig. 9.38 is controlled by a single parameter b, which is equivalent to the volume of water injected per unit fracture area. From the data presented, b ranges from 4–20 mm. If we return to one of the key results from the hydraulic fracture models developed by

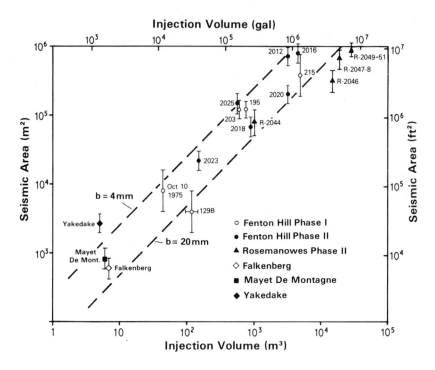

Fig. 9.38 Correlation of seismic area with injection volume for hydraulic fractures in HDR reservoirs worldwide. (Murphy *et al.*, 1983).

Khristianovich and Zheltov and by Geerstsma and de Klerk (see Section 8.8.2), fracture aperture at the wellbore $w(0,t)$ for a circular fracture with no leak off given by equation (8.34):

$$w(0)=w(0,\ t)=2\left(\frac{\mu \dot{q}R}{G}\right)^{0.25} \tag{9.9}$$

where
  μ=fluid viscosity
  G=shear modulus
  R=effective fracture radius
  $\dot{q}$=injection flow rate

So the aperture scales are $R^{0.25}$. Using a value of $R = 400$ m and suitable values for μ of 0.5×10^{-3} Pa-s and G of 30 GPa, $w(0)$ is approximately 1 mm. We would expect $w(0)$ determined from equation (9.9) to be smaller than the empirical parameter b because b has the effect of permeation built into it. Murphy *et al.* (1983) uses a transient, one-dimensional solution for Darcy-like flow into the rock surrounding the fracture walls and the difference between b and $w(0)$ to estimate permeabilities by the equation

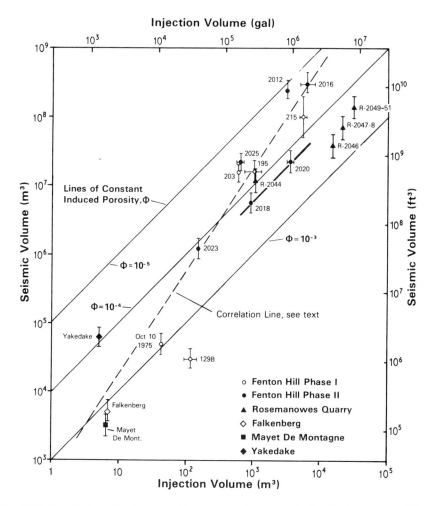

Fig. 9.39 Correlation of seismic and injection volumes for hydraulic fractures in HDR reservoirs worldwide. (Murphy *et al.*, 1983).

$$b-w(0)=\Delta P \left[\frac{\pi k t}{\mu c \phi} \right]^{1/2} \qquad (9.10)$$

Again, using reasonable values for μ and $c\phi$, ranges of k could be estimated for each reservoir. Then estimates would be remarkably close to experimentally measured permeabilities.

A somewhat non-linear correlation is given by Fig. 9.39 which plots V_s against V_{inj}. (A linear correlation would require a 45° slope on a log–log plot). The correlation line drawn is for 2/3 power relationship

$$V_{inj}=aV_s^{2/3} \qquad \text{(where } a=\text{constant)} \qquad (9.11)$$

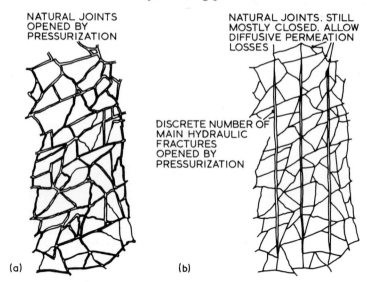

NATURAL JOINTS
OPENED BY
PRESSURIZATION

NATURAL JOINTS. STILL
MOSTLY CLOSED. ALLOW
DIFFUSIVE PERMEATION
LOSSES

DISCRETE NUMBER OF
MAIN HYDRAULIC
FRACTURES
OPENED BY
PRESSURIZATION

(a)   (b)

Fig. 9.40 (a) Volumetric fracturing model (Murphy *et al.*, 1983). (b) Planar fracturing model (Murphy *et al.*, 1983). Joint orientation angles shown are arbitrarily placed.

rather than a simpler model where the formation's porosity corrects the seismic volume to injected volume.

$$V_{inj} = \phi V_s \qquad (9.12)$$

The non-linear correlation of equation (9.11) is consistent with the width of the seismic zone increasing with the square root of time which is exactly the diffusional permeation result of equation (9.10). Although this may be fortuitous it does seem to suggest that the planar fracturing model of Fig. 9.40(b) is correct or at least consistent with these observations.

At this point assuming a direct relationship between microseismicity and fracture location one might hypothesize that hydraulically induced fractures always tend to have their geometry dominated by a planar structure with finite permeation spreading outward in a secondary joint set perpendicular to the main plane. This would produce slab-like, ellipsoidal or tabular, three-dimensional features that have two-dimensional primary flow paths within the main fractures and three dimensional flow in a secondary joint system surrounding these primary paths. This model provides reasonably consistent behavior for the reservoir at Fenton Hill and Rosemanowes both in terms of observed shear-dominated failure modes and the thermal-hydraulic perform-ance of the fracture network that were activated. This is discussed further in Chapter 10.

9.6   Multi-bore reservoirs

One of the most important deductions suggested from the experimental work

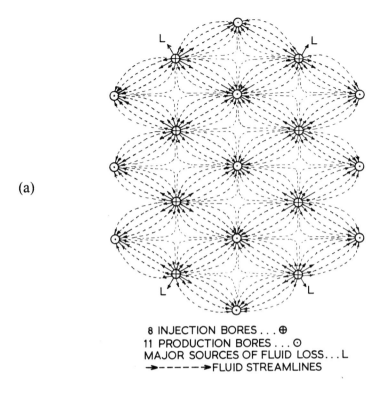

(a)

8 INJECTION BORES . . . ⊕
11 PRODUCTION BORES . . . ⊙
MAJOR SOURCES OF FLUID LOSS . . . L
→ — — — → FLUID STREAMLINES

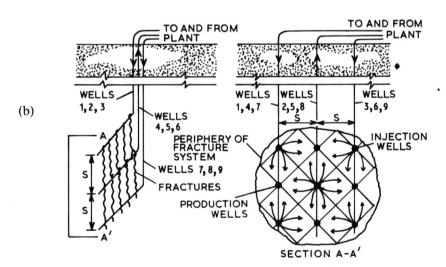

(b)

Fig. 9.41 Multi-bore reservoir concept for HDR exploitation. (a) Hexagonal bore-hole array. The arrangement may be extended outwards indefinitely, with increasing efficiency. (b) 5-spot staggered array.

that has been undertaken at Fenton Hill and Rosemanowes is that the volumetric extent of a fractured reservoir created by a simple pair of bores, as estimated by seismic and other data, is far greater than that part of it which has been usefully employed for the extraction of heat. This implies a low utilization factor for a two-bore reservoir taken in isolation, since only the innermost core of the fractured zone would be put to good use while the outer regions of that zone would be only partially utilized.

The major uncertainty of locating the fractured region causes this efficiency limitation to a two-bore reservoir. Anisotropic stresses and lithologic hetero-geneities can locally alter the orientation and shape of fractured zones in such a complex fashion as to reduce a directional drilling targeting strategy to almost a 'guessing game'. The use of multiple bores in a large HDR field can greatly increase the probability of fracture intersection in such a manner as to optimize heat extraction capacity. In effect, we are attempting to access a three-dimensional fracture system whose dimensions are comparable to the total region of rock that has been stimulated.

If, for instance, a large zone of HDR is subjected to multiple perforation by a number of bores so located that fractured sub-reservoirs formed by various pairs of bores overlap with one another, then the outer regions of many of these sub-reservoirs would then become the inner (and hopefully productive) regions of other sub-reservoirs. Only the outer regions of the whole complex of fractured rock would be unproductive. Thus the effective heat transfer area and reservoir volume would be larger with this network of multiple bores, and would result in a larger overall utilization factor.

Several possible configurations of multiple bores could be devised, although one must keep in mind that these are concepts that have not been tested in practice. One that is suggestive of high efficiency would be a hexagonal pattern of vertical bores with alternating uprisers and injections, as shown in Fig. 9.41(a). Such a system would be completely symmetrical and could be subjected to flow reversal from time to time so as to cool the reservoir at a steadier rate than would be possible with uni-directional flow. Alternatively a staggered 'five-spot' pattern as shown in Fig. 9.41(b) might be effective.

Appendix A   Stress and failure relationships under anisotropic conditions

Consider the situation of the joint at some angle θ to the maximum principle horizontal stress. The normal stress σ_n times the coefficient of friction gives the shear strength of the joint and if this is greater than the shear stress τ_j then the joint is stable. This is the classical Mohr's failure theory, expressed as:

$$\tau_{max} = \sigma_n \tan(\phi + \Omega) \qquad (A.1)$$

where: τ_{max} = shear strength
ϕ = angle of friction
Ω = ramp angle of the irregularities (adjoining angle)
σ_n = normal stress across the joint
$\tan(\phi + \Omega)$ = effective coefficient of friction
σ_n and τ_j can be derived from a Mohr's Circle description

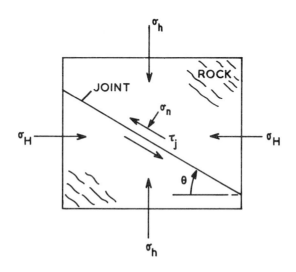

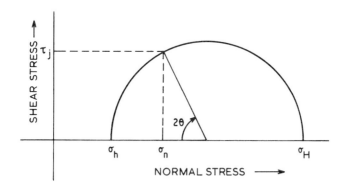

it can be seen that:

$$\tau_j = ((\sigma_H - \sigma_h)/2)\sin 2\theta \qquad (A.2)$$

and

$$\sigma_n = (\sigma_H + \sigma_h)/2 - ((\sigma_H - \sigma_h)/2)\cos 2\theta \qquad (A.3)$$

Therefore, the maximum shear stress τ_{max} can be rewritten using equations (A.1) and (A.3) as:

$$\tau_{max} = \left[\frac{(\sigma_H + \sigma_h)}{2} - \frac{(\sigma_H - \sigma_h)}{2} \cos 2\theta \right] \tan(\phi + \Omega) \qquad (A.4)$$

Now for $\tau_{max} - \tau_j$ to be a maximum, the angle θ can be obtained by combining equations (A.2) and (A.4) and differentiating to give:

$$\theta = (90 - (\phi + \Omega))/2 \text{ for } (\tau_{max} - \tau_j) \text{ maximized} \qquad (A.5)$$

As the joint becomes pressurized σ_H and σ_h are reduced by the fluid pressure, e.g. the effective maximum horizontal normal stress is $\sigma_H = \sigma_H - P$ and the τ_{max} equation becomes

$$\tau_{max} = \left[\frac{(\sigma_H + \sigma_h)}{2} - P - \frac{(\sigma_H - \sigma_h)}{2} \cos 2\theta \right] \tan(\phi + \Omega) \qquad (A.6)$$

where equation (A.5) can still be used to specify θ for a maximum difference between τ_{max} and τ_j.

Chapter 10

Heat mining strategy: thermal and hydraulic reservoir performance

This chapter deals with the theoretical and practical aspects of the heat extraction process itself. Economic aspects of reservoir performance are covered later in Chapter 14. First, basic exploitation strategies are discussed for low and high permeability formations. Quantitative criteria and parameters are defined for assessing reservoir performance. Next the discussion focuses on hydraulic performance: reservoir flow impedance, pumping requirements, flow distribution and water losses are examined. Theory, diagnostic methods, modeling and field results at the Rosemanowes and Fenton Hill sites are covered in detail. Reservoir thermal performance is examined to provide a theoretical basis for properly designing HDR systems for long production lifetimes at commercial heat extraction rates. Geochemical effects on reservoir performance are also discussed. Throughout this chapter, comparisons between theoretical models and field results are made to illustrate both the utility and limitations of using such models to predict performance in the field. The final section considers the future design of HDR reservoirs to improve performance.

10.1 Basic approach

The commercial prospects of a HDR system are of course largely dependent on the costs of forming the reservoir – i.e. of the drilling and stimulation operations. The ideal reservoir should have the greatest possible heat production rate, the longest possible life and the minimum investment and operating costs. Reasonable commercial goals for a single reservoir would be a sustained production for 25 to 30 years at flow rates of 40–100 kg/s (300 000 to 800 000 lb/h) for each injection/recovery pair of wells. At a mean production temperature of 150°C this would yield about 16 to 42 MWth at the wellhead per pair of bores.

It is also important that we should be able to predict with reasonable accuracy the future performance of the reservoir from test data obtained over a short period (days or months). Fortunately there are several theoretical models on which such predictions can be made on the basis of short term field data. These models, however, should be used with caution because some of them which fit the short term data may sometimes predict quite widely different long term performance.

10.1.1 Specification of quantitative performance criteria

Table 10.1 lists the major reservoir performance parameters that could be used to judge the thermal–hydraulic characteristics of any HDR system. Many of the underlying concepts for these parameters have been discussed in Chapter 7. The first three, flow impedance, water loss, and flow distribution, refer to the reservoir's hydraulic characteristics while the last two, fluid geochemistry and thermal drawdown, are direct measures of thermal performance.

As shown by the example given earlier in Section 7.5, there are many components contributing to the overall impedance. The pressure losses discussed in that example could have been subdivided into those that were related to the surface pipework and borehole system and those that were directly tied to the reservoir. The losses of the former can be reduced to acceptable levels by increasing borehole and surface pipework diameters. However, the reservoir losses are tied to the structure of the labyrinth itself, which as we have seen in Chapter 9 can be quite complex. These reservoir-related impedance effects will be covered in detail in Section 10.2.

One important attribute of operating a two-hole HDR system is the thermohydraulic aid provided by having cold (more dense) fluid in the injection well and hot (less dense) fluid in the recovery well for extended depths. The example given in Section 7.5 shows a buoyancy-driven, thermo-syphon boost of some 5.9 MPa (856 psi) for a 3.5 to 4 km deep system at 240–275°C. The magnitude of this effect scales directly with the depth of the system and the average temperature of the reservoir relative to the injection temperature. As pointed out in that example, if the combined borehole/reservoir impedance is low enough then the system could be self-pumping. However, a circulation pump would, even in this case, be necessary for startup purposes. It should also be noted that all of the reservoirs tested so far in the field have required external pumping throughout circulation tests.

Instantaneous water loss rate and total accumulated water loss during reservoir operation is another important parameter since it will determine makeup pump sizes and water requirements. Large water losses could be economically unacceptable and may be related to induced seismic effects (see Chapter 13). In the extreme, excessive water losses will reduce thermal energy recovery, because a substantial fraction of available heat is being swept out of reach as fluid permeates away from the injection point. In most practical applications, water losses would have to be less than about 5% of the injected flow.

Flow distributions in HDR systems are usually determined with a flow-through tracer that follows the distribution of residence times within the

Table 10.1. HDR reservoir performance parameters

Parameter/units	Quantitative representation	Description
(1) FLOW IMPEDANCE I GPa/(m³/s) (SI) psi/gpm (Imperial)	$I = \Delta P / \dot{q}_{out}$ $\Delta P = P_{inlet} - P_{outlet}$ $\dot{q}_{out}$ = production flow rate	Measure of resistance to fluid flow, may depend on average reservoir initiation pressure in some cases. Determines pumping requirements.
(2) WATER LOSS $\dot{q}_{loss}$ m^3/s (SI) gpm (Imperial) Q_{loss} m^3 (SI) gallons (Imperial)	$Q_{loss} = \dot{q}_{loss} dt$ $\dot{q}_{loss}$ = instantaneous loss rate $\dot{q}_{loss} = \dot{q}_{in} - \dot{q}_{out}$ Q_{loss} = total water loss t = time	Instantaneous and integrated water loss to reservoir by permeation and/or fracture growth during production. Fluid loss may be regarded as transient storage within reservoir that may be recoverable under certain operating conditions.
(3) FLOW DISTRIBUTION $<V>$, $\overset{\vee}{V}$, $w_{1/2}$, $[V]$ m^3 (SI) US gallons	$<V>$ and $[V]$ = integral mean and median reservoir volume $\overset{\vee}{V}$ = modal reservoir volume σ^2 = variance of the residence time distribution as a measure of the spread of fluid residence times $w_{1/2}$ = width of the distribution at half peak height as a measure of the degree mixing in major reservoir flow paths.	Measure of reservoir size by mean or modal volume from tracer response. Tracers also give the distribution of fluid residence times within the reservoir. They can be used to identify long and short flow paths and to assess the sweep efficiency of fluid as it passes through the labyrinth.
(4) THERMAL DRAWDOWN $\dfrac{T_{out}(t) - T_{reinjection}}{T_{out}(t=0) - T_{reinjection}}$	Fractional reservoir production fluid temperature decline in the recovery well as a function of time at a specified production flow rate.	Measure of the active heat transfer area and volume of the reservoir
(5) FLUID GEO-CHEMISTRY $C_i(t)$ ppm, mole/liter	Concentration of species i as a function of operating time t. Species i could include dissolved SiO_2, Ca^{+2}, Na^+, K^+, Cl^-, or other ions.	Measure of rock chemical dissolution and solubility effects. Used in geothermometry to estimate reservoir temperatures and to identify mixing effects with indigenous 'pore fluid' originally contained in the rocks' natural porosity.

fractured reservoir. Knowing about this distribution is extremely important for two reasons: first, the overall volume or size of the reservoir can be estimated, and secondly the degree of fluid mixing within the labyrinth can be assessed. These data can then be used to model the behavior of the system.

The term 'sweep efficiency' has been developed for oil and gas reservoir engineering operations to quantify how well fluid flowing from an actual reservoir approaches its ideal limit for some assumed reservoir geometry. To consider this concept further, assume that the ideal fracture system between two wells is a single 'penny-shaped' circular vertical fracture of area $A = \pi R^2_{max}$. The sweep efficiency for this case might be defined in terms of the effective area or fracture radius R_{eff} determined from tracer or thermal measurements of reservoir volume. For a circular fracture, this area could be expressed as an integral of the product of aperture $w(r)$ and the local circumferential perimeter $(2\pi r)$ over all possible radii:

$$\eta_{sweep} = \text{sweep efficiency} = \int_0^{R_{eff}} w(r)2\pi r dr \left/ \int_0^{R_{max}} w(r)2\pi r dr \right. \qquad (10.1)$$

where $R_{eff} < R_{max}$. Thus η_{sweep} is always less than unity. There are many other definitions of sweep efficiency that one could devise depending on how well the flow geometry is known. For example, three-dimensional concepts are equally as appropriate as two-dimensional for defining sweep efficiency. As will be shown later, the overall efficiency of the heat mining process itself can be related to a sweep efficiency concept as applied to the available heat transfer area or volume in a fractured HDR reservoir.

Although it is difficult to say quantitatively what sort of sweep efficiency would be acceptable because of the vagueness of its definition, arbitrarily safe levels would be somewhere about 50% or more.

Thermal drawdown as specified by fluid production temperature decline in the recovery well is a direct measure of the heat transfer performance of the reservoir. At a particular mass circulation rate and initial rock and fluid reinjection temperatures, the measured outlet temperature scales directly with the extracted thermal power level. If referenced to a constant surface reinjection temperature T_{rej}, the instantaneous thermal power $P(t)$ in Watts becomes:

$$P(t) = \dot{m} C_p [T(t) - T_{rej}] \qquad (10.2)$$

where $\dot{m}$ = mass flow rate of fluid (kg/s) and C_p = heat capacity of fluid (J/kg K). As will be shown in later sections of this chapter, $P(t)$ will be used to estimate the effective heat transfer area or volume of an HDR reservoir by empirical matching of predictions from a specific model with adjustable parameters to specific field results for thermal drawdown.

The final parameter is termed fluid geochemistry since it refers to the behavior of specific molecular or ionic species found in the circulating fluid. The concentrations of these species as a function of time may indicate changes in reservoir temperatures since mineral dissolution and reprecipitation rates and solubilities of minerals in the rock are sensitive to temperature. Furthermore, a certain amount of indigenous, highly saline 'pore fluid' is naturally found in the rock's porosity. This tends to diffuse and mix with recirculated reservoir fluid – thus providing some clues regarding fluid circulation patterns throughout the reservoir. Although the resulting models are often complex, fluid geochemistry results can provide useful data to check for consistency when using tracer and thermal drawdown data to model the system.

Again as pointed out earlier, the finite chemical interaction that occurs between the circulating fluid and the host rock may lead to inferior performance probably by one of the following effects:

(1) deposition of supersaturated minerals or compounds (e.g. SiO_2 or silica) in the heat exchanger of a closed cycle system;
(2) deposition of material in the cooler injection bore or entrance region to the fracture introducing unacceptable impedance increases;
(3) removal of material by dissolution in hot regions of the system causing undesired reductions of impedance locally and resulting in channeling.

Before discussing actual field results however, reservoir development strategies for low and high permeability formations designed to meet our quantitative performance goals are considered.

10.1.2 Strategy for low permeability formations

The general criteria for reservoir formations have been established in Chapter 7 – so we can now be specific as to what is required for a commercial-sized HDR system. The objective is to create sufficient rock heat transfer area with low fluid flow impedance to ensure both a high rate of heat extraction and long life. Although it is somewhat difficult to generalize for all cases, an overall impedance of about 1 GPa s/m³ (10 psi/gpm) or less, after buoyancy corrections have been added, is a reasonable goal for a commmercial HDR system. If the reservoir is to last for n years at a known rate of thermal power withdrawal, then the *minimum* volume of rock can be estimated simply by knowledge of the geothermal gradient and the heat capacity and density of the rock. The heat extracted from this minimum volume corresponds to the total thermal power produced.

$$\int_0^n P(t)\mathrm{d}t = \int_{Z_1}^{Z_2} A\rho_r C_{pr}[T_{ri}(Z) - T_{ro}]\mathrm{d}Z \qquad (10.3)$$

$$V_r^{min} = \int_{Z_1}^{Z_2} A\mathrm{d}Z$$

A = area of reservoir in plan view

ρ_r = rock density

C_{pr} = rock heat capacity

n = reservoir lifetime

$T_{ri}(Z)$ = initial rock temperature at a particular depth Z

$\quad T_{ro}$ = reference temperature = minimum acceptable rock temperature for heat mining

$\quad Z$ = depth

However, actual reservoir sizes would have to be considerably larger than calculated from equation (10.3) because the temperature or grade of the energy extracted would no doubt decline as finite drawdown occurred with $T_{ri}(Z)$ decreasing toward T_{ro}. To elaborate on earlier discussions in Section 7.10, the minimum rock temperature T_{ro} will be specified by the mode of heat removal from the reservoir and by the surface system or end use requirements in a particular application. For example, if electric power is to be generated, thermal drawdown in the produced fluid temperature in the recovery well might be restricted to something like 20% of the original design temperature – 160°C if the plant design temperature was 200°C. Depending on how heat is removed from the rock, this will place limits on the amount of cooling allowed in certain parts of the reservoir. For example, Fig. 7.6 shows the effect of altering the flowrate within a given system where low flow rates tend to cause a thermal wave to move within the reservoir as production proceeds.

In order to proceed further with developing a strategy we need to reexamine the field results and the appropriate theory. First, the question of reservoir structure. Is the reservoir predominately single planar fractures discretely oriented in a particular direction, or is it just a mass of more or less randomly distributed joints approximating a volumetric (in the sense of three-dimensional) homogeneous porous system? From the results observed at Fenton Hill, Rosemanowes, and elsewhere, it would be unwise to select either extreme case. The actual situation is probably in between, with near vertical primary joints opening against the least principal horizontal confining stress providing the major conduits and with secondary and possibly tertiary sets of parallel joints hydraulically connected to this primary set. The secondary and tertiary sets would be oriented away from the more favorable direction perpendicular to the least principal stress, and thus would not find it as easy to open under pressurization as the primary joint system but could well fail in shear at much lower pressures (Pine and Batchelor, 1984). This situation becomes more exaggerated as the horizontal stresses become more anisotropic with $S_Y < < S_X$. The two extremes and the more probable situation are diagrammatically depicted in Fig. 10.1.

By assuming that the vertical fracture dominated network structure is favored for HDR in low permeability formations, the optimal reservoir will result by using a wellbore pair inclined from the vertical with injection and recovery wells separated by as large a distance as is hydraulically feasible to ensure good communication. If the dominant fracture system is vertical, then

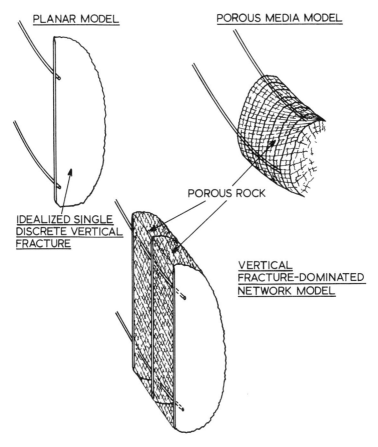

Fig. 10.1 Conceptual schematic of possible HDR fractured reservoir structure models showing extreme cases of single and multiple planar fractures.

the more horizontal the wells become in the active region the better. Of course, drilling horizontally from a vertical position several kilometres above would be impossible. As will be shown in Chapter 11, some compromise is necessary for practical directional drilling. At Fenton Hill, a 35° angle from the vertical was selected for their Phase II reservoir, while at Rosemanowes 30° was used. Since both wellbores will intersect the fracture system at essentially discrete points, there will be a diverging–converging aspect to the flow pattern that should be considered. The lower position in the wellbore pair would normally be selected for injection into the reservoir and the upper for recovery. This would tend to maximize areal sweep through the labyrinth via buoyancy-driven flow. The cold injected fluid would tend initially to sink into the lower portion of the reservoir and then expand and move upward toward the outlet contacting a larger portion of rock volume. The flow would converge as the outlet region is approached. In a qualitative macroscopic sense, the pattern would resemble potential flow between a point source and point sink with buoyancy assistance. If the wells were reversed, short-circuiting between the inlet on the top and the outlet on the bottom would be encouraged by the negative buoyancy effect.

Two options can be pursued for trying to create or open fractures and connect them to form the labyrinth. First is the so-called *targeting strategy*, whereby a single well is drilled and fractured with the fractures mapped seismically and the second well drilled to intersect the fractured region. This first well could be either the lower injection well or the upper recovery well depending on the preferred direction of fracture growth, upward or downward. Obviously, there is a strong incentive to maximize the separation distance between inlet and outlet wells because this will tend to increase the size of the reservoir. However, a good connection with acceptable flow impedance is equally important so one must be somewhat conservative in drilling the second well to ensure that such a connection results. To optimize this strategy a better understanding is required than presently exists of the exact significance of the microseismically defined region.

The second approach is to drill both wells in the hope of optimal orientation, then to fracture out of one or both wells to create the connected labyrinth. The major risks here are that the well direction and separation distance have to be selected before really knowing the orientation and how far usable fractures tend to propagate. As discussed in Chapter 9, this *drill and fracture strategy* has been used for the Phase II system at Fenton Hill and Rosemanowes with only limited success. One can see, however, that as the wellbore pair slants away from the vertical, the probability of intersection with vertical fractures is still quite good even if the principal stresses and/or natural joint pattern are not exactly oriented in expected directions.

Let us return for a moment to our discussion of reservoir structure and re-examine how heat is transferred from the rock to the circulating cooling fluid. By considering a very idealized structure, we can see what actually controls or limits the rate of heat mining. Assume the reservoir to be dominated by vertical fracture connectivity in the form of a circular, penny-shaped crack as a primary system with only secondary flow in parallel joint sets. For now neglect any heat extraction contribution from the secondary system and concentrate on convective circulation within the fracture and thermal conduction through the rock to the rock/fluid interface. As depicted in Fig. 10.2(a), the flow within the crack is two-dimensional, sweeping across the face of the crack, with heat being transferred partly by forced convection. Flow in the crack results from the pressure difference between inlet and outlet and from natural convection due to the buoyancy forces caused by fluid temperature and density differences. In a gravitational field, buoyancy effects will create their own circulation pattern even in the absence of an imposed pressure gradient.

As cold fluid is circulated, a gradient is set up between the surrounding rock and the fluid. Heat transfer in the rock surrounding the fracture is essentially by pure thermal conduction if we neglect the permeation flow effect through the

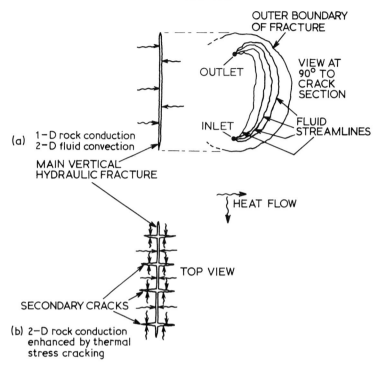

Fig. 10.2 HDR reservoir heat transfer mechanisms. (a) Fracture-dominated flow. (b) Fracture-dominated flow with enhancement by thermal stress cracking.

faces of the crack. Furthermore, if we consider the magnitude of the thermal diffusivity α_r of the rock to be approximately 10^{-6} m²/s, then the thermal penetration depth into the rock will be quite small even after many years of heat extraction. An order of magnitude analysis of the heat transfer resistances provides two unequivocal results:

(1) Thermal conduction through the rock quickly controls the heat extraction rate.
(2) The convective transfer rate within the fracture is so large relative to conduction through the rock, that the fluid may be assumed to be locally isothermal, that is at nearly the same temperature as the rock which is in contact with it.

These two conclusions will be justified quantitatively in Section 10.3.

Given these results, a suitable strategy would consist of creating multiple fracture zones along the inclined set of wellbores with approximate spacings to correspond to thermal penetration depths over the anticipated lifetime of the reservoir. In this analysis we are neglecting any heat that is transferred to fluid that is permeating perpendicular to the fracture walls. If this diffusive loss is less than 5% of the injected flow, which is our goal, then this approximation is ·

reasonable. For a conductively controlled environment, the thermal wave penetration depth δ_t scales as

$$\delta_t = 2\sqrt{\alpha_r t} \qquad (10.4)$$

t = time in seconds (s)
$\alpha_r = \lambda_r/\rho_r C_{pr}$ = rock thermal diffusivity (m²/s)
λ_r = rock thermal conductivity = 3 W/m K for granite
ρ_r = rock density = 2800 kg/m³ for granite
C_{pr} = rock heat capacity = 1000 J/kg K for granite

Thus, using granite properties:

$$\delta_t = 36.8 \text{ m after 10 yr}$$
$$\delta_t = 52.0 \text{ m after 20 yr}$$
$$\delta_t = 63.6 \text{ m after 30 yr}$$

Given that rock thermal conduction controls the rate of heat mining, we have verified that a very large contact area between the working fluid and hot rock is a desired property of the reservoir (item (1) in 5.2.2 and 7.2). With penetration depths of the order of 60 m during the reservoir's 30 yr lifetime, we would need to have either individual fractures of enormously large surface area of order 1 to 2 × 10⁶ m² or a series of smaller, thermally non-interacting fractures that are equivalent to the total area of the single larger fracture. If we approximate the heat capture volume by

$$V_h = (2\delta_t) \times (\text{fracture area}) \qquad (10.5)$$

we can make an estimate of the maximum uniform thermal power extraction rate possible:

$$P(t) \simeq V_h \, \rho_r C_{pr} [T_i - T_{ro}]/t \qquad (10.6)$$

assuming a uniform initial rock temperature T_i of 250°C and a minimum utilization temperature T_{ro} of 125°C, then a 2 × 10⁶ m² granitic HDR fracture system could in principle provide up to 94 MWt of power continuously

$$P(t) = 2 \times 10^6 \ (127.2) \ (2800) \ (1000) \ (125)/30(365)(24)(3600)$$

$$P(t) = 94 \times 10^6 \text{ Wt or 94 MWt per pair of wells} \qquad (10.7)$$

Given this approximate method we can size and space the wellbores for any thermal power load. Figure 10.3 shows three idealized concepts for producing a commercial-sized reservoir capable of producing hot fluid at 40 to 75 kg/s continuously for up to 30 years with no more than 20% drawdown. None of these concepts will completely represent the actual reservoir in terms of structure but they can in many cases be used to model actual field performance as will be seen in Section 10.3. For example, the ideal of a very large single, penny-shaped fracture would be very difficult to create in the field, given the heterogeneous nature of HDR reservoirs filled with joints, varying lithology, and rotating anisotropic stresses. But something between the smaller parallel set of parallel fractures in Fig. 10.3(b) and the fully rubblized region in 10.3(c) is probably not far from reality.

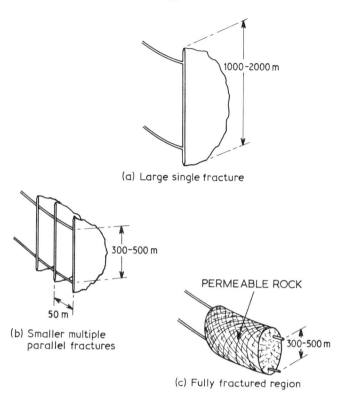

(a) Large single fracture

(b) Smaller multiple parallel fractures

(c) Fully fractured region

Fig. 10.3 HDR reservoir concepts for low permeability formations. Half fractures shown with division at a vertical axis of symmetry. (a) Large single fracture. (b) Smaller multiple parallel fractures. (c) Fully fractured regions.

Individual fracture zones stimulated individually and spaced between 25 and 50 m horizontally along an inclined wellbore pair should result in minimal thermal interference over a 20 to 30 yr period of operation and may represent the optimal strategy given our present state of knowledge. One other point concerning system design needs to be made. As suggested previously, additional cracking induced by thermal stresses might occur and cause the reservoir to *grow* while heat mining is underway. This could, in the proper circumstances, lead to enhanced productivity from the HDR system as shown in Fig. 10.2(b). However, there are so many uncertainties associated with rock fracturing theory and thermal stress cracking phenomena that a more conservative engineering approach is suggested, namely to design the reservoir without anticipating any beneficial effects from thermal cracking.

10.1.3 Strategy for high permeability formations

If *in situ* formation permeabilities are high enough (in the millidarcy range or better), the problem of creating artificial permeability to circulate fluid at low impedance is replaced by that of trying to contain and recover fluid while still maintaining large swept volumes of rock. Although the occurrence of natural high permeability HDR may be rare, the techniques for exploiting them are not very far from currently available technology used for recovery of gas and oil by water-drive or flooding methods. Both production and injection wells would be required and arranged in such a manner as to minimize fluid loss to surrounding permeable formations at the perimeter of the developed field. Figure 10.4 shows a five-spot arrangement with a central injector and four peripheral producers. In these cases, adequate well spacing is essential to simultaneously maximize exposure to hot rock while minimizing water losses. Other well arrangements involving two-spot, four-spot, and similar peripheral flooding techniques employed to enhance oil production with water drive are potentially applicable to high permeability HDR systems (see Ramey *et al.*, 1975, and Slider, 1976, for details).

One would expect that heat recovery from a water-drive system would depend largely on the nature of the permeability within the reservoir. If it is dominated by fractures, then something like the behavior seen for a low-

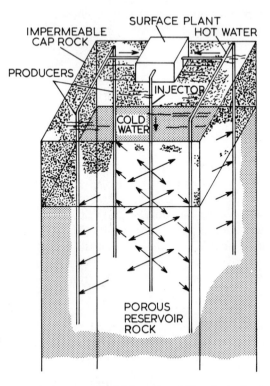

Fig. 10.4 HDR reservoir concept for high permeability formations using a five-spot water drive well arrangement.

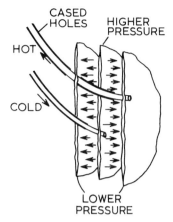

Fig. 10.5 Double parallel fracture concept using water drive for heat mining in high permeability formations. Fracture planes divided at a vertical axis of symmetry (from Tester and Smith, 1977).

permeability fractured HDR system might be observed. If the active reservoir is a layer of high permeability rock that is essentially uniformly porous, one would expect a displacement or breakthrough of the thermal wave or front as the reservoir becomes depleted (see Fig. 7.6).

In another approach to reservoir formation, a parallel network of vertical fractures has been suggested with flow in the rock space between fractures. Figure 10.5 conceptualizes such a system consisting of three parallel fractures of similar area. The centrally-located fracture serves as the fluid injection surface while the two outside fractures receive fluid by permeation through the rock.

The heat extraction performance of either the multi-spot (Fig. 10.4) or the double parallel fracture system (Fig. 10.5) will depend on the distribution and velocity of flow through the rock matrix which in turn depends on the formation permeability (Tester and Smith, 1977). For the parallel fracture system, thermal breakthrough can be ideally related to the lifetime of the reservoir assuming that flow between fractures is essentially one-dimensional (resembling plug flow in a cylindrical tube) and that heat transfer rates between the fluid and rock are very rapid. Breakthrough of the cooler fluid will occur at times t:

$$t > \rho_r C_{pr} A S (1 - \phi) / \dot{m} C_p \qquad (10.6)$$

where

A	= fracture area (m²)
S	= spacing between fractures (m)
ϕ	= matrix porosity
C_p	= heat capacity of water (J/kgK)
$\dot{m}$	= mass flow rate of water (kg/s)
ρ_r, C_{pr}	= rock density and heat capacity (kg/m³, J/kgK)

If we assume that the major flow resistance is in the rock matrix contained between fractures, Darcy's law can be used to estimate pumping power requirements. The pressure drop across the fracture system ΔP is given by

$$\Delta P = \mu \dot{q} S / 2kA \qquad (10.7)$$

where

μ = fluid viscosity (Pa-s)
$\dot{q}$ = volumetric fluid flow rate $(m^3/s) = \dot{m}/\rho$
k = mean formation permeability (m^2 or 10^{12} darcies)
ρ = fluid density (kg/m^3)

Pumping power will scale with the product of ΔP and $\dot{q}$, consequently from equation (10.7) it is proportional to $\dot{q}^2$. Given the thermal breakthrough relation of equation (10.6) and the pumping power proportional to

$$P_{pump} = \Delta P \dot{q} = \mu \dot{q}^2 S / 2kA. \qquad (10.8)$$

an optimum relationship between A and S will result depending on k and prevailing economic conditions. If k is small, large areas and small S would be preferred; if k is large, the reverse is true. An upper limit for ΔP is given by the difference between the formation's fracture extension pressure and the ambient pore pressure. Also fracture area A might have an upper bound set by *in situ* conditions such as lithologic boundaries.

10.1.4 General modeling approach to deal with the non-uniqueness dilemma

Partly because of the inherent structural complexity of these fractured networks and partly because fluid flow and rock failure behavior in fractured-porous media are not completely understood, more than one model of the reservoir may adequately fit all available data. This apparent non-uniqueness bothers many investigators and in fact has led to our conservative dismissal of thermal stress cracking enhancement at this stage of understanding.

Some of the uncertainties regarding fracture and flow behavior include:

(1) uncertain rock failure criteria for *in situ* conditions;
(2) elastic laws for porous, anisotropic rock with heterogeneous structure not understood;
(3) rock elastic moduli vary with pore pressure or effective stress level in a non-linear fashion;
(4) superimposed interactions between thermal and fluid diffusion fields;
(5) flow impedance along roughened fracture face not predictable;
(6) pressure losses due to converging-diverging flow not characterized;
(7) rock-water geochemical interaction kinetics not known.

Given these and other uncertainties, reservoir models have evolved mainly from empirical fits to field data using the simplest theory that appears relevant. For example, water loss has been successfully modeled assuming permeation into a continuum with a constant permeability – even though strong evidence for joint-dominated flow exists.

Geometric simplifications are usually the ones made first to allow for a tractable mathematical description. For example, consider again the models for hydraulic fracturing where rather idealized geometries were assumed for the propagating fracture (see Section 8.8.2).

A non-uniqueness problem will always be an integral part of reservoir engineering and is not necessarily limited to HDR modeling. In fact, many oil and gas reservoirs are more complex geologically and frequently involve multi-component, multi-phase operations – so the number of adjustable parameters to fit a given set of field data can grow to enormous proportions.

10.2 Hydraulic performance

10.2.1 Theory and diagnostic techniques

The three major parameters to describe the hydraulic behavior of a HDR reservoir were first introduced in Table 10.1. They are flow impedance, water loss, and flow distribution.

Although the overall system impedance is defined as the pressure drop divided by the production flow rate, this can be deceptively simple because many factors combine to give the overall value. A buoyancy correction is applied to account for the difference in hydrostatic pressures in the cold injection well versus the hot recovery well. Aside from the obvious pressure losses in the boreholes and surface equipment which can be separated from the reservoir losses, the losses within the reservoir itself have several distinct components. First, there are the entrance and exit losses associated with deceleration as the fluid moves radially outward from the injection point and with acceleration as the fluid moves radially inward converging toward the recovery point. These contributions are termed entrance and exit impedances and can be measured by suitably designed transient pressure-flow measurements. The second category includes losses within the fracture system. These can vary anywhere from a pressure drop linearly proportional to flow velocity associated with Darcy-like, laminar flow between the faces of the fracture, to non-linear, flow-rate-dependent effects associated with non-laminar, induced turbulence mechanisms caused by everything from surface roughness on the fracture faces to geometry dependent effects as fluid passes through the joint network.

The two-dimensional orthogonal joint or FRIP model discussed in Chapter 8 shows how non-linear effects could be observed even if the flow is locally laminar within each joint (Pine and Ledingham, 1983). A few experimental studies have been conducted with flow in laboratory-scale rock fractures to determine appropriate flow laws (Witherspoon and Wang, 1974, and Wilson and Witherspoon, 1970). In many practical situations, a Darcy's law form can be used for flow within a fracture with the linear velocity u given by

$$u=\dot{q}/A_c=(w^2/12\mu)[dP/dx] \tag{10.9}$$

where the effective permeability of the fracture k_f is

$$k_f=w^2/12 \tag{10.10}$$

If there is a jointed network of fractures equation (10.10) is modified by the mean fracture spacing S (metres/fracture)

$$k_f=w^3/12S \tag{10.11}$$

where

w=fracture aperture
A_c=cross-sectional area of the fracture=wH
H=fracture height

If some structural information is known about the reservoir, it may be possible to estimate fracture apertures from impedance measurements. By integrating equation (10.9) for steady one-dimensional flow conditions with an assumed constant aperture, the impedance of the fracture system (I) can be related to aperture by

$$I^{-1}=\dot{q}/\Delta P=w^3H/12\mu L \tag{10.12}$$

where L=length of the flow path.

Thus by knowing fluid viscosity, fracture length and height and impedance, one can estimate w.

One other complicating factor is encountered when dealing with an inclined wellbore system that intersects a series of joints. Measured flows from the recovery bore are a result of the superposition of individual joint flows. Thus spinner surveys are needed to measure contributions from each joint system to characterize the impedance of each joint. There is typically a wide distribution of joint impedances under essentially the same pressurization condition. This might be even more than an order of magnitude say from 1 GPa s/m³ to 10 GPa s/m³.

As the average pressure of the reservoir is increased to approach the effective confining stress, the joints will open and since the impedance scales inversely with w^3 (equation (10.12)) a large decrease in impedance should be observed as reservoir pressures approach the confining stress. Field experiments at Los Alamos using partially choked flow in the recovery well to induce a high back pressure have verified this effect (see Section 10.2.3). One detrimental consequence of higher pressures is larger water loss rates. This is a result both of the higher pressure driving force and the fact that formation permeability increases as the confining stress is approached (Fisher and Tester, 1980).

The water loss rate and integrated total water loss is treated mathematically by a continuum approach with Darcy's law coupled to the fluid continuity equation. For the case of transient diffusion of fluid under a pressure gradient an equation of the form below is used:

$$\nabla\cdot((k/\mu)\nabla P)=\beta\partial P/\partial t \tag{10.13}$$

where

∇P=pressure gradient=$\partial P/\partial x+\partial P/\partial y+\partial P/\partial z$
for three-dimensional diffusion

β=average system compressibility=$\dfrac{-1}{V}\left(\dfrac{\partial V}{\partial P}\right)_T$

For low permeability formations with $k < 10^{-18}$ m² or $k < 10^{-6}$ darcy, k is usually assumed to vary with the cube of porosity, ϕ^3, via the Kozeny–Carmen equation. Fisher and Tester's (1980) analysis of field data has suggested that k, ϕ, and β depend on local pore pressure and effective stress in the form of:

$$k=k_o/(1-C*P)^{3n}$$
$$\phi=\phi_o/(1-C*P)^n \qquad (10.14)$$
$$\beta=\beta_o/(1-C*P)^{n+1}$$

where n and $C*$ are empirical constants and P is the local fluid pressure. The constant $C*$ determines the pressure dependence of k, ϕ and β and can be interpreted as the reciprocal of the sum of the confining stress and the pore pressure.

Even with this complex dependence on pressure, simple trends in water loss rate and accumulated water loss are observed. One can integrate equation (10.13) for the case of one-dimensional diffusion out from the faces of a planar fracture, with k, ϕ, and β assumed constant and the following initial and boundary conditions

for $t=0$ $\quad P=P_o$ at all x (uniform pressure initial condition)

for $t>0$ $\quad \dfrac{-k}{\mu}(\partial P/\partial x)=\dot{q}_{loss}/A$ at $x=0$ (constant flux boundary) $\qquad (10.15)$

for $t>0$ $\quad P=P_o$ at $x\rightarrow\infty$ (far field uniform pressure boundary)

The resulting solution shows that loss rate given by $\dot{q}_{loss}$ is proportional to A and decreases as $1/\sqrt{t}$, where total accumulated loss varies directly with A and increases as $\sqrt{t}$.

A single hydraulic parameter can be used to characterize the one-dimensional portion of fluid loss behavior.

$$\alpha=A\sqrt{k\beta} \qquad (10.16)$$

where A = effective fracture permeating area. Equations (10.14) and 10.16) can be combined to show that α itself is merely a function of A (a structural parameter), n and $C*$ (empirical rock parameters) and k_o and β_o (physical properties of the formation). In practical applications water loss data would be used to obtain values of $C*$ and n for specified k_o, β_o, and ϕ_o properties. Another virtue of α is that it is a measure of system size since it scales with A.

Measurements of flow distributions constitute the final category of hydraulic parameters. As outlined in Table 10.1, these parameters relate to reservoir size and the distribution of fluid residence times within the active reservoir.

Chemical or radioactive tracers are used to characterize flow distribution using techniques similar to those described earlier in Chapter 9 for detection of fracture intersections in open boreholes. Basically, a pulse of tracer containing a radioisotope such as Br^{82} or a dye solution such as sodium fluorescein is placed in the injection well, pumped through the fracture system and monitored as it returns to the recovery well. Tracer concentrations are determined as a function of time and volume throughput on the surface using suitable analytical methods, or downhole for some Br^{82} measurements using the gamma logging tool described in Section 9.2. The resulting tracer concentration versus time or volume data gives the residence time distribution (RTD) for fluid circulating in the reservoir (see Fig. 10.6). Papers by Tester, *et al.* (1982) and Robinson and Tester (1984) describe the techniques used in detail.

For HDR fracture systems, very dispersed flow with long tailing residence times is common as depicted qualitatively in Fig. 10.6. To evaluate these experimental RTDs, which are actually probability distribution functions, several statistical quantities and normalization procedures are used. Integral (first moment) mean volumes $<V>$, modal volumes $\overset{\circ}{V}$, median volumes $[V]$, variancies (σ^2), and the distribution width at one-half the peak height $w_{1/2}$ are typically used to describe the shape of the RTD. These are defined as follows:

$$\text{volume-based RTD function:} \quad E(V) \equiv C_i(V)/m_T$$

$$\text{mean volume:} \quad <V> = \int_0^\infty VE(V)dV \tag{10.17}$$

$$\text{modal volume:} \quad \overset{\circ}{V} = V_i \text{ when } C_i = f(V_i) \text{ is a maximum} \tag{10.18}$$

$$\text{median volume:} \quad [V] \geqslant \int_0^{[V]} E(V)dV = 0.5 \tag{10.19}$$

$$\text{variance:} \quad \sigma^2 = \int_0^\infty (V - <V>)^2 E(V)dV \tag{10.20}$$

$$\text{width at 1/2 height:} \quad w_{1/2} = [V_i \text{ at } C_i^{max}/2^+ - V_i \text{ at} C_i^{max}/2^-] \tag{10.21}$$

where $C_i(V)$ is the tracer concentration at a given cumulative produced volume V since the start of the tracer experiment and m_T is the total mass of tracer injected. Representations of these parameters are shown in Fig. 10.6 where for the skewed distribution shown $\overset{\circ}{V} < [V] < (<V>)$.

The effective residence time equivalent to any of the volumes given above is obtained by dividing that volume by the steady state volumetric flow rate $\dot{q}\,(m^3/s)$.

As seen in Fig. 10.6, flow in the fracture system can be described as well-dispersed. Theoretical estimates and experimental studies have shown that dispersion of fluid in a single, propped hydraulic fracture cannot account for this degree of mixing. Because the reservoir actually may consist of a jointed

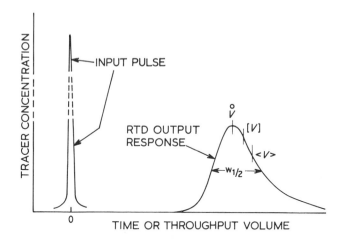

Fig. 10.6 Tracer pulse input and typical response curve or residence time distribution for a fractured HDR reservoir. Statistical parameters qualitatively illustrated.

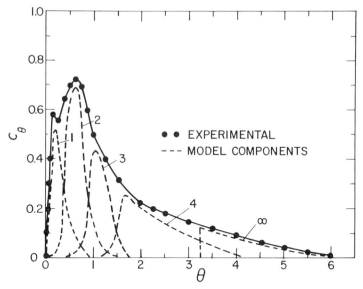

Fig. 10.7 Multiple fracture models as applied to an RTD from a tracer test in the Phase I Fenton Hill reservoir. (From Tester *et al.*, 1982).

network, the observed RTD is caused by the superposition of individual flow paths each having their own RTD as shown in Fig. 10.7.

The median volume of the overall distribution is probably an adequate measure of flow through the major production zones in the reservoir. The integral mean volume is equivalent to the total reservoir volume. However, both mean and median volumes are mathematically biased toward large volumes or residence times because of the integral relationships given in equations (10.17) and (10.19). This bias becomes more pronounced for skewed distributions with long tails. However, the mode of the distribution may be a more accurate measure of main fracture flow volume that is responsible for most of the active heat transfer area of the reservoir. The tail of the distribution is still real and not an artifact, however. It represents fluid passing through the reservoir with long residence times, possibly indicating penetration deep into the fractured labyrinth or possibly matrix-like flow in the rock surrounding the main fractures. In either case, the heat extraction capacity of the system increases as a result of this secondary flow, roughly in proportion to the flow fraction associated with this flow path relative to the main fracture system. However, these secondary flow paths do have higher impedances than the main fractures so their full exploitation is more difficult.

In addition to this statistical or probability-based analysis of the RTD, considerable effort has been expended toward developing analytic and numerical approaches to modeling dispersion in fractured porous media. Interested readers should consult Tester *et al.* (1982) and Robinson and Tester (1984) for details concerning these deterministic methods, because they are only summarized here.

In general, a Fick's law diffusion-type equation is solved for different conditions:

$$\nabla \cdot (D\nabla C_i) - U \cdot \nabla C_i = \partial C_i / \partial t \tag{10.22}$$

where C_i is the tracer concentration, $\nabla C_i = \partial C_i / \partial x + \partial C_i / \partial y + \partial C_i / \partial z$, and D is the dispersion coefficient tensor whose terms can depend on flow velocity and direction and formation properties. As the degree of mixing increases, D increases in magnitude. However, the form of D can be quite complex mathematically and usually forces one to make simplifying assumptions to make solutions possible. An example of a one-dimensional (1-D) model fit to a field-determined tracer curve is shown in Fig. 10.8. In this case only one adjustable parameter is used for the fit, the so-called dispersional Peclet number (Pe):

$$Pe = \bar{U}L/D \tag{10.23}$$

which is a dimensionless group that gives the ratio of the purely convective contribution of the flow $\bar{U}L$ to the dispersive contribution D. Values of Pe were selected to minimize the difference between experimental and predicted values from solving equation (10.22).

A multizone fracture model was also developed that assumed independent, one-dimensional dispersion in M separate zones for this model, as depicted in Fig. 10.7; the predicted tracer concentration results from a superposition of dispersional effects from each zone j:

$$C_i = \sum_{j=1}^{M} \xi_j C_j(X_j, \, \theta_j, \, Pe_j) \tag{10.24}$$

where

$X_j = x/L_j =$ dimensionless distance in fracture j

$\theta_j = \xi_j \, \dot{q} t / V_j =$ dimensionless residence time for fracture j

$Pe_j = \bar{U}_j L_j / D_j =$ Peclet number for fracture j

$\xi_j = \dot{q}_j / \dot{q} =$ flow fraction through fracture j

$V_j =$ volume of fracture j

In most situations the flow fraction ξ_j, fracture lengths L_j, and even individual fracture volumes are known from field measurements, consequently only M adjustable Pe_j parameters remain to fit the data. Optimal fits are usually obtained by minimization of some suitably defined objective function F, evaluated over N data points.

$$F = \sum_{i=0}^{N} \left[\sum_{j=1}^{M} \xi_j C_j(X_j, \, \theta_j, \, Pe_j) - C_i \right]^2 \qquad (10.25)$$

One should note that as M increases, very complex RTDs can be matched easily, thereby reducing the deterministic technique to an exercise in curve fitting rather than true modeling. Consequently, M should be kept to a minimum consistent with observed flow paths.

Other more complex solutions to the convection-dispersion equation are available to account for two- or three-dimensional flow effects as well as velocity and directionally dependent dispersion coefficients. The results, however, are at present beyond the scope of this book (see Tester *et al.*, 1982 for additional information).

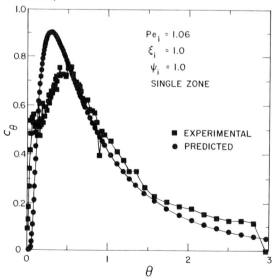

Fig. 10.8 One-dimensional dispersion deterministic model fit to an RTD from a tracer test in the Phase I Fenton Hill reservoir, Pe_j adjusted empirically to give the best fit. (Tester *et al.*, 1982).

Having now established the theoretical basis for characterizing the important hydraulic performance parameters of a reservoir, we can proceed to examine the field results at Rosemanowes and at Fenton Hill.

10.2.2 Field results at Rosemanowes

(a) PHASE I TESTING

As mentioned in Chapter 9, the results of the explosive pretreatment in well RH9D (see Fig. 9.34(a)) were extraordinarily effective in reducing the inlet flow impedance to a very low value. Batchelor, Pearson, and Halladay (1980) discuss the results from pre- and post-shot testing of the reservoir in detail. Following hydraulic fracturing to connect the four-well system, flow testing proceeded. The overall system impedance is shown in Fig. 10.9 as a function of production well pressure. This illustrates the point made earlier that as confining stresses are approached, joints begin to open and flow resistances should drop. The Phase I system at Rosemanowes is now known to be predominantly horizontal and will behave differently from deeper reservoirs. Figure 10.10 plots data from the same experiment but as system efficiency rather than impedance. Batchelor *et al.* (1982) define hydraulic efficiency as

$$\eta_{\mathrm{h}} = f\Delta P_r / \Delta P_i \tag{10.26}$$

where

$\Delta P_r = P(\text{production wellhead}) - P_0$

$\Delta P_i = P(\text{injection wellhead}) - P_0$

$P_0 = \text{reference pressure}$

$f = \text{fraction of injected flow recovered}$

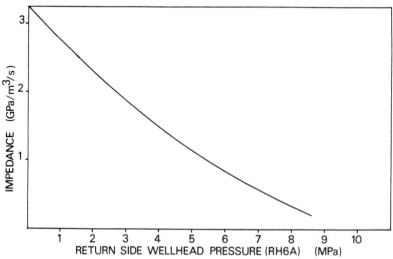

Fig. 10.9 The overall system impedance of the Rosemanowes Phase I reservoir as a function of production well pressure. (Batchelor *et al.*, 1980).

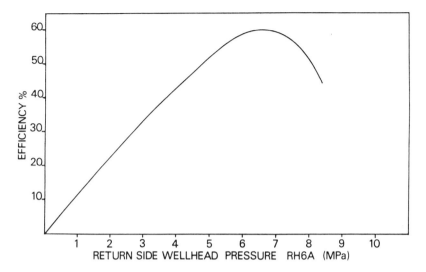

Fig. 10.10 Hydraulic efficiency of the Rosemanowes Phase I reservoir as a function of production well pressure. (Batchelor *et al.*, 1980).

As ΔP_r increases η_h should increase as it does in Fig. 10.10. This is the impedance-dominated region. However, as ΔP_r continues to increase, loss rates climb with f decreasing rapidly. Thus a maximum is reached whereby impedance reduction effects have their optimal impact on system performance.

Tracer tests were also conducted in this system and resulted in a modal volume of 50 m³ of self-propped fracture. If an effective aperture of 1 mm is assumed, then the potential heat transfer area is in excess of 50 000 m² with system impedances as low as 0.4 GPa s/m³. As shown in Fig. 10.11, cumulative water loss from this reservoir varied as the square root of time as predicted by our earlier theoretical discussion in Section 10.2.1 for planar, one-dimensional diffusion or permeation. Assuming an intrinsic rock permeability of 10^{-16} m² (100 microdarcy), a compressibility of 1.95×10^{-11} Pa⁻¹, and a fluid viscosity of 1.3×10^{-3} Pa-s, an effective diffusing area of 280 000 m² is estimated. This may represent the upper limit for potentially available heat transfer area according to Batchelor (1982). Also by extrapolation, the water loss rate would have decreased to 1% of the injected flow after 300 hr of operation.

(b) PHASE II TESTING

The behavior of the hydraulic connection in the deeper Phase II reservoir at Rosemanowes is totally different from that observed in the earlier, shallower system. In three recent project reports (13, 14, 15) Batchelor and his colleagues (1982–1984) reported the chronology of field testing and their interpretation of the results to produce an evolving model of the Phase II reservoir (see also Batchelor, 1984). The explosive pretreatment, hydraulic stimulation, and

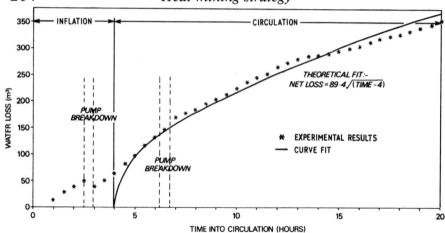

Fig. 10.11 Accumulated water loss during a pressurized circulation test of the Rosemanowes Phase I reservoir. (Batchelor, 1982).

microseismic results have been discussed in Section 9.4.3. The tracer and circulation test results are presented here as well as a summary of their latest interpretation of Rosemanowes Phase II reservoir structure as provided in report 15 and by Pine and Ledingham (1983).

The original objectives of the Phase II reservoir testing at Rosemanowes were to learn how to develop, at intermediate depths and temperatures, a reservoir which, if reproduced at full depth, would be commercially viable. The required reservoir parameters were estimated to be as follows:

(1) an effective heat transfer area of 2×10^6 m²;
(2) an active rock volume in excess of 200 000 000 m³ (or 0.2 km³);
(3) an overall flow impedance between wells of less than 0.1 GPa s/m³ which would allow for 50 to 70 kg/s of flow at differential pressures of 5 to 7.5 MPa.

By 1984, Batchelor and associates felt they had accomplished the first two objectives with 'an inferred heat transfer area of more than twice the required value within a rock volume of double the target size'. The third objective regarding impedance still requires an order of magnitude reduction. However, the water losses have been far too large to consider the present system operable without redrilling and re-stimulation. For example, at an injected rate of 33 liters/s only 24% or 8 liters/s were recovered; at a 17 liters/s injection 35% or 6 liters/s were recovered with the impedance increasing from 1.5 to 1.7 GPa s/m³. With water loss rates now of order 40% of the injected flow, a redrilling effort got underway in the fall of 1984 to improve matters (see Section 15.5 for details).

Nonetheless, much encouraging information has resulted from the latest Rosemanowes tests and has led to new interpretations of HDR reservoirs. For instance, a new mechanism for reservoir growth has been proposed to account for the unexpected downward growth of the fracture system and the larger than expected water losses. Joint shearing and other pressure-related joint dilation effects dominate the hydraulic behavior. This is exaggerated by the misalignment of the large horizontal stress anisotropy with the natural joint sets (see Fig. 9.33).

The Camborne group has made extensive use of its FRIP model to interpret hydraulic test data (see Pine and Ledingham, 1983). In their preliminary and continued interpretation of microseismic results from Phase II, Batchelor *et al.* (1983) have maintained that a low pressure shearing mechanism is responsible for developing very large diffusional areas for water loss. A network of low impedance joints that are non-interconnecting with the recovery well is apparently activated. The large dimensions of the seismic zones shown in Figs. 9.36 and 9.37 provide unequivocal evidence that fluid is permeating into enormous volumes of rock away from the injection bore. They view the water loss as caused both by joint dilation and true permeation into the rock matrix and suggest that future stimulations should try to limit shear failure and enhance tensile jacking to produce maximum joint opening and fracturing to create a region that can be exploited for heat mining as opposed to just parasitic fluid consumption.

Typical tracer results are shown in Fig. 10.12 for the Phase II system. Modal volumes are consistently of the order of 2000 m^3 with somewhere between 30 to 70 hr required for breakthrough. Hydraulic testing, on the other hand, has shown that pressures are transmitted in approximately four hours. The very long tails observed in the RTD may indicate the presence of multiple flow paths and are consistent with results calculated by the FRIP model.

Batchelor *et al.* (1984) conclude their analysis by stating that the present reservoir consists of many parallel flow paths of extremely narrow widths that connect the two wells. Within the stimulated region there are perhaps other much wider paths – but they are not well-connected. Based on FRIP modeling, a typical conceptual model in agreement with the seismic and hydraulic data might have 100 flow paths, each 30–50 μm wide and more than 500 m long. If these zones are truly parallel, then a system height of 500 m and a spacing of a few metres between paths are reasonable. It is this model coupled to the tracer results which suggests heat transfer areas in excess of 2×10^6 m^2.

10.2.3 Field results at Fenton Hill

In order to get properly oriented, we must first refer back to the events discussed in Chapter 9, Section 9.4.1 where the formation of the Phase I fractured reservoir was described. Recall that there are actually two systems, the original and enlarged reservoir (see Fig. 9.22) and five separate large-scale experimental tests: run segments 1–5. The hydraulic arrangement of wellbores GT-2B and EE-1 and the associated fracture systems are given in Fig. 10.13. Since an acceptable connection between EE-2 and EE-3 has only recently been achieved in the Phase II reservoir, long term hydraulic testing results under

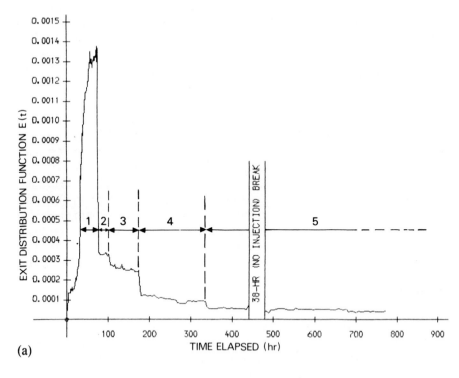

(a)

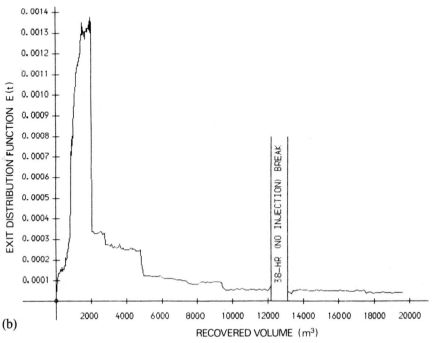

(b)

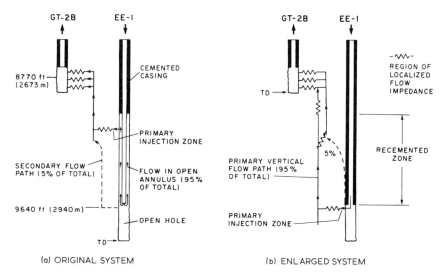

Fig. 10.13 A schematic indicating flow connections and reservoir impedances in the GT-2B/EE-1 Phase I system at Fenton Hill.

circulation conditions have not yet been obtained. So our discussion is limited to the results of testing the Phase I system only. Preliminary results for Phase II are presented in Section 15.5. Primary references used in this section include the review of reservoir testing by Dash *et al.* (1981) as well as the reports from run segment 2 (Tester and Albright, 1979), run segment 4 (Murphy, 1980), and run segment 5 (Zyvoloski, 1981).

(a) FLOW IMPEDANCE

Entrance flows from EE-1 and exit flows into GT-2B during run segment 2 were distributed as shown in Table 10.2 as determined from spinner and tempera-ture surveys. Since impedance is defined in terms of the exit flow, it is possible to assign flow impedances to each of the exit regions, as shown in Table 10.3. Although numerous small, sudden flow changes (the largest was 0.00126 m³/s or 20 gpm) were observed during the test, the overall impedance decreased steadily from 1.7 GPa s/m³ (16 psi/gpm) to 0.326 GPa s/m³ (2.98 psi/gpm) during the 75 day flow test, as shown in Fig. 10.14. This large change is probably due to the extensive cooling of the entire fracture system, with consequent decrease of the fracture-closure stress and partial opening of the fractures. This point will be considered again later.

The so-called 'high back pressure test' or run segment 3 was conducted next.

Fig. 10.12 Residence time distribution expressed as an exit age distribution $E(t)$ as a function of elapsed circulation time (a) and recovered throughput volume (b). (Batchelor *et al.*, 1984).

Table 10.2. (from Dash et al., 1981) Position and magnitude of relative EE-1 injection flows during run segment 2 in the Phase I Fenton Hill reservoir

Cable depth interval (m)	Cable depth interval (ft)	Flow fraction
2073–2179	6800–7150	0.05
2271–2377	7450–7800	0.06
2484–2530	8150–8300	0.01
2606–2652	8550–8700	0.01
2652–2896	8700–9500	0.81
2896–2957	9500–9700	0.06
		Total 1.00

Position and magnitude of GT-2B exit flows, run segment 2

GT-2 laboratory cable depth (m)	GT-2 laboratory cable depth (ft)	Flow fraction	Individual zone flow (gpm)	Individual zone flow (m³/s)
2660–2661	8726–8729	0.38	92	0.0058
2671–2672	8764–8765	0.05	12	0.00075
2686–2688	8812–8820	0.20	48	0.00303
2705–2706	8876–8879	0.12	29	0.00183
2719	8920	0.25	61	0.00385
	Total	1.00	242	0.0153

Table 10.3. Summary of individual fracture zone flow impedances, run segment 2 in the Phase I Fenton Hill reservoir (from Dash et al., 1981)

Fracture depth GT-2B (m)	(ft)	Flow impedance GPa s m⁻³ (psi/gpm) Jan. 30, 1978 Main	Secondary	April 12, 1978 Main	Secondary
2640	8660				12.6 (115)
2660	8728			0.82 (7.5)	
2670	8760			6.6 (60)	
2677	8784				54.1 (495)
2688	8820	2.2 (20)		1.67 (15.3)	
2703	8867				108.2 (990)
2707	8882			2.78 (25.4)	
2719	8920				29.5 (270)
2720	8925	8.7 (80)		1.36 (12.4)	
Fracture set		1.7 (16)		0.34 (3.12)	7.1 (65)
Entire system		1.7 (16)		0.326 (2.98)	

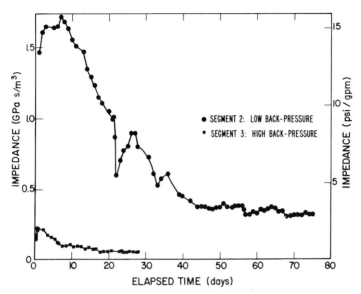

Fig. 10.14 Flow impedance behavior for run segments 2 and 3 in the original Phase I Fenton Hill reservoir corrected for buoyancy effects. (From Dash *et al.*, 1981).

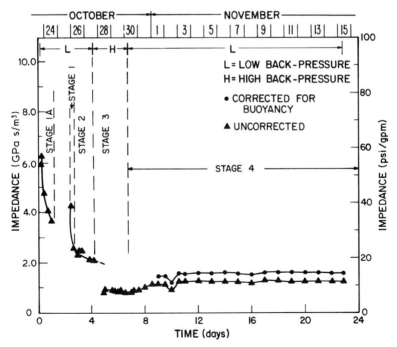

Fig. 10.15 Flow impedance behavior for run segment 4 in the enlarged Phase I Fenton Hill Reservoir.

The reservoir configuration during this test was the same as in run segment 2. However, the GT-2B pressure was raised by valve throttling to 9.65 MPa (1400 psi) in order to observe the effect of the increased backpressure on heat extraction and flow impedance. As expected, a marked decrease in flow impedance was observed, as seen in Fig. 10.14, and the impedance decreased further during the run as more heat was removed and water loss accumulated. The pressurization of GT-2B changed the impedance from 0.82 GPa s/m³ (7.5 psi/gpm) to 0.22 GPa s/m³ (2.0 psi/gpm), based on a low back-pressure measurement made just before the high back-pressure test. A further decrease by a factor of four during the high back-pressure run may be attributed to increased pressurization and cooling.

Run segment 4 involved testing the enlarged fracture system. Following recementing and fracturing, the primary injection zone in EE-1 was moved from approximately 2760 m (9050 ft) to 2940 m (9640 ft) depth, about 300 m below the major production regions in GT-2B (see Fig. 10.13, enlarged system). The measured and buoyancy-corrected flow impedances are shown in Fig. 10.15. After some early transient behavior caused by variable pressure conditions within the reservoir, the flow impedance approached a value somewhat less than 2 GPa s/m³ which is nearly the same as that observed near the end of run segment 2.

The major circulation test (run segment 5) ran for 286 days on the enlarged system. Again the results for impedance were very similar to those observed earlier in segment 4.

After an initial transient decline, the impedance during this run segment remained essentially constant at 1.69 GPa s/m³ (15.5 psi/gpm). Despite the evidence for area and volume growth cited earlier, no evidence of impedance change due to cooling of the rock was observed. The mean impedance values determined by an alternative method using a shut-in technique were as follows for each section of the system (see Fig. 10.13 for positions):

Entrance impedance = 0.13 ± 0.03 GPa s/m³ (1.2 psi/gpm)
Main fracture impedance = 0.29 ± 0.03 GPa s/m³ (2.7 psi/gpm)
Exit impedance = 1.14 ± 0.05 GPa s/m³ (10.4 psi/gpm)
Total impedance = 1.56 ± 0.05 GPa s/m³ (14.3 psi/gpm)

At the end of run segment 5, well GT-2B was shut-in and the entire system pressurized to approximately 15 MPa in order to facilitate readjustment of the reservoir rocks to the new state of stresses, which had been generated by cooling the reservoir. The experiment has been named the 'Stress Unlocking Experiment' or SUE. Numerous microseismic signals were observed indicating that such readjustment was taking place, and changes in fracture impedance were seen in subsequent flow and shut-in experiments as follows:

Entrance impedance = 0.06 GPa s/m³ (0.55 psi/gpm)
Main fracture impedance = 0.34 GPa s/m³ (3.1 psi/gpm)
Exit impedance = 0.50 GPa s/m³ (4.6 psi/gpm)
Total impedance = 0.90 GPa s/m³ (8.2 psi/gpm)

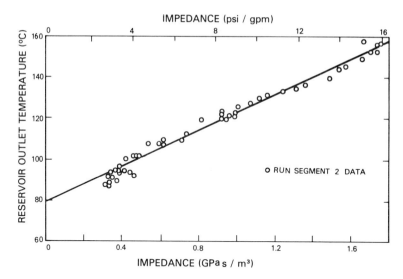

Fig. 10.16 Fracture system specific flow impedance as a function of fluid outlet temperature during the run segment 2 test of the original Phase I Fenton Hill reservoir. (From Dash *et al.*, 1981).

The changes in entrance and exit impedance are statistically significant, and the main effect of SUE is seen to be a reduction of the exit impedance by a factor of two. The system was not operated after SUE for a time long enough to show that this reduction was permanent.

Theoretically, as shown earlier, fracture flow impedance should decrease as $1/w^3$ in both laminar and turbulent flow. Aperture may be increased in several ways: (1) by pressurization of the fracture, (2) cooling of the surrounding rock, (3) dissolution of minerals lining the crack by chemical treatment of the fluid, and (4) by geometric changes resulting from relative displacement of one fracture face with respect to the other. Run segments 2 and 3 were especially useful in demonstrating the correlation between impedance and pressure and temperature as depicted in Figs. 10.16 and 10.17. The impedance changes observed after SUE were probably due to additional 'self-propping' caused by slippage along the fracture faces near the exit or by other pressure-induced geometric changes. Microseismic events were mapped during SUE using fixed geophones in GT-2B and EE-2. Hodographic methods were used to locate events as described in Chapter 9 (see Fig. 10.18). The plan and elevation maps of event hypocenters given in the figure represent the more likely of two sets of solutions. The vertical structure of the more or less planar event cloud shows an aseismic region centered in the section of the reservoir that was substantially cooled during these five circulation tests.

The concentration of impedance near the exit, shown in all the low-back pressure flow experiments, may be desirable when the system impedance is

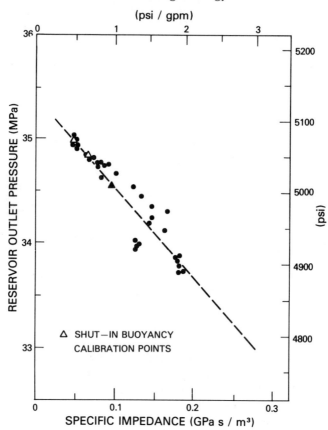

Fig. 10.17 Fracture system specific flow impedance reservoir outlet pressure in GT-2B for the high backpressure test, run segment 3, in the original Phase I Fenton Hill reservoir. (From Dash *et al.*, 1981).

reduced by multiple fractures. In this mode of reservoir exploitation, the possibility of a low impedance 'runaway' fracture cooling and taking too much of the flow is minimized since the exit impedance concentration throttles the flow until reservoir cooling has been extensive. Eventually, the problem of flow control in the individual fractures may arise, and methods of flow control near the fracture entrance may be required. This point will be discussed in more detail in Section 10.5.2. after the thermal performance characteristics of the system have been described in Section 10.3.

(b) WATER LOSS

The water loss rate data of each experiment contain many transients due to operational shutdowns, pump limitations, and various leaks. Consequently, the cumulative water loss is best suited for comparisons since it tends to smooth

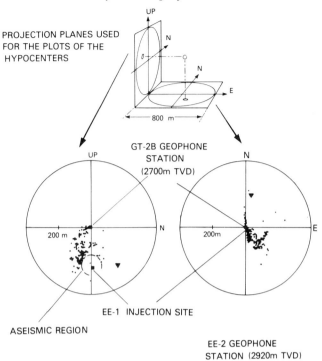

Fig. 10.18 Microseismic event maps located by a two-station method during the stress unlocking experiment (SUE) Elevation (left) and plan (right) shows an aseismic region around the EE-1 injection site as shown on the north-south plane in elevation. (From Dash *et al.*, 1981).

out the transients. Data for run segments 2, 3, and 5 are presented in Fig. 10.19. Run segment 4, only 23 days long, was excluded from this comparison because of the disparate permeable flow conditions under which it was conducted – four separate stages of wellhead pressure conditions were imposed (see Fig. 10.15). The first stage was actually during a massive hydraulic fracturing (MHF) treatment, so the water losses were necessarily high. The succeeding stages consisted of alternating sequences of high, then low back pressure, so that comparison with the other run segments, in which pressure conditions were far less variable, is impractical. Run segments 2 and 5, both conducted under normal, low back-pressure conditions, can be compared directly. Water loss for run segment 5 is approximately 40% higher than that of run segment 2 at comparable times after the beginning of heat extraction. However, because the operating pressure was 10% higher during run segment 5, the water loss for run segment 2 should be scaled up by 10% as in curve 2 of the figure, in order to be directly comparable to run segment 5. Then it is seen that the run segment 5 water loss is only 30% higher than run segment 2, despite a several-fold increase in heat-transfer area and volume. An obvious conclusion is that the heat-

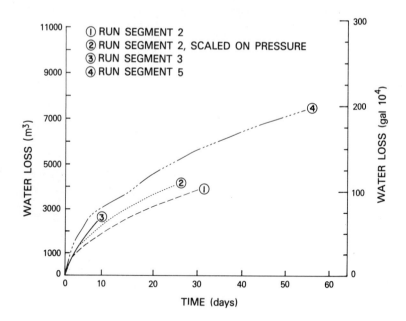

Fig. 10.19 Cumulative water losses during testing of the Phase I Fenton Hill reservoir. (From Dash *et al.*, 1981).

exchange system may only utilize a small portion of a much larger fracture system that controls water loss. More on this later.

At the end of run segment 2, the actual water loss rate was only 1×10^{-4} m³/s, about 1% of the total flow rate circulated through the reservoir. However, this extremely low loss rate was due to the gradual decrease in surface pressure after 30 days. The loss rate extrapolated to the end of the experiment and by model fits to the first 30 days is 7×10^{-4} m³/s or about 7% of the total flow. At the end of run segment 5, the loss rate was 6×10^{-4} m³/s, about 10% of the circulated flow rate. The run segment 3 losses are slightly larger than those of segment 2, which is consistent with the fact that both EE-1 and GT-2B were pressurized in run segment 3. At the end of run segment 3, the loss rate was 1.3×10^{-3} m³/s, about 14% of the circulating rate. Loss rates below 10% are probably commercially acceptable.

The pressure-dependent, water-loss diffusion model discussed earlier was used to obtain theoretical fits to the data. Empirical values of α (equation (10.16)) and C^* (equation (10.14)), obtained from numerical modeling for each run segment are shown in Table 10.4.

Similar to the water losses, α for run segment 5 is 30% higher than that of run segment 2 and probably reflects the addition of the lower half of the reservoir. The value of C^* for segment 5 was determined mainly by only one flow transient. The range of C^{*-1} indicated reflects a lack of sensitivity to this parameter.

Table 10.4. Empirical values of α and C^* obtained from numerical modeling

Run segment	$\alpha(m^3MPa^{-1/2})$	$C^{*-1}(MPa)$
2	1.4×10^{-6}	9.3
3	$1.3 \times 10^{-6} - 2.8 \times 10^{-6}$	9.3
5	1.9×10^{-6}	$13.3 < C^{*-1} < 20.0$

(c) FLOW DISTRIBUTION

Two tracers were used in the Phase I reservoir. One is a visible dye, sodium fluorescein, which is monitored in the produced fluid at the surface with a UV spectrophotometer. The other, radioactive Br^{82} (half life of 36 hr) in the form of ammonium bromide (NH_4Br^{82}), is monitored in the produced fluid at the surface and downhole with the gamma counter described in Section 9.3. The radioactive tracer is not temperature sensitive and therefore does not undergo thermal decomposition as does sodium fluorescein in the higher temperature portions of the reservoir. Flow conditions and preliminary results of tracer tests from the Phase I system are presented in Table 10.5. In this table, the definitions of modal volume, integral mean volume, median volume, and variance are the same as those described earlier. The wellbore volumes are subtracted from the total volume produced to give the true fracture volumes. The integrated mean volume is obtained by integrating the tracer concentration–time curve at the reservoir exit. As described below, the modal volume is considered the most reliable indicator of reservoir volume change. Large changes in the modal volume are observed after the hydraulic fracturing of the system between run segments 3 and 4 and during the SUE, which followed run segment 5. SUE was conducted on December 9 and 10, 1980, and the volume change was evaluated with a tracer experiment on December 12, 1980.

The integral mean volumes show a regular increase during segment 2; however, the integral mean is strongly affected by the volume of fluid produced during a given experiment. During run segments 2 and 4, the length of time a given experiment could be run was determined largely by the limit of detection of sodium fluorescein. The highly increased sensitivity of the method for analyzing Br^{82} over that for sodium fluorescein is responsible for longer tails on the Br-tracer experiments. Dye-tracer experiments end when the dye concentrations in the produced fluid can no longer be detected which is typically <1400 m^3 total produced volume at GT-2B. Bromine-tracer experiments, on the other hand, have continued to 4140 m^3 without completely reaching background levels. Integration of the long tails of the concentration–time curves biases both the integral mean volume and the variance to higher values. Variances of the distributions, calculated for the fully integrated distribution, show the effects of the long tails; however, the flow distribution associated with low back-pressure operation does not change drastically, even after the

Table 10.5. Summary of fluorescein dye and Br^{82} tracer experiments in the Phase I Fenton Hill EE-1/GT-2B fracture system. (Adapted from Tester *et al.* (1982) and Robinson and Tester (1984).)

Experiment	Elapsed days	Injection wellhead pressure P_{EE-1}, MPa	Production wellhead pressure P_{GT-2}, MPa	Total flow Q, $10^3 m^3$ [a]	Production rate q_{GT-2}, $10^{-3}m^3/s$ [b]	Tracer recovery %	Water loss q_{loss}, $10^{-3}m^3/s$	Average fluid production temp. °C	Mean $<V>$, m^3	Median $[V]$, m^3	Mode $\hat{V}$, m^3	90% Trimmed mean $\|V\|$, m^3	Variance σ^2	Variance σ^2 (90%)	Dispersional Peclet number $1/Pe^{-1}$
ORIGINAL RESERVOIR															
Segment 2 (75 day, LBP)															
Phase 1-1 (2/9/78)	8	8.8	1.1	8.0	7.25	69	1.30	150	34.4	25.6	11.4 ± 1.1	29.3	0.65	0.44	0.591
Phase 1-2 (3/1/78)	28	8.5	1.1	26.1	13.1	65	0.60	110	37.5	28.9	17.0 ± 1.5	30.9	0.62	0.43	0.942
Phase 1-3 (3/23/78)	50	6.6	1.1	57.2	13.9	71	0.75	95	54.7	45.1	22.7 ± 2.3	46.3	0.51	0.42	0.944
Phase 1-4 (4/7/78)	65	5.9	1.1	75.7	15.5	>65	0.18	90	56.2	48.4	26.5 ± 2.7	49.2	0.47	0.39	1.120
Segment 3 (28 days, HBP)															
Phase 1-5 (9/28/78)[c]	10	9.3	9.7	9.0	7.7	60	3.17	111	33.1	24.7	$3.8 \,^{+5.7}_{-1.9}$	26.6	0.75	0.56	0.358
Phase 1-6 (10/13/78)[d]	25	9.3	9.7	26.3	9.3	74	3.17	98	56.5	46.5	11.4 ± 1.5	48.1	0.45	0.36	0.306
Phase 1-7 (10/16/78)[g]	28	9.3	1.7	~30.0	15.7	33[e]	4.28	—	49.6	40.5	20.8 ± 1.9	41.8	0.49	0.38	0.347
ENLARGED RESERVOIR															
Segment 4 (23 day, LBP and HBP)															
215-1 (10/26/79)	—	17.2	1.1	1.3	6.4	13(25)[f]	8.70	153	207	192	136 ± 19	184	0.26	0.22	0.195
215-2 (10/29/79)	0	17.2	10.3	6.1	8.1	18	14.00	154	230	211	144 ± 19	209	0.17	0.12	0.360
215-3 (11/2/79)	2	9.3	1.1	9.5	6.6	27	1.13	153	262	216	121 ± 11	243	0.38	0.34	0.281
215-4 (11/12/79)	12	9.3	1.1	16.2	6.4	25	1.27	153	283	236	129 ± 11	263	0.32	0.28	0.310
Segment 5 (282 day, LBP)															
217-A1 (4/16/80)	38	9.8	1.3	13.1	6.2	>57	0.90	158	404	341	155 ± 10	500	0.45	0.44	0.760
217-A2 (5/9/80-Br)[f,h]	61	9.5	1.3	26.5	5.9	88.4	0.50	158	1311[m]	1280	161 ± 4	941	0.53	0.52	—
217-A3 (9/3/80-Br)	178	8.8	1.3	74.0	5.7	81.4	0.38	154	1845[m]	1750	178 ± 4	1274	0.56	0.54	—
PreSUE 217-A4 (12/2/80-Br)	268	8.5	1.1	121.0	5.1	—	2.38[j]	149	581[n]	541	187 ± 10	525	0.40	0.38	—
PostSUE 217-A5 (12/12/80-Br)	278	8.4	1.3	126.0	8.1	66.1	0.90[j]	149	2173[n]	1880	266 ± 4	1009	0.46	0.42	—

a 1 gal. = 3.785 liter = $3.785 \times 10^{-3} m^3$
b 1 gpm = 6.31×10^{-2} liters/s = 6.31×10^{-5} m^3/s
c includes diffusional formation loss and leaks
d volume correction added $7.344 \ m^3$ (1940 gal.) to eliminate negative volumes (early arrivals)
e volume correction added $3.937 \ m^3$ (1040 gal.) to eliminate negative volumes (early arrivals)
f tracer used Br^{82} in place of Na-fluorescein
g 10/26/78 test actually during exp. 190 at LBP

h 1 millicurie (mCi) strength of feed at injection
i experiment terminated before tail of distribution
j includes annulus leak
k SUE-Stress Unlocking Experiment
l $Pe^{-1} = D/UL =$ inverse dispersional Peclet number for single 1-D zone fit
m calculated with assumed exponentially decaying tail
n exp. terminated before tail of distribution

NB dates are in the form: month/day/year

hydraulic-fracturing episodes. This fact is most clearly shown in Fig. 10.20 where normalized tracer concentrations from run segment 5 are plotted against produced volume (the volume produced from the time of reservoir injection). Because of the inconsistency in the calculated integral mean volumes, the concentration normalization is performed relative to the modal volume, $\overset{\circ}{V}$:

$$C_\theta = C_i \bigg/ \left[1/\overset{\circ}{V} \int_0^\infty C_i \mathrm{d}V \right] \tag{10.27}$$

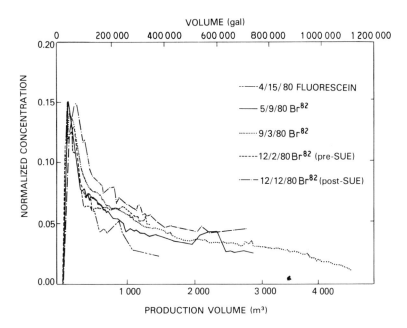

Fig. 10.20 Variation of normalized tracer concentrations with produced volume of fluid uncorrected for tracer recirculation effects in the Phase I Fenton Hill reservoir. (From Tester *et al.*, 1982).

Robinson and Tester (1984) established that it was necessary to correct the Br^{82} tracer data for the effects of recirculation of produced fluid containing small amounts of tracer. The true RTD or $E(t)$ curve was obtained by deconvoluting the concentration–time data of Fig. 10.20 using the convolution integral:

$$C_{out}(t) = \int_0^t C_{in}(t - t^*) \, E(t^*) \, \mathrm{d}t^* \tag{10.28}$$

where the outlet response of the system $C_{out}(t)$ to an arbitrary input C_{in} is the sum of all contributions of tracer injected previously. For the case of complete recirculation, $C_{out}(t)$ is the measured outlet response, while $C_{in}(t)$ is initially the pulse concentration at $t = 0$ followed by the recirculated concentration of tracer. $E(t)$ is then calculated by numerical integration of equation (10.28). Figure 10.21 compares the corrected and uncorrected RTD for experiment 217-A2 (9 May 1980). The really significant difference among the Phase I tracer experiments shown in Fig. 10.20 is the drastic increase in modal volume due to the SUE experiment. However, the mode also increases regularly with time due to heat-extraction throughout segments 2–5. The volume of the system increases systematically even though the shape of the distribution remains fairly unchanged, even after the SUE. This increase in modal volume is shown in a chronological manner in Fig. 10.22. The correlation between this growth and heat extraction and pressurization effects will be covered in Section 10.3.4 following the discussion of reservoir thermal performance.

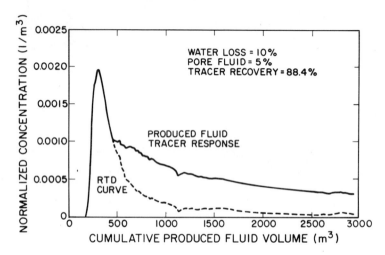

Fig. 10.21 Fenton Hill Br^{82} tracer results showing uncorrected data and the RTD curve calculated by subtracting the effect of fluid recirculation. Wellbore plug flow delay volume of 161 m³ included in the response curves.

As mentioned in the theoretical section on flow distribution effects, the overall RTD is really a superposition of effects from individual fracture zones. For several of the Br^{82} tests, tracer responses from separate zones were monitored with downhole gamma logging. By combining these logs with spinner data from the production hole, individual RTDs for each zone were resolved (Robinson and Tester, 1984).

As can be seen from the Br^{82} test 217-A2 conducted on 9 May 1980 during run segment 5, given in Fig. 10.23, dispersion within each production zone is large and indicates that observed dispersion may result primarily from mixing of various residence times in a highly complex network of interconnected fractures and flow paths. For these systems, the deterministic approach using the convective-dispersion equation (10.22) for flow in a continuum is clearly inappropriate even though it may give reasonable fits to the observed RTD.

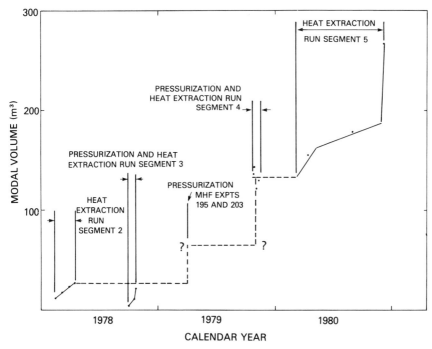

Fig. 10.22 Historical summary of modal volume growth resulting from fracturing operations and heat extraction from the Phase I Fenton Hill reservoir.

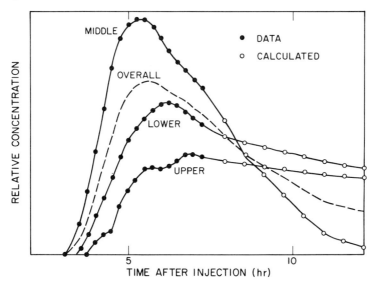

Fig. 10.23 Individual fracture zone tracer concentration-time curves obtained by Br[82] pulse injection followed by gamma logging in the production wellbore in the enlarged Phase I Fenton Hill reservoir. (From Robinson and Tester, 1984).

10.2.4 Summary of hydraulic performance

As a major result of field testing at Rosemanowes and Fenton Hill many of the desired features of commercial HDR systems seem within reach. These include the following.

(1) *Flow impedance.* Fracture flow impedances of 1 GPa s/m^3 (10 psi/gpm) or less were observed for wellbore separations of up to 300 m. Proper pressurization may also be used to reduce impedances even further. Impedance levels also appear to be very stable in time indicating that plugging does not occur.

(2) *Water loss.* Diffusional models of fluid loss by permeation provide adequate predictions of observed reservoir behavior. Aside from the early testing of the Phase II system at Rosemanowes, loss rates have been acceptably low, approximately 10% of the total circulated flow or less. Proper reservoir management has been used successfully at Rosemanowes to stabilize losses to unproductive joint systems.

(3) *Flow distribution.* Techniques for determining flow characteristics including reservoir volume have been refined with tracer methods. Flows within the Rosemanowes and Fenton Hill reservoirs are well-mixed and highly dispersed, indicative of a complex jointed system. There is no apparent short-circuiting of flow from inlet to outlet and reasonable sweep efficiencies should be expected. These findings all will have a positive effect on thermal performance as will be discussed in the next section.

10.3   Thermal performance

The basic concepts have been introduced earlier, so we can proceed with a description of model development and general design criteria for commercial-sized HDR reservoirs. Following this, field results on thermal performance will be summarized and then theory will be compared to practice in a general synopsis of heat mining tactics.

10.3.1   General design criteria

The two extreme reservoir structures (planar fractures versus a porous zone) for low-permeability HDR systems should have very different thermal drawdown characteristics. In the case of planar fractures, there must be a balance between the rate of convective transport by the circulating fluid to the rate of conduction through the rock walls.

To start with, consider a very simple, 'zeroth order' fracture model depicted in Fig. 10.24 as a single perfectly rectangular fracture of constant aperture w in which all fluid is uniformly distributed in one-dimensional plug flow from the bottom to the top of the fracture. Heat transfer in this geometry can be treated analytically for the case of uniform flow of fluid $\dot{q}\,(\text{m}^3/\text{s})$ at a fixed temperature

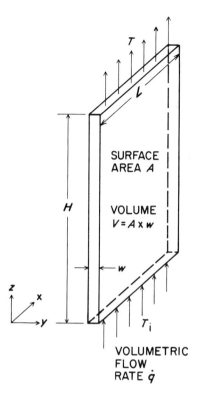

Fig. 10.24 One-dimensional plug flow in a rectangular fracture: simplified HDR reservoir heat transfer model.

corresponding to the temperature T_i of fluid as it leaves the injection well and enters the formation. The rock, extending outward in the $+$ and $-y$ directions, is initially at a fixed temperature T_∞. For all times greater than zero, the rock temperature at an infinite y distance on either side of the fracture remains at T_∞. A closed-form solution to this problem exists (Arpaci, 1966), as cited by Murphy *et al.* (1981), where the equation governing the rate of heat transfer is,

$$\alpha_r \left(\frac{\partial^2 T}{\partial y^2} \right) = \frac{\partial T}{\partial t} \tag{10.29a}$$

which can be solved to give the temperature of the rock as a function of position y and z and time t, subject to the initial and boundary conditions cited above:

$$T(y,z,t) - T_i = (T_\infty - T_i)\, \mathrm{erf} \left[\frac{y/2 + \lambda_r z/\rho C_p \bar{U} w}{(\lambda_r/\rho_r C_{pr} t)^{1/2}} \right] \tag{10.29b}$$

This expression can then be simplified to give the outlet water temperature appearing in the recovery well (T_{out}) with $z = H$ and $y = 0$ as

$$T_{\mathrm{out}} - T_i = (T_\infty - T_i)\, \mathrm{erf}\, [\lambda_r A/\rho C_p \dot{q} (\alpha_r t)^{1/2}] \tag{10.30}$$

where
  λ_r=rock thermal conductivity (W/m K)
  α_r=rock thermal diffusivity=$\lambda_r/\rho_r C_{pr}$ (m²/s)
  ρ=fluid density (kg/m³)
  C_p=fluid heat capacity (J/kg K)
  A=area of fracture on one side, for example if the fracture were
       rectangular $A=L \times H$ or if circular $A=\pi R^2$, (m²)
  w=fracture aperture (m)
  $\bar{U}$=fluid velocity in z (vertical direction)=$\dot{q}/Lw$ (m/s)

Equation (10.30) can be modified to give the actual thermal drawdown as a ratio of thermal power extracted at time t to the initial thermal power:

$$P(t)/P(t=0)=\mathrm{erf}[(\lambda_r \rho_r C_{pr}/t)^{1/2}\, A/\dot{m}C_p]\qquad(10.31)$$

where

$$\dot{m}=\text{mass flow through the fracture}=\rho\dot{q}\ \text{(kg/s)}$$

Thus, for a given set of rock and fluid thermophysical properties, the single parameter that will determine the drawdown rate is $\dot{m}/A$ or $\dot{m}/R^2$ for a circular fracture.

This parameter can be regarded as the mass loading per unit of area of the fracture. As seen in Fig. 10.25, the drawdown rate over a period from 0 to 30 years can be shown to scale directly with the magnitude of $\dot{m}/R^2$ for an ideal circular fracture. As $\dot{m}$ increases or R decreases the drawdown rate will increase.

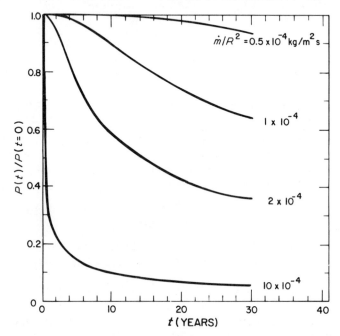

Fig. 10.25 Parametric thermal power drawdown curves for a single idealized circular fracture of radius R.

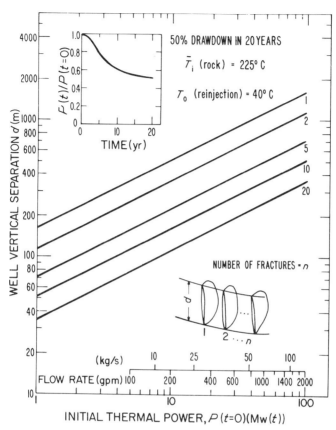

Fig. 10.26 Conceptual design nomograph for low permeability formations with a specified level of thermal drawdown (50% in 20 yr of operation). Wellbore separation distances that define single or multiple fracture sizes are shown as a function of initial thermal power level. Flow rates given in kg/s and gal/min (gpm).

A few simple calculations using equation (10.31) quickly reveal that very large surface areas are required for low drawdown rates with wellbore flows in excess of 40 kg/s – the lower limit for commercial-sized systems. In designing HDR systems, one uses this zeroth order model to size single large fractures or smaller multiple fractures to achieve desired drawdown rates at given initial power extraction levels. Figure 10.26 shows the required wellbore separation distance as a function of initial thermal power or mass flow rate and the number of fractures involved. For this case, it is conservatively assumed that wellbore separation distance defines the outer boundary of the fracture and that ideal uniform flow occurs. A 50% drawdown level after 20 years of continuous operation is assumed to specify $\dot{m}/R^2$ overall at about 1.8×10^{-4} kg/m²s, where drawdown here means the fractional loss of fluid production temperature, for example:

$$\frac{T(t=20 \text{ yr}) - T(\text{reinjection})}{T(t=0 \text{ yr}) - T(\text{reinjection})}.$$

The fractures are spaced at 50 m or larger intervals to minimize thermal interference.

Equation (10.29) can also be used to estimate the recovery time required to restore the reservoir to some fraction of its initial condition. The point concerning the long recovery times required was first discussed in Chapter 5 and can now be mathematically justified. Wunder and Murphy (1978) treated a number of cases of thermal recovery from a partially drawndown single fracture reservoir. They were able to correlate the recovery behavior to a single dimensionless parameter which scales the actual time t to the thermal-hydraulic properties of the fracture-rock-fluid system, where

$$\theta \equiv [\dot{q} \rho C_p / A (\lambda_r \rho_r C_{pr})^{1/2}]^2 t \tag{10.32}$$

As an example, consider the recovery that occurs for a dimensionless time θ of 2.0 where the drawdown, $(T_{out} - T_i)/(T_\infty - T_i) = 0.3$ from equation (10.30). After $\theta = 5.0$, the reservoir rock temperature has recovered to 80% of its original value. If we assume the drawdown period was actually 20 years, then the ratio of dimensionless times can be used to calculate the actual recovery time, t_r:

$$t_r \text{ (for 80\% recovery)} = (5/2)^2 \, 20 = 125 \text{ yr.}$$

As the present recovery is increased toward 100%, the recovery period required gets extremely large – thereby justifying the practicable non-renewability of heat mining in HDRs.

The other extreme of a highly jointed fracture network that resembles a porous medium can also be treated theoretically. Gringarten *et al.* (1975), Bodvarsson (1974), Wunder and Murphy (1978), and Zyvoloski (1981) have considered this problem from different viewpoints and we only summarize their results here. Thermal breakthrough times for an ideal porous medium with plug flow were given earlier in equation (10.6). By defining new dimensionless variables for a fracture length of L (m) and a fracture height of H (m):

for time: $t_d = C_p^2/(\lambda_r \rho_r C_{pr}) \, (\dot{m}^*/H)^2 t$

for distance: $x_{ed} = (C_p/\lambda_r) \, (\dot{m}^*/H) \, (L/2n)$

where

  $n =$ number of fractures
  $\dot{m}^* =$ mass flow rate divided by effective fracture aperture $= \dot{m}/<w>$

Then at ideal volumetric breakthrough

$$t_d = 2 \, x_{ed}$$

Zyvoloski has shown that the upper limit of fracture spacing to approximate volumetric heat mining occurs when $x_{ed} = 0.5$ as seen in Fig. 10.27. Solving for L/n

$$S = (L/n)_{max} = \sqrt{(2x_{ed}\alpha_r t)} = \sqrt{(\alpha_r t)} \tag{10.33}$$

or one-half the thermal wave penetration depth discussed earlier. Using typical granite properties as we did before,

$S = 26$ m at 20 years.

Figure 10.27 can be used for preliminary designs of HDR reservoirs where multiple interacting fractures are anticipated.

In a recent study, Pruess and Bodvarsson (1984) examined the movement of thermal fronts through fractured hydrothermal reservoirs that are influenced by long term reinjection. Cold fluid is injected into the reservoir, heated by the rock adjacent to the fractures, and finally appears in the production well. This process amounts to a partially recharged reservoir operating under waterdrive conditions and therefore is very similar to heat mining from an HDR. Pruess and Bodvarsson admit to the ambiguities inherent in specifying the reservoir's geometry as well as to the nonuniqueness aspects of modeling performance, but nonetheless, they have provided insight into how to interpret and predict thermal breakthrough effects. Their approach is a mixture of numerical and analytical modeling, incorporating the basic result of the zeroth order model given in equation (10.29) where the fluid temperature is a function of the position in the fracture and time for a given set of thermal physical properties and fluid flow rates. They treat the simple case of one-dimensional linear flow as given in Fig. 10.24 to define a thermal breakthrough time as the sum of two effects: the first accounts for heat content of fluid and rock within the fracture

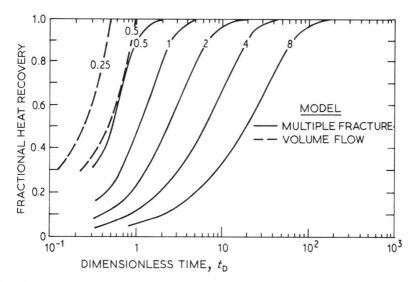

Fig. 10.27 Fractional heat recovery as a function of dimensionless time t_d (equation (10.30)) for multiple fracture and volume flow model of an idealized HDR reservoir. Lines show various values of dimensionless fracture spacing x_{ed} given by the equations on p. 284. (After Gringarten *et al.*, 1975)

aperture and the second for heat extracted from the rock surrounding the fracture faces. For most cases of interest in fracture-stimulated, low permeability rock, only the second effect will be important, as the first is roughly equivalent to the mean residence time within the fracture. In simplified mathematical form, a thermal breakthrough time τ_f was defined by Pruess and Bodvarsson (1984) as

$$\tau_f = \tau^* + \tau_h \tag{10.34}$$

where

$$\tau^* = \frac{\overline{\rho C_p}}{\rho C_p}\left[\frac{\tau}{\phi_f}\right]$$

$$\tau_h = \frac{\Psi}{\rho^2 \zeta_f^2}\left[\frac{\tau^2}{\phi_f^2 w^2}\right]$$

with $\overline{\rho C_p}$ representing the average density-heat capacity product for the rock and water contained within a fracture of aperture w and porosity ϕ_f and mean residence time $\tau = <V>/\dot{q}$. The terms ψ and ζ_f refer to parameters resulting from the zeroth order model of equation (10.29) which models transient one-dimensional conduction in the rock surrounding one-dimensional linear flow contained in a rectangular fracture:

$$\Psi = \text{thermal physical property group} = \lambda_r \rho_r C_{pr}/C_p^2$$

$$\zeta_f = \sqrt{\lambda_r \rho_r C_{pr}}\, x_f/w\phi_f \rho C_p \bar{U}\sqrt{t-\tau^*} = \text{error function argument}$$

$$\zeta_f = \sqrt{\Psi}/[(\dot{m}/A)\sqrt{t-t^*}] \tag{10.35}$$

where $\dot{m} = $ mass flow rate $= \rho \bar{U} w H \phi_f$ and $A = $ fracture area $= x_f H$.

The second time term τ_h in equation (10.34) is a measure of the effectiveness of the heat mining process itself. It reflects the ratio of the fluid residence time τ to the thermal conduction penetration time into the rock surrounding the fracture which in turn is a function of mass loading (flow rate $\dot{m}$ per unit area A) and the thermal physical properties of the rock and fluid given by the term Ψ.

A given level of thermal drawdown or breakthrough corresponds to a certain magnitude of the term ζ_f at the fracture outlet where $x_f = L$. In this case $\zeta_f(x_f = L,t)$ is given by:

$$\frac{T_\infty - T_{out}(t)}{T_\infty - T_{in}} = \text{erfc}[\zeta_f(x_f = L,t)] \tag{10.36}$$

For example if $\zeta_f = 0.81$, the drawdown is approximately 25% with:

$$T_{out} \cong 0.75\, T_\infty + 0.25\, T_{in}$$

Thus if one knows the degree of drawdown at various times in the reservoir's production period, equation (10.36) can be used to estimate ζ_f which should vary as $t^{-1/2}$. This was verified by Pruess and Bodvarsson for the run segment 2 data from Fenton Hill shown in Fig. 10.32. A reasonable correlation of ζ_f with $t^{-1/2}$ provides a basis for predicting reservoir performance into the future.

10.3.2 Theoretical model development

Modeling of thermal drawdown is used to determine the effective heat transfer area and to predict HDR reservoir lifetime. To make solutions possible mathematically, major assumptions concerning flow and structure are commonly made as we did in the development of the zeroth-order model.

As Murphy *et al.* (1981) point out for the Phase I Fenton Hill system depicted in Fig. 9.22, 'Inclusion of heat transfer effects in the dipping natural joints providing the outlets to GT-2B was not attempted, as this would have required a fully three-dimensional simulation for which the computing expenses were not justified considering the geometric uncertainties. Instead, the heat transfer area associated with the joints, expected to be small, was simply absorbed into each hydraulic fracture, and the total swept area was sought in terms of equivalent circular fractures. However, the influence of the joints on flow patterns and residence times did have to be considered; for example, a joint providing a fracture flow outlet very close to the inlet could result in short circuiting in which most of the injected flow could follow the shortest path to this outlet, thus bypassing much of the fracture area potentially available for heat extraction. To account for these effects, more complicated models with multiple flow outlets were also considered.'

Essentially, their model treated two-dimensional fluid flow within the fracture coupled to one-dimensional heat conduction through the rock perpendicular to the fracture surface. Because each hydraulic fracture itself is much more permeable than the surrounding rock, fluid (essentially water with minor amounts of dissolved minerals) was assumed to be confined inside each fracture, and it was also assumed that heat from the rock was transported to the fluid only by means of thermal conduction in the rock. In fact, a small amount of fluid did penetrate the surrounding rock; to correct for this, fluid loss was assumed to occur uniformly over the fracture area. This permeation effect was approximated by assuming that, on the average, heat was removed from the reservoir by all of the produced fluid flow rate plus half the difference between the injected and produced fluid flow rates. The possible enhancement of the heat transfer area because of thermal stress cracking which would be expected to produce secondary flow paths (Harlow and Pracht, 1972) is not explicitly considered in this model.

Similar to the coordinate system selected for the zeroth-order model (Fig. 10.24), the horizontal coordinate in the fracture plane is taken as x and the vertical coordinate as z. Solid rock heat conduction takes place along the y coordinate, perpendicular to the x–z fracture plane. Coordinate y is measured from the fracture midplane into the rock, and modified coordinate y is measured from the rock wall of the fracture. The transport equations expressing conservation of momentum, mass, and energy are written below in nondimensional form for a variable fracture aperture and temperature-dependent water properties. The fracture aperture w is so small (typically about 1 mm) in comparison with the

x and z directions of fracture extent that fluid motion in the y direction is negligible. The y component of the momentum equation is therefore neglected, and simplified momentum equations for the fracture fluid are obtained by integrating across the aperture; i.e., the x and z velocities, u and v, that follow are average velocities, averaged with respect to y. Furthermore, only velocity gradients taken with respect to y are important, so the Navier–Stokes equations can be approximated with the normal boundary layer formulation.

Buoyancy effects in the vertical z direction are treated with the usual Boussinesq approximation as

$$\beta_T \rho g(T - T_i^*) \tag{10.37}$$

where $T_i^* = $ reference fluid temperature at injection point and $\beta_T = $ volumetric thermal expansion coefficient for water. Viscous laminar flow within the fracture in the x–z plane is modeled with Darcy's law with the fracture permeability k_f given by $w^2/12$. Further, by neglecting viscous dissipation, flow work, and kinetic and potential energy terms, all of which can be shown to be small (Murphy *et al.*, 1981 and Harlow and Pracht, 1972) the momentum and continuity equations for flow in the fracture can be written as

$$-\partial P/\partial x = \mu u/k_f \quad x\text{-direction horizontal}$$
$$-\partial P/\partial z = \mu v/k_f - \beta_T \rho g(T - T_i^*) \quad z\text{-direction vertical} \tag{10.38}$$
$$\partial/\partial x((wk_f/\mu)\partial P/\partial x) + \partial/\partial z((wk_f/\mu)\partial P/\partial z) - \beta_T \rho g(T - T_i^*) = 0$$

The importance of buoyancy or natural convection effects relative to forced convection is usually indicated by a dimensionless group K_b given by the ratio of Grashof (Gr) to Reynolds (Re) numbers:

$$K_b = Gr/Re$$

with

$$Gr = w^{*3}\rho^2 g\beta_T(T - T_i^*)/\mu^2 \quad \text{and} \quad Re = \rho \bar{U} w^*/\mu$$

where

$\bar{U} = $ characteristic mean velocity $= q'/rw^*$
$w^* = $ maximum fracture aperture
$r = $ characteristic fracture length $=$ radius for circular fractures

and fluid density ρ and viscosity μ are evaluated at an appropriate average reservoir temperature. If K_b is much less than unity, then buoyancy effects will be small.

The next step in developing the model is to couple the heat transfer between the rock and circulating fluid. This is done with an energy balance where heat transferred from the rock to the fluid at the fracture surface is treated as a source term:

$$\rho C_p \left[u\frac{\partial T}{\partial x} + v\frac{\partial T}{\partial z} \right] = \frac{-2\lambda_r}{w}\left(\frac{\partial T_r}{\partial y} \right)_{y=0} \tag{10.39}$$

where the subscript r refers to rock properties, the factor of 2 on the right-hand side accounts for both faces of the fracture. For all practical cases of interest, the transient conduction of heat within the rock surrounding the fracture can be treated as one-dimensional with the following form of Fourier's law:

$$\frac{\partial T_r}{\partial t} = \alpha_r \frac{\partial^2 T_r}{\partial y^2} \qquad (10.40)$$

This can easily be seen to be true if one examines the thermal penetration depth for 30 years of operation in a granite system and compares it to a typical fracture radius. If

$$\delta_t^2/R^2 \ll 1$$

then the one-dimensional form is correct. If R is of order 500 m or larger and δ_t from equation (10.4) is of order 60 m for 30 years, then

$$\delta_t^2/R^2 = 60^2/500^2 = 3600/250\,000 \ll 1$$

and the one-dimensional form is acceptable. A transient form for heat transfer in the rock, as given in equation (10.40), must be utilized, since no viable heat extraction concept based upon rock conduction, as this one is, could rest on the steady state alternative; that is, heat fluxes which are limited to the usually quite small regional geological heat flows. Instead, this concept requires the depletion of stored heat in the rock, and in the absence of additional creation of heat transfer area, caused perhaps by thermal stresses and contraction, eventually results in thermal decline. Equation (10.40) is subject to the initial and boundary conditions.

Initial conditions $t = 0$ $T_r(\text{all } y, t = 0) = T_{ri}(x,y,z)$
prescribed temperature field in rock   (10.41)

Boundary conditions $t > 0$ $T_r(y = 0, t) = T(x,y,t)$
fluid and rock temperature equal
at fracture wall   (10.42)

$T_r(y \rightarrow \pm \infty, t) = T_{ri}(x,y,z)$
semi-infinite relaxed boundary   (10.43)

$T_{ri}(x,y,z)$ represents the initial temperature in the rock, normally this would only be a function of z (depth) given by the geothermal temperature gradient. The second condition (10.42) implies that heat transfer between the rock and fluid at the interface is rapid compared to all other processes including heat transfer through the fluid across the fracture aperture.

The fracture aperture w itself can also be a specified function of location (x,z) in the fracture plane and of temperature to account for the effect of thermal contraction as heat is mined.

Equations (10.38 through 10.43) comprise a set that can be solved for u, v, P, T and T_r. They are, however, coupled with equation (10.39) and non-linear because of the temperature effects on buoyancy and viscosity, so that numerical methods provide the most general solutions, although, as discussed below, an analytical solution can be obtained for simplified geometries. Harlow and Pracht (1972) and McFarland and Murphy (1976) have presented numerical solutions that indicate that when the fracture inlet and outlet positions are some distance from the fracture periphery, as would normally be the case with typical drilling and fracturing operations, the buoyancy effect can improve sweep efficiencies across the fracture face. Increased heat production results if the outlet lies vertically above the inlet. For this reason, fluid should always be injected in the deepest fracture-well intersection. However, when the buoyant effect is small, $K_b < 1$, the density differences between the cold fluid entering a fracture and the other fluid within it do not significantly affect fracture flow patterns. The streamlines instead exhibit a pattern similar to a two-dimensional potential flow solution in which streamlines radiate out from the inlet and then converge upon the outlet. There is little tendency for the inlet fluid to flow first downward before flowing up towards the outlet. These effects are illustrated by the streamline patterns in Fig. 10.28 (a) and (b). Figure 10.29 shows a typical result from a performance simulation for a large single fracture of 1000 m radius where both production temperature and thermal power histories are plotted.

The Los Alamos team has modified this two-dimensional convection/one-dimensional conduction single fracture model to treat more complex geometries. As discussed by Dash *et al.* (1981) these include the following.

(1) *Multiple independent fractures model*: where the total flow is divided into parts that enter separate non-thermally interacting fractures with two-dimensional flow and one-dimensional rock conduction. The numerical solution approach is identical to the presentation just discussed. Fluid production temperatures and power levels are obtained by weighted summation over individual fracture contributions. This type of approach could easily be used to model the early conception of the Phase I reservoir shown in Fig. 10.30(a).

(2) *Multiple interacting fractures model*: one-dimensional flow in each fracture is assumed following the depiction for the rectangular fracture in Fig. 10. 24. Consequently fluid mechanics considerations do not directly enter into the heat mining process; that is, the sweep efficiency is implicitly assumed to be 100%. However, a rigorous treatment of two-dimensional rock conduction has been incorporated for the rock space between fractures shown in Fig. 10.30(b).

Also, both Los Alamos and Camborne are currently developing three-dimensional numerical codes for heat performance modeling to provide predictive capability for complex jointed rock networks. Although these models are inherently more complex and require advanced high-speed computers for generating acceptable results, the basic physical concepts presented in this section are still followed. For instance, a finite element model has been developed by Zyvoloski (1983) that allows for two-dimensional flow in a set of fractures with arbitrary geometry and spacing and thermal interactions and fluid permeation between fractures. Zyvoloski's model not only provides a

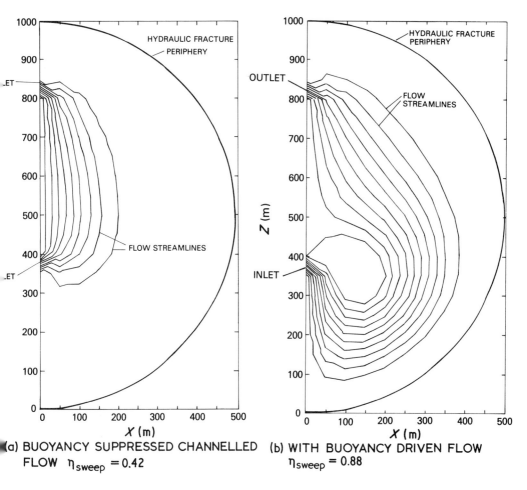

Fig. 10.28 Flow streamlines resulting from the two-dimensional convection/one-dimensional conduction numerical model of a vertical, 500 m radius, planar fracture with elliptical cross-section. (Adapted from McFarland and Murphy, 1976).

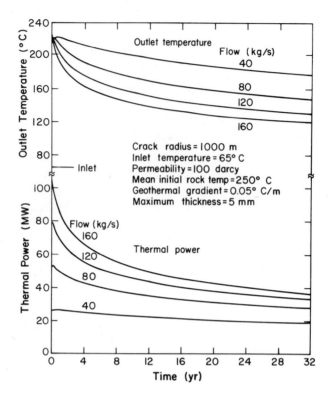

Fig. 10.29 Thermal drawdown and power production predictions for the two-dimensional convection/one-dimensional conduction model of a vertical, 1000 m radius, planar fracture of elliptical cross section. (From McFarland and Murphy, 1976).

prediction of thermal performance and of water loss rates but also can be used to solve the generalized convective-dispersion equation to estimate tracer response curves (see equation (10.22)).

10.3.3 Field results at Fenton Hill

The discussion that appears in this section is based primarily on a summary account of testing by Dash *et al.* (1981).

During run segments 2 through to 5, the temperature of the water exiting the reservoirs was measured with a thermistor surveying tool positioned in GT-2B. This tool has a resolution of 0.5°C. Combined with the thermistor is a flow rate sensor; that is, a 'spinner,' so that both temperature and flow rates of the water exiting the reservoir and entering GT-2B via various natural fractures or joints could be monitored.

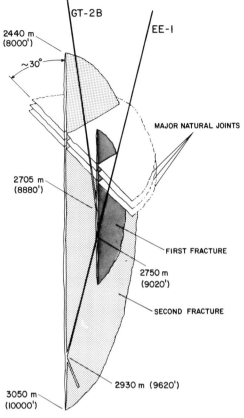

GT-2B

EE-1

2440 m
(8000')

~30°

MAJOR NATURAL JOINTS

2705 m
(8880')

FIRST FRACTURE

2750 m
(9020')

SECOND FRACTURE

2930 m (9620')

3050 m
(10000')

(a) EARLY CONCEPT

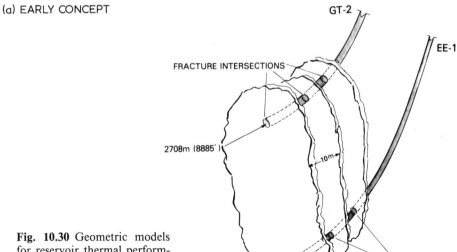

GT-2

EE-1

FRACTURE INTERSECTIONS

2708m (8885')

10m

FRACTURE INTERSECTIONS

2930m (9620')

(b) LATER CONCEPT

Fig. 10.30 Geometric models for reservoir thermal performance prediction. (a) Multiple independent fractures model (early concept). (b) Multiple interacting fractures model (later concept).

Cumulative thermal energy extracted from the Phase I reservoir during run segments 2 through to 5 is summarized in Fig. 10.31. The upper curve depicts total energy produced at the surface. Measured GT-2B production temperatures were used and a constant 25°C reinjection temperature was assumed for EE-1. The lower curve represents energy extracted from flow through the reservoir and excludes contributions from the wells. Based upon wellbore heat-transmission calculations, an inlet reservoir temperature in EE-1 of about 65°C was used when measured downhole temperatures were unavailable. The outlet reservoir temperature was taken from GT-2B temperature logs measured at 2590 m (8500 ft). Average energy extraction rates at the surface were 3.1 MWt for run segment 2, 2.1 MWt for run segment 3, 2.8 MWt for run segment 4, and 2.3 MWt during run segment 5.

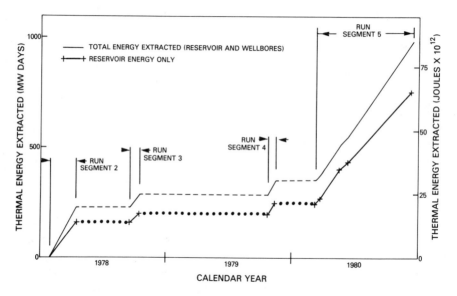

Fig. 10.31 Cumulative thermal energy extracted from the Phase I Fenton Hill reservoirs, run segments 2 through to 5. (Dash *et al.*, 1981).

Although run segment 2 produced thermal energy at the greatest rate, thermal drawdown was quite severe. After 75 days of operation the GT-2B reservoir temperature had dropped from 175 to 85°C. Temperatures during run segment 3 ranged from 135 to 98.5°C. Finally during run segment 5, the initial reservoir temperature of 156°C climbed to about 158°C after 60 days, then dropped to about 149°C by the experiment's end (286 days).

The rate of decline, or drawdown, of these temperatures, when analyzed with the heat-transfer models described earlier, permits estimates of the effective heat-transfer areas of the reservoirs. We say *effective* because some parts of the total area are either inaccessible to, or inefficiently swept by, the water flow because of fluid dynamic and structural constraints.

The most recently developed model, which we call the multiple-interacting fracture model, is based upon the geometry of Fig. 10.30(b) and assumes that the fractures are parallel rectangles and that flow is distributed uniformly along the bottom of each fracture and uniformly withdrawn from the top of each fracture while heat conduction with rock contained between fractures is rigorously treated. In contrast, the older model (the multiple independent-fractures model), based upon Fig. 10.30(a), assumes that the fractures are circular (but other assumptions are permissible) and allows proper local positioning of the inlet and outlets; i.e., the point-like intersection of the injection well with the fracture can be modeled, as can the intersection of the main hydraulic fractures and the slanting joints that provide the connections to GT-2B. However, as was cautioned earlier, while the fluid dynamic effects of the joint/outlets can be faithfully modeled, the heat-transfer effect of the joints cannot; the area of the joints must be lumped with the main fractures. In view of this more faithful representation of inlet and outlets, and the fact that a complete two-dimensional solution to the Navier–Stokes fluid dynamic equations is incorporated, the multiple independent-fractures model results in a more realistic assessment of the effect of fluid dynamics and sweep efficiencies upon heat extraction. The penalty, however, is that in the present two-dimensional version of the code, thermal interaction as the temperature waves in the rock between fractures overlap cannot be realistically represented, as it is with the multiple interacting-fracture model.

In run segment 5, it was concluded that both models could adequately predict the overall thermal drawdown observed. This agreement between models suggests that the more complicated three-dimensional models may not be appropriate given our present state of knowledge concerning reservoir geometry.

A re-analysis of earlier thermal modeling of run segments 2–5 has indicated that growth of the heat-exchange area occurs not only from pressurization and hydraulic fracturing, as might be expected, but also as a consequence of thermal cooling and thermal stress cracking. Reservoir growth due to thermal cracking during heat extraction was predicted as early as 1972 by Harlow and Pracht, but this is the first time we have been able to detect such growth in the actual drawdown behavior. A more sophisticated model for thermal stress cracking enhancement is given by Harlow and Demuth (1979).

The first application of the multiple independent fracture model was to the original reservoir, when only the smaller hydraulic fracture schematically shown to the right in Fig. 10.30(a) was active. This reservoir was tested extensively during the 75 day run segment 2 test (see Tester and Albright, 1979). Based upon spinner and temperature surveys in the production well, the depths of the intersections of the production well with the slanting joints were estimated as well as the flow rates communicated by each joint. In the thermal performance calculations, the actual temporal variations of production and injection flow rates were utilized. The fracture inlet temperature was estimated with a separate wellbore heat-transmission calculation. With this information, estimates of the thermal drawdown were calculated with the model for various trial values of fracture radii and vertical positions of the fracture inlet. It could not be assumed that the inlet was located at the center of the fracture because the earth stresses increase with depth, so that during its creation the fracture probably grew preferentially in the upward direction. A fracture radius of 60 m with an inlet located 25 m above the fracture bottom resulted in a good fit to the measurements, and, as shown in Fig. 10.32, the computed thermal behavior was in good agreement with the measured temperature. The temperature shown is the mixed mean reservoir-outlet temperature. That is, the mean outlet temperature is taken as the combination of the joint outlet temperatures measured in the production well, averaged, or weighted, by the flow-rate fraction in each joint. *The time dependence of this mean outlet temperature is the only direct indicator of the overall thermal performance of the reservoir.* It is eventually what all predictive models must be correlated with or tested against.

A radius of 60 m, as indicated by the fit to the data in Fig. 10.32, implies a total fracture area (on one side) of 11 000 m²; however, because of hydrodynamic flow sweep inefficiencies, the effective heat exchange area was only 8000 m² during run segment 2. With the independent fracture model, it was not possible to obtain an indication of thermal growth from the drawdown data of Fig. 10.32. The very low rate of drawdown in the later period of the test prevented resolution of any potential area increase.

Run segment 3 was the next test conducted in October of 1978. Its purpose was to examine reservoir behavior under conditions of high mean-fracture pressure. The test duration was short, less than one month, but modeling the thermal drawdown with the independent fractures model suggested that the effective heat transfer area was nearly the same as in the earlier low-back-pressure test. However, flow rate (spinner) surveys in GT-2B indicated that because of the the higher pressure level, most of the flow was entering the well at positions that averaged 25 m deeper than during run segment 2 at low back pressure. In effect, the reservoir flow paths were shortened by 25 m out of the original 100 m separation distance. A reduction of at least 25% in heat-transfer area would have been expected because this vertical shortening would also result in horizontal contraction of the streamlines, and yet the area estimated from actual drawdown was about the same at 8000 m². While pressurization did indeed result in partial short circuiting of the streamlines, it also resulted in a notable decrease in impedance as described in Section 10.2.3, which apparently improved fluid sweep through the remaining area.

The Phase I reservoir was enlarged during the massive hydraulic fracturing operations of 1979 as discussed in Chapter 9. With the multiple independent

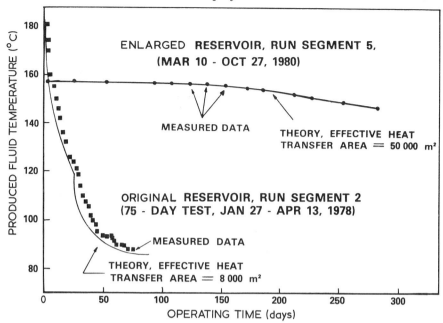

Fig. 10.32 Computer model prediction using a single fracture version of the independent fractures model with field data for run segment 2 and a dual independent fracture model for run segment 5 of the original Phase I Fenton Hill reservoir. (Tester and Albright, 1979 and Zyvoloski, 1981).

fractures model, the enlarged reservoir is portrayed as two fractures, the original one operative in run segments 2 and 3, and a new and larger one shown to the left in Fig. 10.30(a). The enlarged reservoir was evaluated during run segments 4 and 5. Based on run segment 4 results, the old fracture had an effective heat-transfer area of 15 000 m² and the new fracture had an effective area of at least 30 000 m². Because the heat-extraction period was only 23 days, far too short to result in significant depletion of the new fracture, only a minimum area could be specified. The area determined in run segment 4 for the original smaller fracture was almost twice that determined in run segment 2. This trend of increasing area is now attributed to thermal stress cracking enhancement.

Improved estimates of the total effective heat transfer area of both fractures were obtained in run segment 5, during which the thermal drawdown was about 8°C. The drawdown rate suggested an active heat transfer area of 50 000 m². The mean outlet temperature actually increased slightly during the early portion of run segment 5. This temporary increase is due to transport of deeper, hotter water into the production well as qualitatively depicted in Fig. 7.6, as well as to some interaction of the fractures. For simplicity the effect was neglected in the multiple independent fractures model as it is fairly small, less than 2°C.

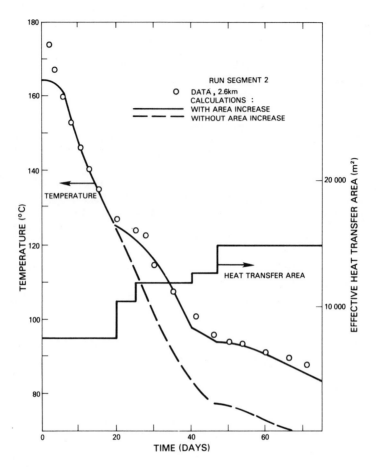

Fig. 10.33 Computer model predictions using the multiple interacting fracture model with programmed increases in heat-transfer area to fit data from run segment 2 of the original Phase I Fenton Hill reservoir. (From Dash *et al.*, 1981.

Fig. 10.34 Heat-transfer area growth determined by model fits to drawdown data and wellbore temperature logs in the Phase I Fenton Hill reservoirs during run segments 2 through to 5. As indicated by the question marks, the area increase due to the MHF experiments (195 and 203), is uncertain. The heat-transfer area was not measured until the later stages of run segment 4. Unfortunately, the first stages of run segment 4 were fracturing operations in their own right, about 3 MPa (450 psi) lower in pressure, but seven times the injection volume of MHF experiments 203 and 195. Consequently, the area increase measured in run segment 4 is due to the combined effects of all the fracturing operations, and cannot be individually ascribed to the separate operations. This uncertainty will also be apparent in the multiple-fracture modeling and the reservoir-tracer experiment.

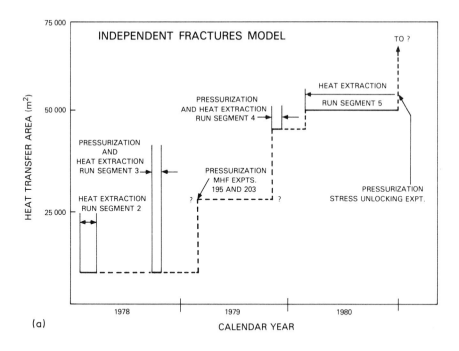

(a)

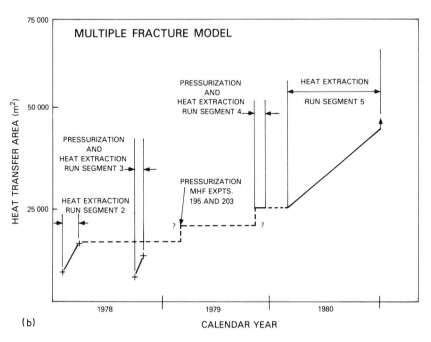

(b)

Unlike the constant area predicted by the independent fractures model when applied to run segment 2 data, the multiple interacting fracture model required a major increase in active heat transfer area to properly fit the GT-2B drawdown data. Areas increased from approximately 7 500 m² at day 0 to 15 000 m² at day 75 as shown in Fig. 10.33. Similar results were also observed with the run segment 3 data on the original fracture system as well as with the enlarged system during run segments 4 and 5.

Figure 10.34 shows the history of effective heat transfer area increases predicted by both models over the entire set of Phase I tests. For both models, general increases from 8 000 m² to 50 000 m² are indicated even though there are differences in detail. This increasing trend is supported by other independent measurements, including tracer measurements, pressure-transient fracture inflation and venting data, water loss rates by permeation, and microseismic event mapping.

10.3.4 Theory versus practice

Table 10.6 lists a number of fracture area estimates obtained from different sets of data. Areas from fits to the thermal drawdown are consistently less than all other estimates which suggests that a large portion of the reservoir not yet active in the heat mining process may be possibly stimulated to increase reservoir productivity. In fact, the enlargement of the original Phase I fracture system is consistent with such a stimulation hypothesis.

The extremely large areas based upon fracture pressurization, inflation, and propagation are obtained from total injected volumes divided by the assumed effective aperture of the fractures. The actual area associated with the inflation volume could be larger, as some of the water must be in small-scale porosity with smaller apertures. Most of this area, however, cannot be expected to participate in heat exchange.

The diffusion area is obtained mainly from water-loss data using techniques discussed in Section 10.2.1. The microsiesmic area is the vertical projection of the locus of seismicity given by methods and data described in Sections 9.3.1 and 9.4.1. The magnitude of the locus projection is multiplied by n, an assumed number of fractures in the microseismic volume. For $n = 3$, as suggested by the multiple interacting-fractures model, the total mapped microseismic areas are very close to the diffusion area.

The volume of water vented at high flow rates after the system has been pressurized for a long time is also a measure of system size. The higher vent rates are assumed to come from a large low-impedance system before the decreasing internal pressure closes the fractures. The vent areas in Table 10.5 are for a fracture system with an assumed 4 mm aperture, obtained from the heat-exchange areas and modal tracer volumes.

The fracture inflation and vent areas were obtained from volume measure-

Table 10.6. Comparison of fracture area estimates for the Phase I Fenton Hill Reservoir. (From Dash et al., 1981)

Segment	Source of area estimate – area (m²)				
	Fracture inflation volume[a]	Micro seismic maps	Diffusional water loss	Fracture vent[b] properties	Thermal drawdown modeling[c]
Original system 1		>20 000		40 000	7 500→15 000
2	5×10^6		~250 000		6 000→12 000
3	5×10^6				
Enlarged system 4	20×10^6	$75\,000 \times n$[d]	~350 000	>250 000	>21 000
5		$90\,000 \times n$			37 000→50 000

[a] 1 mm aperture fracture assumed
[b] 4 mm aperture fracture assumed
[c] → indicates growth during heat extraction
[d] n = number of fractures in the volume of rock defined by all microseismic events

ments and aperture estimates. In each case a maximum estimate of the aperture was used to minimize calculated areas. For the inflation areas, a nominal 1 mm was chosen to be consistent with the hydraulic fracturing models introduced in Chapter 8. Furthermore, the larger, uncooled fracture system into which water is forced during inflation and propagation, as opposed to circulation, should have a smaller aperture. Pressure-transient testing of the fracture system effective in heat exchange gives an estimate of 1 to 2 mm for the aperture of the active heat-transfer region, whereas run segment 2 tracer modal volumes (Table 10.4) and heat-exchange areas (Table 10.5) provide an aperture estimate of 1 – 5 mm for the heat-exchange regions. Consequently, a nominal 1 mm was chosen for the larger and, essentially, still undeveloped reservoir during pressurization and inflation. Apertures associated with the low-impedance vent volume should be the same as or smaller than the apertures corresponding to the internally depleted, active heat-transfer regions of the reservoirs. The largest aperture estimates for the depleted reservoir were obtained at the end of run segment 5 when the modal volume and heat-exchange areas gave apertures of about 4 mm.

Temperature logs in both the EE-1 and GT-2B wellbores have revealed certain detailed structural aspects of the system not specifically seen in other data. Given the complex nature of the injection borehole profile as depicted for EE-1 in the temperature log of Fig. 9.14, multiple fracture connections are certainly present and all of them possess different hydraulic characteristics. Another facet of the temperature log analysis in the injection well is that it can be used to crudely map the thermal drawdown region if the wellbore should pass parallel to the major heat extraction zone and not far from it horizontally. As seen in Figs 9.22 and 10.30(a), a substantial portion of the lower part of EE-1 is close enough to the main fractures to experience thermal drawdown effects. One might hypothesize that the microseismic event region determined during the stress unlocking experiment (SUE) would provide a boundary around a thermally-depleted aseismic region. Figure 10.35 superimposes a post-run segment 5 EE-1 temperature log on the observed planar microseismic event region given by Fig. 10.18 for SUE. As can be seen, the aseismic zone extending from approximately 2850 to 2950 m coincides closely with the region of highest thermal drawdown.

A more detailed analysis of the SUE results (see Dash *et al.*, 1981; and Murphy *et al.*, 1981) suggests that thermally induced stress relief in the extensively cooled regions of the reservoir caused the region to become aseismic during heat mining. If this interpretation is correct, a new fracture mapping technique will result to define the extent and location of these portions of an HDR from which thermal energy is being extracted. Periodically during operation, the production well could be shut-in with injection continued. The mappable microseismic region would then define the thermally-depleted aseismic zone and the adjacent band of microseismic activity which could be correlated to effective heat transfer area.

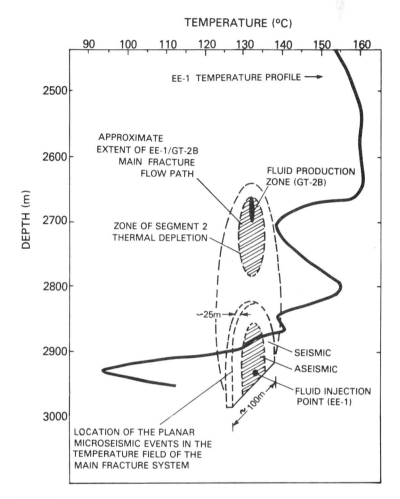

Fig. 10.35 Recovery temperature survey taken in the EE-1 injection hole on 10 July 1981, seven months after heat extraction ended, superimposed on the mapping results from microseismic measurements during the stress unlocking experiment (SUE). (From Dash *et al.*, 1981).

A complete review of the tracer-test data from the Fenton Hill segments 2 through to 5 has revealed pertinent information regarding the growth of the reservoir due to heat extraction and pressurization effects. The reservoir growth due to heat extraction is, to be precise, really a thermal-contraction effect – as the rock surrounding the fractures shrinks, the fractures, and consequently, the measured volumes, expand. Ultimately, we would like to correlate measured tracer volumes with effective heat-transfer surface. In addition, the interpretation of tracer volume changes could be used to develop

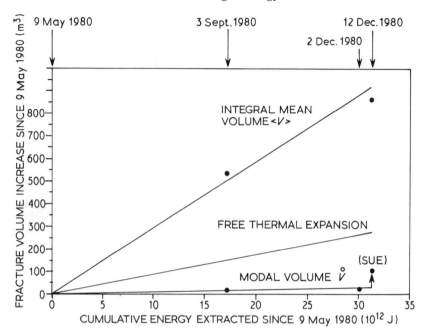

Fig. 10.36 Increase in tracer modal volume as a function of thermal energy extracted for the Phase I Fenton Hill reservoir.

improved methods of reservoir operation – for example, remedial pressurization for stress relief such as SUE, or a huff-puff operation mode in contrast to our normal (stress-constrained) continuous mode of extracting heat (see Section 10.4.5).

Figure 10.36 is a linear plot of modal-volume increase ($\Delta \mathring{V}$) as a function of net thermal energy extracted from the reservoir (ΔE). Essentially identical linear behavior is observed for the low back-pressure experiment, run segment 2, of the original reservoir and the low back-pressure experiment, run segment 5, of the enlarged system. In spite of nonlinear coupled effects of thermal contraction, pore and fracture inflation due to sustained pressurization, and local irreversibilities resulting in fracture propagation, a simple correlation between V and E exists. Furthermore, this simple relationship persists even in the presence of the confining stresses surrounding the active reservoir, which indicates a constrained behavior. The slope of the line at low back pressure is only about 10% of what would be expected for free thermal expansion of the rock ($\Delta \mathring{V} = [(\beta_T)_r/(\rho C_p)_r] \Delta E$) in a stress-free environment. Values of $(\beta_T)_r = 24 \times 10^{-6}$ K^{-1}, $C_{pr} = 1000$ J/kgK and $\rho_r = 2700$ kg/m^3 were used to represent the granite matrix. In practice the region between the low-pressure modal volume and the free thermal volume lines defines an envelope of reservoir operating conditions. As stresses are relieved, for example during SUE, or the high back-pressure test of the original reservoir (run segment 3), or the high-pressure, hydraulic-fracturing stage at the beginning of run segment 4, one moves away from the normally constrained condition toward the free thermal expansion line. An even stranger result appears if the change in integral mean reservoir volume ($\Delta <V>$) is plotted as a function of ΔE. Again a straight line correlation results, but it is above the free thermal expansion line. The only possible explanation for this is the activation of a secondary fracture system by the induced thermal stresses – thus providing direct evidence for the first time of thermal stress cracking. The growth in heat transfer area that was estimated during the Phase I tests using the multiple interacting fractures model (Fig. 10.34(b)) is consistent with a thermal stress cracking enhancement mechanism.

Perhaps the most promising aspect of the tracer test is its potential for estimating the effective heat-transfer surface area of a reservoir. This becomes clear when the modal volumes from Table 10.5 are compared to the corresponding estimated heat-transfer area from thermal performance modeling as shown in Fig. 10.37 for both the Fenton Hill Phase I reservoir and the Phase I reservoir at Rosemanowes (Batchelor, 1982). The similarities of the growth of area and volume are quite striking. This can be quantified by considering the

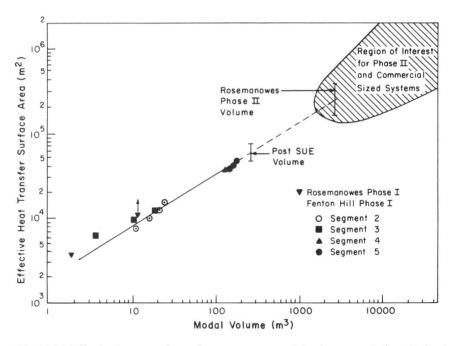

Fig. 10.37 Effective heat transfer surface area versus modal volume correlation obtained using thermal drawdown and tracer measurements. HDR reservoirs tested at Fenton Hill and Rosemanowes. Fenton Hill Post SUE and Rosemanowes Phase II volume points are located by extrapolation of the area versus V line.

relationships between area, volume, and aperture (or effective fracture opening). The volume V, is simply the product of the area A, and the mean aperture w:

$$V = Aw \qquad (10.44)$$

During heat extraction and/or pressurization, the area and aperture can both vary; therefore the volume is a function of two variables rather than one. For constant aperture, the tracer volumes should scale directly with heat-transfer area. A suggestion of this behavior is seen in the data from run segment 2, shown in Fig. 10.37, where the area versus modal-volume curve has an intercept that corresponds to a constant 1.7 mm aperture. Subsequent pressurization has apparently increased the aperture. Because only minimum estimates of the heat-transfer area are available for two of the measured volumes in run segments 4 and 5, not all of the tracer data can be used in this figure. It is likely that the data from run segment 5 would also fall upon a line of constant aperture, in which case, the mean fracture aperture would be roughly 4 mm. There is insufficient information available to provide upper bounds for the heat-transfer area from run segment 5, so the correlation between heat-transfer area and modal volume is inferred by analogy with run segment 2. Further development of this empirical correlation could provide a direct and independent method of estimating reservoir heat transfer area without requiring thermal drawdown, which for commercial-sized systems, may take 10 to 20 years to observe.

10.4 Geochemical effects

10.4.1 General phenomena

The major effects of geochemical phenomena in HDR systems arise from two sources: the chemical interaction of water with rock in the reservoir which can result in the dissolution and reprecipitation of certain minerals within the system during recirculation, and the displacement of indigenous fluid contained naturally in the pores and voids of the reservoir rock. Both of these effects can potentially contribute to operational problems, such as fouling or scaling on heat exchange surfaces, pipes, or even flow passages within the fractured labyrinth. In most cases, fouling caused by mineral deposition will lower performance either by reducing the heat transfer coefficient of surface heat exchangers or by increasing the flow impedance within the system, thereby increasing pumping power requirements. Even in the absence of solids precipitation, corrosion of metal surfaces is a possibility, particularly if the dissolved salt concentrations become high and if dissolved oxygen levels are not controlled.

A side benefit of monitoring the fluid chemistry is the possibility of utilizing transient geochemical behavior to size the HDR reservoir itself. Because rock dissolution rates and mineral solubility are usually strong functions of temperature, observed changes in fluid composition should be related to both the temperature and flow distribution within the reservoir. Using this information with some sort of appropriate geometric and kinetic model may provide an alternative diagnostic method for predicting reservoir performance. Admittedly, these methods will still have to deal with the general non-uniqueness problem associated with the inadequate description of flow within the HDR, but nonetheless should provide an independent check on the results of the thermal, hydraulic, and seismic models.

The evidence for pore fluid displacement comes from comparison of field data taken at Fenton Hill to what would be expected for dissolution of the reservoir rock. First, dissolved silica (SiO_2) concentrations typically exceeded their solubility limit based on the maximum operating temperature of the reservoir. In most cases, silica concentrations approached saturation limits at the original, undisturbed rock temperature that would be encountered deep in the rock matrix outside of the region defined by the thermal penetration depth over the fracture face. Secondly, the concentration–time behavior of many other cations and anions, including Na^+, K^+, Li^+, and Cl^-, was kinematically similar and could not be related to the dissolution of any single mineral indigenous to the reservoir. And thirdly, observed changes in the oxygen and hydrogen isotope ratios were indicative of pore fluid mixing rather than rock dissolution (Grigsby, 1983).

The effect of pore fluid migration and displacement can be mathematically modeled by a mixing term that supplies a constant composition saline fluid at a specified flow rate ($\dot{q}_2$) to the recirculating reservoir fluid (Grigsby and Tester 1978; Grigsby, 1983; Robinson, 1982). Another effect that must be properly accounted for is water lost to permeation during pressurized recirculation. In real field operations, a surface make-up water supply is provided. The dilution effect of this make-up stream can easily be included because its composition and flow rate is known.

The last effect to be modeled is the actual dissolution rate of the minerals comprising the rock matrix within the reservoir. Based on several years of study, as summarized by Robinson (1982) and Grigsby (1983), the major effect observed in the granite reservoirs at Fenton Hill is quartz dissolution into an undersaturated aqueous solution. This process is temperature sensitive as both the rate constant k^* and saturation concentration C^∞ of silica are exponential functions of temperature. Another problem is the strong dependence of dissolution rate on the amount of active rock surface area exposed to the circulating fluid. Even with these complexities, dissolution rates can be correlated by a very simple empirical relationship:

$$\frac{dC}{dt} = k^* a^* (C^\infty - C) \tag{10.45}$$

This so-called rate law is first order and linearly dependent on a concentration driving force ($C^\infty - C$) that expresses the degree of undersaturation. Both k^* and

C^{α} can be expressed by conventional formulae given by Robinson (1982) and Grigsby (1983):

$$k^* = f^* k_o^* \exp\left[-E_{act}/RT\right] \tag{10.46}$$

$$\log_{10} C^{\alpha} = 5.19 - 1309/T \qquad \text{(for } T \leqslant 250°C\text{)} \tag{10.47}$$

where

k_o^* = pre-exponential factor = $10^{0.433}$ (m/s)
a^* = ratio of rock surface area to fluid volume (m^{-1})
E_{act} = activation energy for dissolution = 78.3 ± 3.8 kJ/mol
T = absolute temperature (K)
C^{α} = solubility of quartz at temperature T (mg/liter)
C = concentration of silica in solution (mg/liter)
f^* = surface area enhancement factor ($\geqslant 1$)

The rate of granite dissolution was found to have a similar temperature dependence as that for pure quartz but was about a factor of 100 faster due to the presence of fines which increased the effective surface area for dissolution. The mathematical formulation of the general geochemical model for a recirculating HDR reservoir is described in detail by Grigsby (1983). The basic approach used is one of a dynamic material balance with water loss, make-up water, pore fluid displacement, and chemical dissolution treated as source or sink terms. Even if the solution became super-saturated, no reprecipitation was permitted as it was not observed in the field.

The general approach uses a modified form of the convection-dispersion equation introduced earlier (equation (10.22)):

$$\nabla \cdot D\nabla C - U \cdot \nabla C + k^* a^* (C^{\alpha} - C) + \frac{\dot{q}_2}{V}(C^{\alpha}) - \frac{\dot{q}_L}{V} C = \frac{\partial C}{\partial t} \tag{10.48}$$

where the additional terms account for reaction ($k^* a^* (C^{\alpha} - C)$), fluid loss ($\dot{q}_L C/V$) and pore fluid displacement $\dot{q}_2 C^{\alpha}/V$. Note that both fluid loss and pore fluid displacement are treated as a continuous sink or source that leaves or enters the reservoir uniformly at all positions. In some cases, it may be more appropriate to treat them discretely (see Grigsby, 1983).

Equation (10.48) can be greatly simplified by making assumptions regarding the nature of the flow and temperature fields within the reservoir and whether or not dissolution and/or pore fluid displacement is involved. For example, Robinson (1982) neglected dispersion and used the zeroth order, one-dimensional linear-plug flow model to illustrate several key aspects of the coupling of thermal drawdown to the concentration buildup of silica. Grigsby (1983) treated dispersion using a one-dimensional axial disperson model to provide a tractable approach for predicting Fenton Hill reservoir performance. In both cases, effective reservoir heat transfer areas can be estimated by fitting concentration-time data to the appropriate mathematical model.

10.4.2 Field results at Fenton Hill

Two general types of geochemical monitoring have been conducted during the drilling, fracturing, and field testing of the Phase I and II systems at Fenton Hill. The first is passive reconnaissance to observe changes in solution composition behavior that might indicate scaling or deposition of solids within the system or corrosion of piping. The second is active in the sense that dynamic changes were expected and were studied to characterize and hopefully size the Fenton Hill reservoirs.

The passive monitoring at Fenton Hill showed that continuous recirculation of water during heat extraction eventually resulted in a steady state composition that corresponded to a balance of dilution by make-up water and addition of material by displacement of pore fluid from the secondary flow paths. In addition, for silica, active dissolution of reservoir rock in the hotter, longer residence time flow paths provided a source of material. The steady state dissolved $SiO_{2(aq)}$ concentration and ratios of Na^+, K^+, and Ca^{+2} ion concentrations observed at Fenton Hill corresponded to equilibrium saturation or solubilities at temperatures near the maximum or virgin rock temperatures found in the reservoir.

Total dissolved solids levels never exceeded 3000 ppm with the major components typically ranging from (Murphy *et al.*, 1980 and Zyvoloski, 1981)

Species concentration

	ppm
$SiO_{2(aq)}$	220–240
Na^+	350–450
K^+	40–60
Ca^{+2}	25–50
Cl^-	350–600
HCO_3	300–475
SO_4	100–250

No evidence of solids deposition or excessive corrosion was observed on well casing, piping, or heat exchange surfaces in the Fenton Hill system during any of the runs, totalling about 1 year of operation. Although this is encouraging, rock temperatures encountered were limited to 200°C which kept the silica saturation level at about 265 ppm corresponding to solubility of quartz at that temperature. In the Phase II system at Fenton Hill, rock temperatures as high as 325°C may be encountered. This could increase the silica content in the produced fluid to approximately 800 ppm. Thus, on the cold side of the heat exchanger and in the reinjection bore, supersaturation levels will be much higher and will increase the tendency for deposition. Assuming that amorphous silica is the first solid phase to deposit, the percent supersaturation increases from 56% to 370% for a reinjection temperature of 40°C.

Aside from the practical aspects of managing the chemistry of circulating fluid, its dynamic interaction with the reservoir rock and matrix pore fluid provides useful information for sizing the system and assessing its lifetime. Earlier we mentioned that heat transfer areas could be estimated by correlating the rate of dissolved silica buildup in the reservoir to a kinetic/transport model. Robinson (1982) and Grigsby (1983) did this and found that dissolution in the main fracture system at Fenton Hill could not account for the observed buildup during tests where fresh water, low in silica, was continuously injected into a fracture of approximately 50 000 m^2 in area. Robinson (1982) had to use a factor of 100 increase in the rate constant (equation (10.46)) to provide a large enough rate to fit the field data. This is consistent with laboratory testing on crushed granite core material which showed a net dissolution rate about 100 times faster than pure quartz. However, later study revealed that small particles resulting from the crushing and grinding process increased the effective surface area of the granite samples by about the same factor to completely account for the apparent rate increase.

An alternative mechanism to account for the observed rate has been proposed which suggests an encouraging future for HDRs similar in configuration to the Fenton Hill reservoir. Figures 10.20 and 10.22 showed that the residence time distribution for both the original and enlarged Fenton Hill systems have long tails indicative of about 20 to 30% of the fluid having long residence time paths. Two effects could cause the observed silica buildup: the first is displacement of pore fluid presumably providing dissolved silica at a saturated level of approximately 265 ppm and the second is active dissolution in the hotter regions of the reservoir defined by the long residence time paths. Grigsby (1983) used the similar behavior of Li$^+$, Na$^+$, K$^+$, and B ion concentrations to estimate the pore fluid displacement rate. Even after accounting for the maximum amount of silica that could be provided by displacement, additional dissolution in the hotter region of the reservoir at the normal quartz dissolution rate was required to properly fit the field results.

When this result is superimposed with the fact that reservoir growth in terms of modal and integral mean volumes and heat transfer area was observed, it may be hypothesized that mining heat itself may in fact stimulate and enhance the reservoir's productive lifetime.

10.5 Future considerations

It should be clear from the preceding evidence presented in Chapters 9 and 10 that many of the required features of HDR systems have been demonstrated in the field tests at Fenton Hill and Rosemanowes. In essence, one could argue that the technical feasibility of the heat mining concept from HDR has been established. However, an attitude of cautious optimism is appropriate because there are still important requirements for *commercial-sized* systems that have yet to be demonstrated or developed. The central focus is to properly stimulate large volumes of low permeability rock and to interlink injection and production wells to optimize performance. This involves developments in several specific areas:

(1) Improved diagnostic methods to map the structure of HDR reservoirs in order that wells can be interlinked better and that reservoir performance may be extrapolated.
(2) Development of downhole flow control devices or techniques for multiple fractured reservoirs.
(3) Ability to adjust impedances within the reservoir and wellbores artificially.
(4) Ability to reduce water losses.
(5) Development of advanced reservoir concepts to maximize heat extraction capacity of the HDR reservoir to help lower costs.

10.5.1 Improved reservoir mapping diagnostics

Based on the current 'state of the art', HDR mapping methods are evolving into two frontier areas: improved downhole microseismic mapping and chemical reacting tracers that are sensitive to the temperature-field within the reservoir.

By developing a better understanding of microseismic source mechanisms and improving event location resolution with better hardware and advanced hodographic and triangulation analysis techniques, the future holds great promise for correlating microseismic maps with reservoir thermal hydraulic performance. The SUE (Murphy *et al.* (1981)) results and the latest developments at Los Alamos (Fehler *et al.*, 1984) and at Camborne (Batchelor *et al.* 1982–84; Batchelor, 1983 and 1984; and Pine and Batchelor, 1984) are extremely encouraging.

Chemically reacting tracers are being developed as a direct extension of the non-reactive or inert tracer work discussed here and practiced extensively at Fenton Hill and Rosemanowes. Robinson, *et al.* (1984) have proposed the use of easily detected reactive organic compounds that are selected because they react vigorously over a temperature range that would be encountered between the injection and recovery points in an HDR reservoir. These materials would be injected as tracers with their composition monitored in the recovery bore as a function of time and volume throughput as is done for the inert tracers. Now, however, the tracers' concentration-time or -volume dependence will depend both on the distribution of flow path residence times as well as the structure of the temperature field along these flow paths.

The following is a simplified example to illustrate the concept of chemically reacting tracers. Returning to the idealized, rectangular fracture model shown in Fig. 10.24, Robinson has calculated the conversion of reactant and temperature decline expected for flow in a commercial-sized HDR system

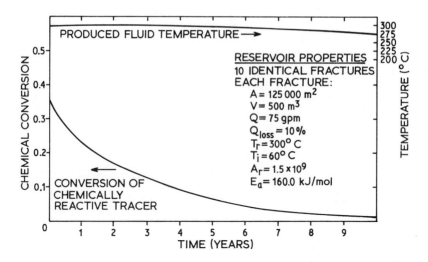

Fig. 10.38 Comparison of the produced fluid temperature and outlet conversion of a chemically reactive tracer as a function of time for a commercial-sized HDR reservoir.

consisting of ten identical fractures, each with an area of 125 000 m² with a total flow rate of 45 kg/s (750 gpm). The results given in Fig. 10.38 show that measurable differences in conversion could be detected in less than one year of operating while measurable thermal drawdown requires at least 5 years if not 10 years for accurate modeling.

Although the analysis and eventual deconvolution of the RTD for flow through a network of fractures with a complex temperature field presents a great challenge, chemically-reacting tracers should provide an enormously more sensitive tool for modeling thermal performance before drawdown appears in the recovery well. In addition, other modes of system operation such as huff-puff can effectively use reactive tracers for sizing purposes.

In a separate development, Kruger *et al.* at Stanford have suggested radon as a natural tracer as it may provide structural information about the fracture surface area in contact with circulating fluid in the reservoir. (Tester and Albright, 1979). The basic idea is to measure radon concentration changes caused by circulation of fluids through the fractured labyrinth. Radon is the radioactive noble gas produced naturally from the decay of radium dispersed in the earth's crust. The principle isotope of radon is Rn^{222} that is an alpha-particle emitter with a 3.83 day half-life. This isotope is produced from Ra^{226} (1600 yr half-life) which is part of the natural uranium series decay originating from U^{238} (4.5 × 10⁹ yr half life). Consequently radon is produced essentially 'forever' in the reservoir but decays with its characteristic 3.83-day half-life when separated from its radium parent radionuclide in the formation. Thus, it introduces a 'time element' in tracer studies. This property of radon contrasts sharply with the stable gas components of geothermal fluids, such as CO_2, H_2S, and O_2 isotopes.

Radon concentration in geofluids depends on several reservoir parameters, primarily the concentration and distribution of radium in the reservoir formation, the conditions for emanation and diffusion into the pore and circulating fluids, and the transport properties of the convecting fluids from the reservoir emanation sites to the producing wellhead. These parameters, in turn, are related to different geologic factors. Radium is found rather uniformly distributed in sedimentary and igneous rocks at an average concentration of about 1 pg/g. Its distribution depends on the local thermodynamic and hydrochemical history of the formation. Since radium is a chemical homolog of the alkaline earth elements calcium, strontium, and barium, it can become redistributed with these elements in hydrothermal regions.

The emanation of radon in rock matrices depends on the chemical, mineral, structural, and thermodynamic properties of the rock. The recoil energy of Rn^{222} on alpha decay of Ra^{226} is 86 keV, sufficient to migrate about $1\mu m$ in rock. The emanation from the rock is thus strongly dependent on the surface area exposed, the porosity, and the composition of the cementing materials in the rock. It is also dependent on the pressure, temperature, and density of the pore fluid in the formation to some extent. The transport of radon depends on its solubility in the convection fluid and the hydrodynamic properties of the reservoir, such as permeability-thickness, reservoir pressure gradients, and flow rate.

Two general types of information may be obtained with radon as an internal reservoir tracer. Under steady flow conditions, changes in reservoir properties will result in changes in radon concentration in the produced geofluids. Under steady emanation conditions, changes in flow regime will also result in changes in radon concentration. The challenge of successfully applying radon transient analysis to geothermal reservoir engineering rests with the ability to show the relationships between changes in reservoir or flow properties and changes in radon concentration. Sufficient data are needed to separate the changes due to each effect.

Experiments at Fenton Hill and Rosemanowes have yielded very encouraging preliminary results regarding the potential of using radon as an active tracer for reservoir sizing. However, the actual mechanism of radon transport in multiply fractured systems will depend on fracture aperture and length distributions within the reservoir. Therefore, deconvolution of the data into a *unique* structural model seems unlikely.

10.5.2 Flow control into multiple fracture systems from two boreholes

In this case, it is easy to define the potential problem in that fractures with widely varying impedances have been observed both at Fenton Hill and Rosemanowes. The fractures with the lowest impedance will tend to accept the majority of injected flow which could lead to poor overall sweep efficiency and inferior heat mining. Two potential solutions exist. The first involves partially choking the flow into each fracture zone with a restricted aperture or orifice system located downhole to offset the differences in natural impedance among fracture zones. These restricted orifices could be placed on the inlet or outlet side. Of course, a complex set of plumbing hardware is required, especially if the number of zones to be controlled becomes large. Furthermore, this hardware will have to perform reliably for long periods. Interestingly enough, systems of this type already exist for downhole flow control in off-shore oil and gas producing systems with multiply-deviated wells. The continued successful operation of these systems is encouraging because it suggests that many potential design failures may have been already worked out. The second approach involves actually trying to alter the entrance or exit impedance into each fracture zone. Because methods for doing this may also improve fluid sweep efficiencies and therefore may enhance reservoir production capacity, they are discussed separately in the next section.

10.5.3   Impedance adjustment

We have already discussed several methods of physically altering flow impedance in the entrance and exit region of the fracture to borehole connection where pressure losses can be high due to accelerating and decelerating flows. These techniques include explosive fracturing under development by Batchelor and his team at Camborne, pressure-propping as examined in the high back pressure tests at Fenton Hill and Rosemanowes, and solid particle propping as practiced during the hydraulic stimulation of many oil and gas and some geothermal reservoirs. Another method is available that takes advantage of the heterogeneous nature of the mineral assemblage that comprises the rock matrix. Chemicals that dissolve specific mineral components of the rock have been used to stimulate oil and gas reservoirs for several decades. Since these commonly involve the use of reactive acids such as hydrofluoric (HF) or hydrochloric (HCl), the general technique has been come to be known as 'acidizing' and when accompanying hydraulic fracturing as 'acid fracturing.' For a primary reference on this subject, interested readers should consult the monograph by Williams, Gidley, and Schechter (1979). In crystalline rock formations of low permeability, two major treatment design choices exist. One can try to selectively dissolve a portion of the silica-based minerals themselves using HF or strong basic solutions of sodium hydroxide (NaOH) or sodium carbonate (Na_2CO_3) or one can try to remove materials that commonly are found in sealed natural joints. Frequently precipitated carbonate minerals are found in these sealed joints which could be dissolved with HCl or similiar acids. Some preliminary tests were conducted at Los Alamos in the early stages of the project before an adequate connection was made between EE-1 and GT-2. This involved injecting an aqueous solution of 1 Normal

Na_2CO_3 into the reservoir to selectively dissolve quartz (SiO_2) at temperatures of 150 to 200°C. Although several tons of SiO_2 were removed, no observable change in the impedance was measured – apparently the 10–20 m separation distance between the non-communicating fractures initiated from EE-1 and GT-2 was too large to be penetrated effectively.

However, the promise of chemical stimulation for impedance control is still very much alive. In fact, it is one of the few methods that has the potential of deep penetration into the fracture system away from the exit and entrance regions.

10.5.4 Downhole pumping to reduce water losses

The major cause of finite water losses in low-permeability HDR systems is the higher-than-ambient operating pressures required to circulate fluid through the fractured labyrinth. In some cases with self- or particle-propped fractures or selective chemical dissolution to induce its own propping, it may be possible to operate at considerably lower pressures within the reservoir. By placing a downhole pump in the recovery bore, the necessary head could be supplied to overcome all losses in the well casings, surface equipment, and the reservoir; a lower average downhole pressure would result. Since most HDR reservoirs will be deeply sited at depths greater than 3 km, sufficient hydrostatic pressure exists at the borehole to avoid unwanted flashing. The pump could be set at a depth of less than 1000 m in the recovery hole if the net pressure drop due to impedance across the fracture zone is less than 10 MPa (1450 psi). If the normal ambient pressure is hydrostatic, then water losses would be essentially eliminated. A recent experiment at Rosemanowes has demonstrated this during field testing of their Phase II system. (Batchelor, *et al.,* 1984). One economic disadvantage that has to be dealt with is the larger well casing diameter required to accommodate the downhole pump. Another aspect has to do with the high temperature capability of the downhole pump unit itself. As a part of the development projects for natural hydrothermal systems, pumps capable of operating at temperatures approaching 200°C are being designed and tested. For higher temperature operation, downhole pumps are only in the developmental stages with a few prototypes tested (see Tester, 1982).

10.5.5 Advanced heat mining techniques

(a) FURTHER IMPROVEMENTS TO THE FENTON HILL RESERVOIRS

The summary of heat extraction test results from Fenton Hill presented earlier indicated that the Phase I reservoir was of modest size, representing about 50 000 m^2 of effective heat-transfer area. However, other indications such as geochemical, microseismic, water losses, and venting volume measurements suggest that the reservoir is potentially much larger. In particular, the microseismic data imply that hydraulic communication was achieved at distances very far from the injection well. Roughly speaking, a circle drawn around the microseismic epicenters measured during run segment 4 has an area of about 225 000 m², about five times the effective heat-transfer area. For three fractures, according to the multiple, interacting fractures model, Table 10.6 suggests a maximum microseismic area of 270 000 m². Furthermore, the microseismic data suggest that this larger potential reservoir is not planar, but highly jointed and multiply fractured, so that the potential reservoir, if sufficiently exploited, would represent a volumetric rather than an areal source of heat. For the same level of power production, a volumetric source results in less thermal decline than an areal source, which is severely limited by the requirement to conduct heat in the low-conductivity rock for large distances perpendicular to the areal feature.

The explanation for the large difference in reservoir sizes provided by heat-transfer results and the other indications, such as geochemistry and microseismicity, etc. is provided by fluid mechanics and flow patterns in reservoirs. Even if a reservoir were physically large, should fluid-dynamic short circuiting occur, then the effective heat-transfer size of the reservoir would be much smaller. The most important criterion is the separation between reservoir inlet and outlet. For continuous flow circulation, in a low-buoyancy mode of heat extraction, the effective heat-transfer area scales roughly with the square of the inlet-to-outlet spacing for individual fractures and is proportional to the cube of the separation distance for volumetric, multiply-fractured reservoirs. As discussed below, these guidelines must be modified when buoyancy or natural convection effects are present, or for a cyclic (huff-puff) mode of heat extraction. However, during run segments 2 through to 5 at Fenton Hill, these conditions did not exist, so for practical purposes wellbore separation distance controlled the reservoir's heat extraction capability. In the first reservoir, before recementing and enlargement, the separation was of the order of 100 m which is consistent with an observed initial heat-transfer area of only 8000 m². In the second enlarged reservoir, the separation was 300 m with an effective area of approximately 50 000 m². For either reservoir, fluid dynamics dominated so that heat production was limited by the separation of inlet and outlet which effectively defined the active reservoir region.

In connection with this conclusion regarding fluid-dynamic limitations to heat production, it must be remembered that the design of the Phase II reservoirs, currently under development at Fenton Hill and Rosemanowes is based largely upon Phase I technology and experience. Of particular importance is the fact that the vertical spacing of the Phase II wellbores is 370 m, only 20% larger than that of Phase I. Consequently, heat production in the Phase II fracture will most likely be subject to the same fluid-dynamic limitations.

Fluid streamline patterns could be improved without redrilling, by changing the mode of heat transfer. One means of doing this is to resort to cyclic (huff-puff) operation as originally suggested by Rex (see Smith 1979), in which water is injected while the production well is shut-in. If the pressure and flow-rate conditions are appropriate, water can be forced to the reservoir extremities where the rock temperatures are high. The heated water is then withdrawn by venting the production well in the 'puff' phase of the cycle. A secondary fracture growth effect may be caused by the periodic higher pressures associated with cyclic operation.

A second means of improving the flow streamline pattern is to promote the effects of high buoyancy as depicted in Fig. 10.28(b). For the cases shown, computed results are for a vertically oriented, circular fracture in which the inlet and outlet locations are separated by 400 m, nearly the same value as in the Phase II reservoir. The fracture is 0.5 km in radius. Figure 10.28(a) shows a case where the flow impedance is so high that buoyant, or natural convection, effects are entirely suppressed. The streamlines flow directly from inlet to outlet and bypass much of the area potentially available. In fact, only 40% of the total area is used effectively for heat transfer. In Fig. 10.28(b), the impedance is low enough that buoyancy is important. The cold entering fluid first flows downward due to its greater density, then eventually turns and flows to the outlet. In so doing, almost 90% of the total fracture area, more than twice that of the first case, is effective in heat transfer.

Further testing of the Fenton Hill Phase I system to explore the potential of cyclic huff-puff and buoyantly-driven modes of heat mining has been recommended by the Los Alamos team.

(b) OTHER ADVANCED CONCEPTS

The general application of special reservoir operating techniques such as cyclic huff-puff operation may require different wellbore fracture system designs to optimize performance. As discussed by Smith (1980), both single-bore and two-bore systems are feasible. One major concern in the single-bore system is the cyclic nature of the thermal stresses on the steel casing and cement caused by the requirements to heat up or cool down the bore on each half cycle. Also, because of the finite rates of heat transfer through the bore casing walls, a single-hole system loses efficiency. The two-hole concept avoids both problems because one hole is used for injection and the other for recovery during each half cycle. One other limitation of the huff-puff system results from the progressive nature of the advancing cold wave in the reservoir. As time progresses, deeper and deeper penetration away from the injection bore will be required to reach hot zones of rock. This will require either higher rates of injection or longer injection periods or a combination of both. The same is true of the recovery period for either a single- or two-bore system. Certain adjustments to ranges of half cycle time can be made to accommodate base or peaking load requirements.

In the Rosemanowes Phase I reservoir, a hybrid flow-through huff-puff concept was tested (Batchelor, Pearson, and Halladay, 1980). Using an episodally-choked flow, fluid was pumped into the reservoir and vented with the production well partly shut-in. The analysis of Batchelor *et al.* (1980) defined the system's hydraulic performance efficiency in terms of water loss rates, specific impedances and specific acceptances. This early study has laid the groundwork for future evaluations of huff-puff operation where the dependence of parasitic losses can be easily related to measured reservoir parameters.

Smith (1980) also reviews several other possible HDR reservoir stimulation concepts besides hydraulic fracturing that have been proposed. These include (1) the use of downhole heat exchangers in a solution-mined hot salt dome or limestone cavity (Jacoby and Paul, 1975), (2) the use of nuclear explosives to create large rubblized regions of permeable rock (Kennedy, 1964; Burnham and Stewart, 1975), and (3) the use of forced circulation through well characterized but unproductive fractures in a hydrothermal system (Bodvarsson and Reistad, 1975). Although all three ideas are technically feasible, they each have some disadvantages over the hydraulic fracturing approach for low permeability formations. The first and third concepts require a suitable geologic resource, so their application is more site-specific, and many of the factors that control heat extraction rates in hydraulically stimulated HDRs also are important to both of these alternatives. The second case introduces obvious social and political obstacles in that the peaceful uses of nuclear explosives are widely regarded as socially unacceptable forms of alternative energy stimulation. In addition, expected ground shocks from such detonations would create a hazard for nearby population centers. Anyone in the free world who would advocate their use, even on the strong technical grounds that they might provide an HDR reservoir with potentially near-ideal volumetric heat extraction characteristics, is probably doomed to failure. The Soviets have apparently been pursuing some sort of technical effort in this area but, apparently, nothing has been published.

In another conceptual design study by the MAGES (International Energy Agency, 1979) group, 2nd generation HDR systems employing large diameter shaft access to the reservoir with multiply deviated sidebores have been considered. However, drilling and mining costs may be too high for these systems to ever be more than a dream.

Chapter 11

Gaining access to the resource: current and future drilling practice

As we have stated in previous chapters, and will now justify, drilling to gain access to hot dry rock (HDR) at depths in excess of 3 km is a difficult and costly process. In fact, drilling-related costs are almost always likely to be the single largest cost component of any heat mining project. Drilling is a risky process because of our inability to predict what lies in front of a penetrating drill bit. It is also an art as much as a science and it involves an extraordinary infrastructure of specialized disciplines that have evolved in the drilling industry during its long history (about 100 years).

In this chapter, the major features of drilling technology for geothermal HDR wells are discussed in an economic context. Current practice for rotary drilling and completing both hydrothermal and HDR wells are summarized and the experience at Fenton Hill in America and at Rosemanowes in England is extensively reviewed. This is followed by a brief discussion of the impact of drilling costs on the total development costs for a HDR geothermal power plant and of various economic tradeoffs. The final section deals with some advanced drilling concepts that may have potential for improving penetration rates and reducing costs.

11.1 Current rotary drilling practice for hydrothermal systems

Because natural hydrothermal fields have been under commercial development for almost 80 years, with over 3000 wells drilled in more than 12 countries, a drilling technology is already well established. The necessary hardware, support services, and labor forces are available in countries like the United States, Italy, Japan, Iceland, and New Zealand. Furthermore, because geothermal well drilling practice closely resembles that of oil and gas well drilling (with some modifications to downhole hardware and techniques), any country that is active in petroleum exploration and production should be able to adapt quickly to the geothermal development business.

Only the major elements of hydrothermal well drilling technology are discussed here with respect to their influence upon HDR drilling technology, as

(a)

(b)

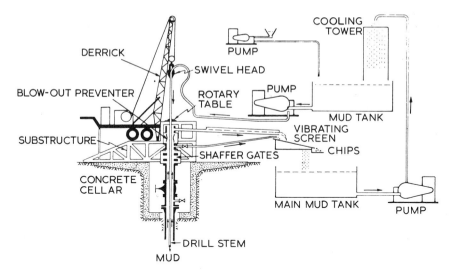

Fig. 11.2 Schematic of typical geothermal drilling rig and pipe assembly. (Craig, 1961, p. 123).

the relevant drilling literature is quite extensive. In fact, previous publications of both authors (Armstead, 1983; Milora and Tester, 1976; Tester, 1982) summarize the technology and cost trends associated with geothermal drilling and make detailed reference to this extensive literature. Recent summaries by Maurer (1975), Kestin *et al.* (1980), and Greene and Goodman (1982) are also particularly informative.

The basic process of rotary drilling involves mechanically induced crushing at the drill bit to chip and spall the rock, producing cuttings that are removed from the hole with suitable pumped fluids (air, water, foams, or drilling muds). As shown in Fig. 11.1, special serrated tricone bits that contain either teeth of hardened steel or 'buttons' of ultrahard material such as tungsten carbide are rotated at the end of a pipe or drill string over the surface of the rock by a mechanical drive, usually from the surface by means of a large diesel engine or electric motor as shown in Fig. 11.2. The drill string is rotated by means of a square- or hexagonal-shaped pipe or Kelly that passes through a bushing which rotates on a rotary table, usually at speeds of 30 to 300 rpm. Normally, drilling fluid is pumped down the hollow drill string using positive displacement 'mud pumps' which are also powered from the surface. A hose and swivel

---

Fig. 11.1 Tricone roller bits. (a) hardened steel tooth bit for softer formations. (b) Button bit with tungsten carbide inserts for hard rock formations.

arrangement allows for the continuous pumping of fluid while the drill string is rotated. Fluid leaves the string through jets in the bit and is recirculated up the annular space between the outside of the drill string and the inside of the open hole. This fluid normally performs three functions: it cools and lubricates bearings within the bit, washes away rock cuttings from the hole to aid penetration, and it helps to stabilize the hole from collapse. In geothermal applications, the cooling and lubrication function becomes more difficult because of the hot environment and coaxial heat exchange effect arising from the temperature difference between the downgoing and upcoming fluid legs in the hole. Surface cooling is frequently used to reduce the inlet temperature of the fluid before it reaches the bit face.

When more stabilization of the formation is required than can be provided by the hydrostatic pressure of the fluid against the borehole walls, specially-formulated drilling muds are used that create a filter-cake effect on the walls. For temperatures up to about 200°C, water-based muds using bentonite or other clay-based materials with special additives, such as chrome lignite/chrome ligno-sulphonate, are used. As higher temperatures are approached, substantial polymerization or gelling can cause severe degradation of the mud properties. Above 260°C, serious problems arise with almost any natural or artificial rheological fluid, including the oil-based muds. Fortunately at a large number of hydrothermal sites, hole stabilization is not a serious problem, at least in the hotter sections of the hole, and air or water can be used as the drilling fluid.

Although drill rigs come in all sizes with many portable power system arrangements, most geothermal drilling would use rigs having 1500–3000 hp of live diesel power or an equivalent 1100–2250 kilowatt (kW) diesel-electric unit. Power levels of this magnitude are necessary to turn the rotary table and to pull the loads associated with removing and replacing long and heavy drill strings from and into the hole as worn bits have to be replaced, and for lowering steel casing into place for cementing. Holes deeper than 4.5 km (15000 ft) would require proportionally higher power levels than indicated above.

In addition to the drilling platform and mud pumps that will be sized according to hydraulic requirements, usually in the range of 500 to 1200 hp, (375 to 900 kWe) an entire fluid circulating system for screening and settling out rock chips and cooling and storing the drilling fluid is included with the rig set-up.

The drill string itself usually consists of steel pipe sections or joints approximately 10 m (30 ft) in length that are coupled together in groups of three to form a 'stand' of pipe. Near the bit are placed several very heavy walled drill collar sections to stiffen the string and to maintain a straight drilling direction as well as to increase the load on the bit. These heavy collars bring the neutral point in the drill string, where there is a transition from axial tension to compression, as close to the bit as possible. This helps to minimize bending which would cause the drill pipe to rub on the wall of the hole. This lower section of the drill string is commonly called the bottom hole assembly (BHA), and in addition to the drill collars and bit it may contain side wall reamers and stabilizers, shock absorbers, and a non-magnetic section of pipe to allow for magnetic hole surveys.

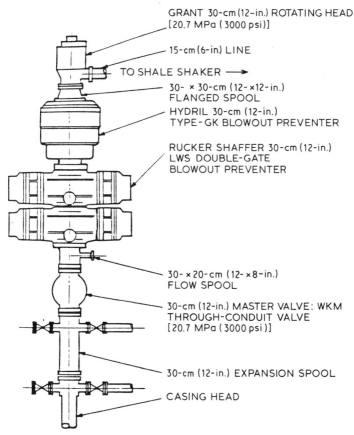

GRANT 30-cm (12-in.) ROTATING HEAD [20.7 MPa (3000 psi)]

15-cm (6-in.) LINE

TO SHALE SHAKER ⟶

30- × 30-cm (12-×12-in.) FLANGED SPOOL

HYDRIL 30-cm (12-in.) TYPE-GK BLOWOUT PREVENTER

RUCKER SHAFFER 30-cm (12-in.) LWS DOUBLE-GATE BLOWOUT PREVENTER

30-×20-cm (12-×8-in.) FLOW SPOOL

30-cm (12-in.) MASTER VALVE: WKM THROUGH-CONDUIT VALVE [20.7 MPa (3000 psi)]

30-cm (12-in.) EXPANSION SPOOL

CASING HEAD

Fig. 11.3 Typical blowout preventer. (From Rieke and Chillingar, 1982).

Blowout preventers (BOP) are used to control pressures and prevent unwanted surges of fluid from blasting out of the well. A typical BOP is shown in Fig. 11.3. Many other designs are available, but the basic concept of forcing elastomer rams (shown as the Rucker Shaffer doublegates) to seal against the drill pipe in the event of fluid surges remains the same.

A normal hole drilling procedure would consist of the following steps:

(1) *Site preparation.* Securing the necessary water supply, construction of concrete pad and cellar to support the weight of the rig, road and

temporary housing construction if needed, and installation of communication and power transmission lines.

(2) *Rig mobilization.* Delivery and assembly of the rig on site, including power units for rig operation, derrick and platform, mud pumps, settling and holding tanks, pipe racks, air compressors and fuel tanks.

(3) *Installation of the conductor or surface casing.* Large diameter, 1 m or more (36 inch or more) steel casing cemented from the surface usually to a depth of 30 m (100 ft) or more to stabilize the initial drilling process.

(4) *Drilling and installation of anchor casing.* Intermediate diameter casing, up to 0.6 m (24 inch) cemented in place after drilling first 300 to 800 m (1000 to 2600 ft) of hole to stabilize the upper section.

(5) *Drilling.* The main portion of the hole is drilled into the production region to its total depth (TD). Several intermediate casing operations may be involved here if hole stabilization is required. These will lead to a reduction of bore diameter, but must allow the BHA to pass inside each casing string. The drilling operation itself will involve assembly of the BHA with new bits and lowering the string into the hole as pipe joints are threaded in place until the current hole bottom is reached. The Kelly would be placed above the last joint and pass through the Kelly bushing. Weight on the entire drilling string would be adjusted to the proper level, the mud pumps started, and the rotary table would begin to rotate to initiate drilling. Penetration rates would be monitored and the rotational speed and string weight adjusted periodically. When bit wear becomes excessive, causing the hole to go too far off-gauge and reducing penetration rates too much, the drill string would be removed from the hole and the bit replaced. This so-called 'tripping' process is repeated throughout the drilling operation. It is very time-consuming, especially for deep holes, but is necessary because of finite bit lifetime. At 3000 m (10 000 ft) it takes about 10 hours while at 10 000 m (30 000 ft) it may take longer than 24 hours. As bit life may be as short as 24 hours or less for very hard rock formations, tripping time greatly reduces the overall penetration rate. Logging, surveying, coring, and other operations may take place periodically between drill string runs.

(6) *Installation of production casing.* This involves the assembly of production liner and casing, lowering into the hole with proper centralization, and cementing in place. For very deep holes, cementing will be done in stages. A wellhead valve unit (or 'Christmas tree') must also be installed.

(7) *Rig demobilization.* Disassembly and removal of rig from site.

Fig. 11.4 (a) Standard casing arrangement and open hole production zone completion for a well at the Geysers dry steam field in California. (From Rieke and Chillinger, 1982).
(b) Casing arrangement and slotted liner production zone completion for a well at the Wairakei field in New Zealand. (From Craig, 1961, p. 123).

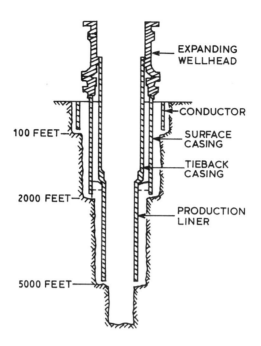

(a)

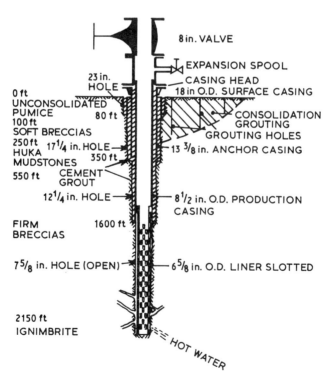

(b)

Lining the hole with steel casing that is properly cemented in place is essential to ensure hole stability and long-term production. Usually several strings of different diameter casing are telescoped into the bore with large diameters near the surface and smaller diameters lower down in the hole as shown in Fig. 11.4. In the production region uncemented slotted or perforated sections of pipe are frequently used to stabilize the hole against sloughing while maintaining good hydraulic communication with the reservoir fluids (see Fig. 11.4(b)). In some cases, however, where the formation is stable, an open hole completion is possible as shown in Fig. 11.4(a) for a well at the Geysers.

One of the most technically challenging parts of the completion procedure is proper cementing of the casing in place. This is partly because of the generally high temperature of the well which degrades the strength and permeability properties of most cements and partly because of the thermal stresses induced if temperature cycling takes place. Rieke and Chillingar (1982) present the details for proper casing and cementing design to optimize performance in geothermal environments. But as temperatures rise above 250°C, even the best high temperature cements have problems. Common practice today is to use a Portland (API Class G, H, or J) cement stabilized with 30–50% silica flour (200 to 400 mesh size) with appropriate curing retarders, water loss agents, and friction reducers to improve its 'pumpability.'

Directional drilling is used in some hydrothermal applications to improve production and/or to reduce surface piping costs. For example, if fluid production comes mainly from major vertical faults that are horizontally spaced, then a deviated well, slanted off-vertical, will intersect more fractures and be more productive. Directional drilling has also been used extensively in petroleum production. This has created a large technology base that has been applied to directional applications in geothermal systems. Because HDR wells will almost always require careful directional control, we have deferred discussion of directional drilling until Section 11.3.

11.2　Current rotary drilling practice for HDR wells

Before proceeding to discuss the field experience gained by the Americans at Fenton Hill and by the British at Rosemanowes, it is important that the general requirements and hardware for HDR drilling be outlined. Similarities and differences to current rotary drilling practice for wells in natural hydrothermal reservoirs are also discussed.

11.2.1　General requirements

As mentioned previously in Chapters 9 and 10, HDR reservoirs in deep-seated crustal pockets at depths of 3 to 6 km in low-permeability crystalline basement rock are of primary interest. Consequently, because minimum stresses will normally lie in the horizontal plane at those depths, wells slanted from a vertical position are preferred. Furthermore, they should be directed more or less perpendicular in plan view to the least principal stress direction ($S_Y = S_3$). In the production region of hot rock, these deviated bores might be inclined 25–35° from vertical and run parallel to each other with the maximum vertical separation possible between them to maximize heat transfer area and volume and still maintain low flow impedances. This geometric arrangement of boreholes will require careful directional control of azimuth and dip (inclination) to maintain desired trajectories. Directional drilling techniques will be required to achieve the desired ±3° to 5° of azimuth and ±1° of inclination control. The casing design and cementing procedures must be able to meet the thermal contraction and expansion stresses that might occur during thermal cycling. In addition, casing bursting strength must be sufficient to withstand normal operating pressures and even higher fracturing pressures unless a smaller fracturing tube is used. Provisions for long lengths of production and recovery sections (1000–1500 m) of the wellbores must be made to permit proper communication and control of fluid flow into and out of the fractures. The completed interior hole diameter must be large enough to permit entry of surveying and logging tools, hydraulic flow control equipment, fracturing hardware, and even drill bits for side tracking or redrilling if required.

11.2.2 Drilling and well completion hardware and techniques

Apart from the rotary power and hoisting capacity requirements which will determine the size of the rig, the rig itself and its ancilliary equipment will be determined by what is available on a competitive bidding basis. Figures 11.5 and 11.6 are photographs of the large rigs used in Phase II drilling at Fenton Hill and Rosemanowes. In some cases, special drill pipe with hardbanding to reduce wear in the highly deviated holes of these HDR systems has to be used. The string itself is frequently subjected to severe stresses that can cause permanent damage (see Fig. 11.7).

A summary of temperature limitations of a large number of downhole equipment and hardware items is presented in Fig. 11.8. Major pieces of equipment or hardware with special design for HDR well environments are discussed below in the following sections.

(a) DRILLING BITS

Based on early experience at Fenton Hill in GT-2 (Pettitt, 1975), tricone hard rock mining bits manufactured by Smith Tool Company, the Security Division of Dresser Industries and Hughes Tool Company, were almost exclusively used for the crystalline rock sections in GT-2, EE-1, EE-2, and EE-3 at Fenton Hill and RH11 and RH12 at Rosemanowes. Some typical bits are shown in Fig. 11.9. Although bit lifetime and penetration rate were quite variable and

Fig. 11.5 (a) Raising the derrick on the Brinkerhoff-Signal Rig no. 56 in a snowstorm at Fenton Hill in New Mexico on 5 April 1979. (b) Rig in place drilling EE-2, summer of 1979. (Photos courtesy of R. Pettitt, Los Alamos National Laboratory.)

Fig. 11.6 (a) Raising the derrick on the National 80UE Rig on location at the Rosemanowes quarry in Cornwall on 17 August 1981. (b) Rig in place drilling RH-11, fall of 1981. (Photos courtesy of A.S. Batchelor, Camborne School of Mines.)

Fig. 11.7 Bent drill pipe resulting from parted drillstring that broke at 1387 m (4549 ft) in EE-2 at Fenton Hill. (Courtesy of R. Pettitt, Los Alamos National Laboratory.)

strongly dependent on the hole condition and drilling parameters such as weight on bit and rotation speed, reasonable performance was obtained and no temperature limitations were encountered in drilling any of the Phase II holes. Typical bit performance data are given in Table 11.1 for drilling EE-2.

(b) DOWNHOLE OR BOTTOM HOLE ASSEMBLIES (BHA)

A typical assembly for straight or 'locked-in' rotary drilling is shown in Fig. 11.10(a) while a deviating or sidetracking assembly using a downhole motor for directional drilling is shown in Fig. 11.10(b). In the locked-in BHA, reamers and stabilizers are used with stiff drill collars to maintain direction. In the deviating assembly, a 30 ft section of non-magnetic collar, and a 2° Bent

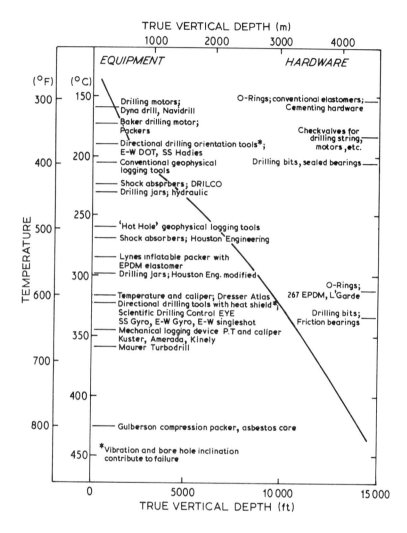

Fig. 11.8 Temperature limitations of downhole hardware and equipment. (From Helmick *et al.*, 1982.) The line represents the conditions for the Fenton Hill reservoir.

orienting sub are used to align the bit face in the desired direction. Water circulated by the rig mud pumps is used to power the downhole drill motor which for the case shown was a positive displacement type motor. If the deviating BHA fails, one can resort to an oriented whipstock for sidetracking with rotary drilling from the surface (see Fig. 11.11).

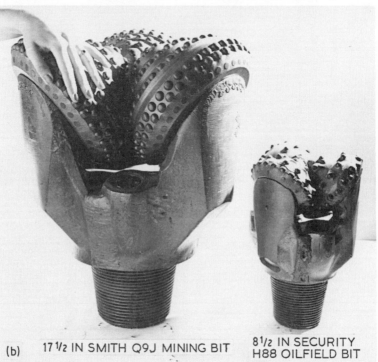

(b) 17½ IN SMITH Q9J MINING BIT   8½ IN SECURITY
                                          H88 OILFIELD BIT

Table 11.1. Drill bit performance in the Fenton Hill Phase II hole EE-2
(Helmich *et al.*, 1982)
Surface-driven rotary bits

Downhole motor/turbodrill driven bits							
Bit no.	Size	Make	Type	End of run depth	Footage	Operating hours	Deviation
17	12–1/4	Sec	H-100	2 999	358	48.75	0.5°
RR28	12–1/4	STC	F-9	4 855	493	53.25	6°
34	12–1/4	Sec	H-100	5 892	345	24.00	7.75°
35	12–1/4	HTC	J-55	6 492	600	22.75	5.5°
45	12–1/4	HTC	J-55	7 689	505	21.75	6°
47	12–1/4	HTC	J-55	8 204	461	23.75	15°
55	12–1/4	STC	F-5	8 975	308	23.75	—
104	8–3/4	STC	7GA	12 521	214	13.50	35°
105	8–3/4	STC	7GA	12 728	207	13.50	34°
117	8–3/4	STC	7GA	13 955	298	22.00	35°
138	8–3/4	STC	7GA	14 189	233	23.00	34.5°
137	8–3/4	STC	7GA	14 501	312	33.50	35°
140	8–3/4	STC	7GA	14 962	462	42.00	35°
142	8–3/4	STC	7GA	15 273	307	22.50	35°

Downhole motor/turbodrill driven bits							
Bit no.	Size	Make	Type	Depth out	Footage	Hours	Deviation
41	12–1/4	Sec	H-100	6 818	100	12.25	4.75°
52	12–1/4	STC	Q9JL	8 545	131	3.00	13.75°
59	12–1/4	STC	Q9JL	9 188	106	3.00	—
60	12–1/4	STC	Q9JL	9 311	123	3.00	13.5°
74	12–1/4	STC	Q9JL	10 035	118	5.50	—

Sec = Security Division of Dresser Industries
STC = Smith Tool Company
HTC = Hughes Tool Company

(c) DOWNHOLE MOTORS AND TURBODRILLS

The use of fluid- or mud-driven downhole motors to rotate a drill bit has certain advantages for directional control. The bent sub-assembly as shown in Fig. 11.10(b) is much easier to place and remove from the hole than a whipstock which must be locked or cemented in place. To simulate surface-driven rotary conditions, high-torque, low-speed downhole motors are required when using

Fig. 11.9 Typical tricone tungsten carbide insert button bits for hard rock drilling. (a) Tricone $9\frac{5}{8}$ in tungsten carbide insert roller bits used at Fenton Hill. (Pettitt, 1975). (b) $17\frac{1}{2}$ in and $8\frac{1}{2}$ in tungsten carbide insert roller bits used at Rosemanowes. (Batchelor and Beswick, 1972.)

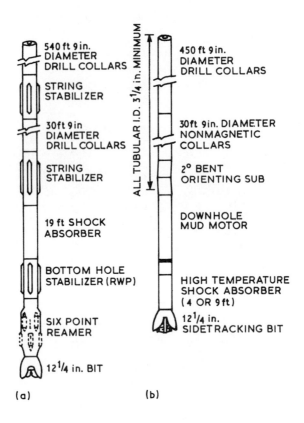

Fig. 11.10 Typical bottomhole assemblies (BHA). (Helmick *et al.*, 1982). (a) For straight or locked-in rotary drilling. (b) For deviating, side-tracking directional drilling.

tungsten carbide button bits in drilling hard rock. Furthermore, high bit loads or string weights are also needed to enhance penetration rates because a crushing mechanism is a key element in the drilling of crystalline rock.

As shown in Fig. 11.12, there are two basic approaches to downhole-powered drills: (1) downhole motor using a fluid driven positive displacement mechanism designed on the Moineau principle and (2) a turbine-drive drill which rotates as fluid is expanded across a set of multistage turbine rotor and stator blades. Several manufacturers have produced commercial units of both types. The ones used at Fenton Hill and Rosemanowes are described in Table 11.2. As can be seen, only the Maurer turbodrill shown in Fig. 11.12 is capable of high temperature operation at temperatures up to 360°C. The major limitation of the Moineau-type motors is that they use elastomeric seals which cannot withstand temperatures much above 175°C. Given that some cooling occurs by circulating the drilling fluid, directional drilling with motors of this

Fig. 11.11 Photograph of a $10\frac{3}{4}$ in diameter whipstock successfully used in sidetracking EE-2 at Fenton Hill. (From Rowley and Carden, 1982).

type has been achieved for short periods at formation temperatures above 200°C.

While most of the experience at Fenton Hill was with Dyna-drills and Baker-motors, Batchelor *et al.* (1982a) report excellent performance with Drilex-motors which are based on a Soviet design and licensed to Drilex in Aberdeen, Scotland. Much of the directional drilling at Fenton Hill used surface-powered, rotary-driven, angle-building assemblies with short downhole drill motor runs for course correction. According to Helmick *et al.* (1982) and Rowley and Carden (1982), this procedure coupled with the bearing and seal problems that were encountered with the motors and turbodrill, frequently led to short bit life and severe wear as shown in Fig. 11.13. Rotational overspeed under insufficient loading (15 000 to 20 000 lb) was a major cause of poor performance. In contrast, at Rosemanowes, the Drilex system was capable of handling bit loads of 40 000 to 60 000 lb at rotational speeds comparable to what is used for optimal drilling penetration under surface-driven rotary conditions. The Camborne group claims that the Drilex motor runs had performance similar to normal rotary drilling during the RH11/RH12 campaign. However, it must be borne in mind that the Rosemanowes reservoir temperature never exceeds 100°C.

The Maurer turbodrill or one of similar design holds great promise for directional drilling in deeper, hotter environments. The Maurer drill was specifi-

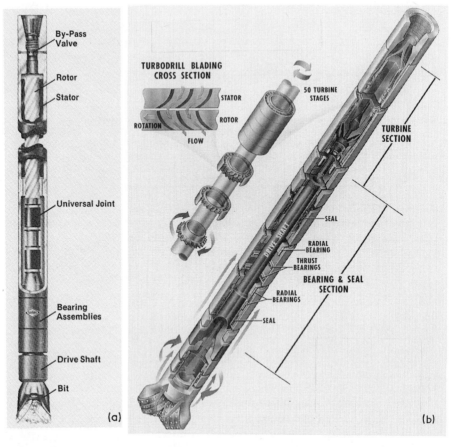

(a) Navi-Drill   (b) Maurer turbodrill

Fig. 11.12 Down hole directional drills. (a) Navi-Drill system powered by a positive displacement, Moineau-type motor. (Courtesy of the Christensen Division of the Norton Company). (b) High temperature downhole turbodrill developed by Maurer Engineering, Inc., of Houston, Texas and Los Alamos National Laboratory for hard rock directional drilling up to 360°C.

cally designed for high temperature service with special bearings and seals and an improved turbine design over earlier models (Maurer, 1975). Given that only prototypes were tested in EE-2 and EE-3, the performance was quite good, with failures being attributed to relatively minor malfunctions rather than serious design errors. Two sizes were used at Fenton Hill: a 19.7 cm ($7\frac{3}{4}$ in.) diameter drill and a 13.7 cm ($5\frac{3}{8}$ in.) diameter drill. The larger one was used in the 31.1 cm ($12\frac{1}{4}$ in.) diameter sections of the hole while the smaller one in the 22.2 cm ($8\frac{3}{4}$ in.) diameter sections. Performance in EE-2 and EE-3 is summarized in Table 11.2. In the deeper sections of both holes, beyond the temperature capability of the Moineau-type designs, the Maurer turbodrill played a key role in controlling the inclination of the wells. However, given the valve and bearing failures and excessive bit wear that was observed periodically, more development work was required to correct these deficiencies before commercialization. Maurer has done this and their turbodrill is now a commercial product.

(d) ORIENTING, SURVEYING, AND STEERING TOOLS

All directional drilling operations require knowledge of hole azimuth and inclination. This can be done in a separate logging operation using a single- or multi-shot magnetic survey or surface-readout gyroscopic survey tool. Alternatively, a continuous surface readout can be used on a steering tool that is mounted inside the drill string in the bottom hole assembly near the bit (see Fig. 11.14). As shown in Table 11.2, three different steering tools were used with downhole motors and turbodrill in EE-2 and EE-3. At Rosemanowes, the 'EYE' tool was used exclusively. Periodically, between bit runs, single or multi-shot magnetic surveys would be run to check hole direction. All of these devices have temperature limits shown in Fig. 11.8 and may require a heat shielding or active cooling in an insulated dewar to increase downhole service life.

(e) CORING

Coring in hard rock is difficult, time-consuming and costly. In all cases it requires a separate coring bit run with a core barrel to retrieve the core that is cut. This requires an extra round trip of the drill string to replace the BHA and at least one day of rig time. Two types of bits have been used for coring in HDR wells (see Fig. 11.15). Diamond impregnated core bits are very expensive and have not performed as well as the specially designed hybrid quadracone bit which uses tungsten carbide inserts on the cones and diamond-impregnated Stratapax studs. However, even the Stratapax core bit had its problems in the deep, hot holes at Fenton Hill with core recoveries varying from 0 to 100% and averaging 65% (14.4 m cut, 9.3 m recovered) for the seven core runs in EE-2. At Rosemanowes, 26 m of core was cut with 20 m recovered in ten runs for a 75% overall recovery from RH11 and RH12.

(f) HIGH TEMPERATURE PACKERS

Most conventional packers used for hydraulic isolation and/or fracturing operations in oil field service do not require high temperature capability. One notable exception is the type of packer used in a steam flooding enhanced oil recovery operation. The Guiberson Division of Dresser Industries, under contract to Los Alamos, modified their steam injection, compression-type

Table 11.2. A summary of directional drilling equipment. Performance for Phase II Drilling at Rosemanowes and Fenton Hill Data Sources: Batchelor and Beswick (1982)—RH11/RH12, Helmich *et al.* (1982)—EE-2, Rowley and Cardin (1982)—EE-3

Downhole motors and turbodrills

Hole use	Type	Diameter	Temperature rating	Length	Supplier	Symbol
EE-2	Positive displacement ($\frac{3}{4}$ lobe)	17.1 cm ($6\frac{3}{4}$ in.)	175°C (350°F)	6.0 m (19.9 ft.)	Baker Service Tools Houston, Texas	BPDM
EE-2/EE-3	Positive displacement ($\frac{1}{4}$ lobe)	19.7 cm ($7\frac{3}{4}$ in.)	155°C (310°F)	6.0 m (19.9 ft.)	Dyna-Drill, Smith International, Irvine California	DDPDM
EE-3	Positive displacement ($\frac{1}{4}$ lobe)	20.3 cm (8 in.)	160°C (320°F)	—	Navi-Drill, Christensen Tool, Houston, Texas	NPDM
RH11/RH12	Positive displacement ($\frac{9}{10}$ lobe)	17.2 cm ($6\frac{3}{4}$ in.) 19.7 cm ($7\frac{3}{4}$ in.)	155°C (310°F)	—	Drilex, Aberdeen, Scotland	DXPDM
EE-2/EE-3	Turbine	19.7 cm ($7\frac{3}{4}$ in.) 13.7 cm ($5\frac{3}{8}$ in)	320°C (610°F)	6.3 m (20.7 ft)	Maurer Engineering, Inc., Houston, Texas	ME17 ME15

Performance comparison

Hole use	Type symbol	Runs drilled	Average length driller per run (m)	Average run period (min)	Average penetration rate (m/hr)
EE-2/EE-3	BPDM	8	22.4	10.9	2.0
EE-2/EE-3	DDPDM	14	17.6	3.1	5.7
EE-3	NPDM	9	19.0	4.0	4.5
RH11/RH12	DXPDM	27	60.4	11.3	5.4
EE-2/EE-3	ME17	35	16.2	3.0	5.5
EE-2/EE-3	ME15	5	18.1	2.1	8.6

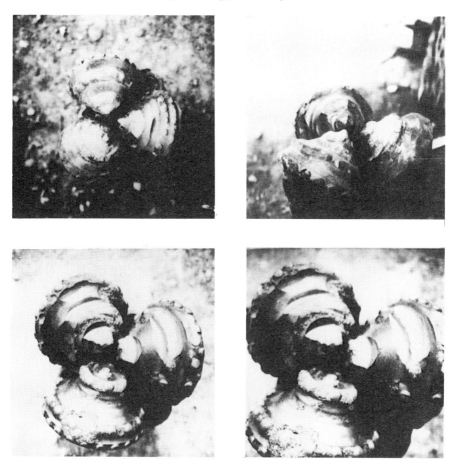

Fig. 11.13 Views of four $8\frac{3}{4}$ in. tungsten carbide insert bits with severe wear due to turbodrill overspeed with insufficient bit loading. (From Rowley and Carden, 1982).

packer that has an asbestos core for hot hole service in HDR wells. As seen in Fig. 11.8, the temperature capability of this packer is well over 400°C based on bench-scale testing. In a separate development, Los Alamos contracted with Lynes to construct a high temperature inflation packer using a thermally stable EPDM type polymer base for the packer element. (see Fig. 11.16). In addition to high temperature stability, the packer must hydraulically isolate a section of the borehole to allow for high injection pressures above the surrounding conditions. For example, pressures may be as high as 350 to 700 bar (5000 to 10 000 psi) above the hydrostatic head during hydraulic fracturing treatments.

For HDR applications not using explosive initiation, packers that seal against a rough open borehole wall rather than against the smoother steel casing

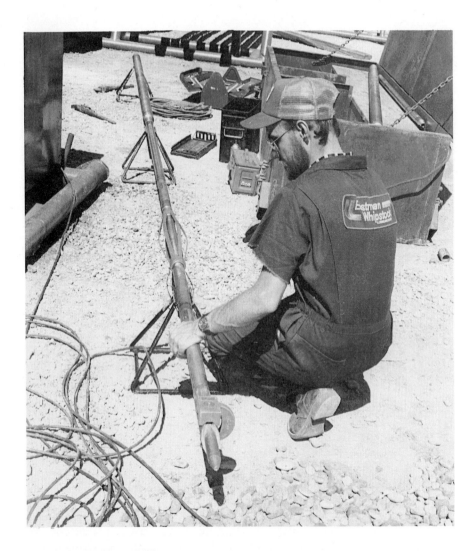

Fig. 11.14 Eastman Whipstock direction drilling steering tool being prepared for use at Fenton Hill. (Courtesy of R. Pettitt, Los Alamos National Laboratory.)

are needed for fracturing. An inflatable packer which uses fluid pressure to expand its soft elastomeric element against the borehole is potentially better suited for open hole operation than the compression-type which uses rotation of the drill string to activate a mechanical setting mechanism to seal the packer against the wall. In either case, the potential for slippage and leakage around the element exists as the isolated section is pressurized. This can be minimized by

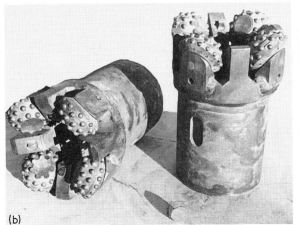

Fig. 11.15 Coring bits used in HDR wells at Fenton Hill and Rosemanowes. (From Batchelor and Beswick, 1982). (a) Christensen $8\frac{15}{32} \times 4$ in. diamond impregnated corebit. (b) Used hybrid 4-cone Stratapax Smith Oil Tool corebits. (c) and (d) Security six-cone tungsten-carbide coring bits.

using the packers in a straddle-arrangement where multiple-packer elements are placed above and below the zone to be isolated. Tests at Fenton Hill with open hole compression and inflation packers have only seen limited success: casing packers on the other hand have worked better.

Fig. 11.16 High temperature Lynes inflation elastomer-based packers made of EPDM being prepared for a fracturing string assembly. (Courtesy of R. Pettitt, Los Alamos National Laboratory.)

The alternative solution to removable packers, which have difficulty performing in hot wellbores, is cementing steel liners in place. This was necessary in some situations at Fenton Hill for fracturing and circulation tests. Typically a 100–300 m section of steel casing containing a so-called polished bore receptable (PBR) at the top end is cemented in place. A smaller diameter piping string is attached to the PBR to isolate the section of the hole above the liner. Pressurization or circulation can then take place into the zone below the steel liner. Some regard this method of borehole isolation as more foolproof than using packers. Unfortunately, the cemented liner is more or less permanent, as a very costly 'wash-over' or milling operation would be required to remove it.

11.2.3   Fenton Hill experience at high temperatures in granite

The early history at Fenton Hill involved drilling three holes: GT-1, an exploratory hole in Barley Canyon about 2.5 km from Fenton Hill, that just penetrated into the granite at approximately 700 m depth (see Fig. 9.20); GT-2 and its sidetracked sections GT-2A and GT-2B, the production hole of the Phase I reservoir, and EE-1 the injection hole of the Phase I reservoir (see Fig. 9.21). An accurate account of the drilling operations for these holes is given by Pettitt (1975, 1977) and is not reported here other than to say that a considerable amount of technology development was motivated by the failure of hardware (including casing cement) or of improper techniques for drilling and servicing HDR wells in hot granite. Of more importance to the development of commercial-sized HDR systems are the drilling results from Phase II which involved EE-2 and EE-3 as deep, inclined holes vertically separated to define the reservoir depicted in Fig. 11.17. A detailed drilling plan for EE-2 and EE-3 was formulated. Helmick *et al.* (1982) and Rowley and Carden (1982) summarize the plan's strategic points as:

(1) Drill the 800 m (2500 ft) section of volcanic and sedimentary rocks while drilling the lost-circulation zones without returns of cuttings as far as feasible and run and cement the 34.0 cm (13° in.) casing about 50 m (150 ft) into the granite. After cementing the bottom section, this casing string would be tensioned to guard against thermal buckling of the casing upon heat-up during production.

(2) The kickoff point (KOP) for directional drilling of the 31.1 cm ($12\frac{1}{4}$ in.) diameter borehole was to be at about 2000 m (6500 ft) where the inclination would be raised and the azimuthal angle started in a northeasterly direction.

(3) Directional drilling of the 31.1 cm ($12\frac{1}{4}$ in.) diameter hole would continue until an inclination angle of $35° \pm 1°$ has been established and an azimuth angle of N55° $\pm 5°$E, at a MD of about 3124 m (10 250 ft), [a true vertical depth of 3109 m (10 200 ft)] at a vertical distance of 370 ± 15 m (1200

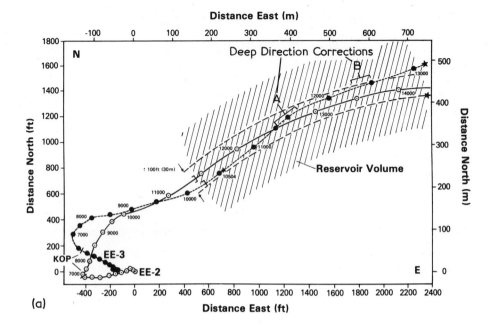

(a)

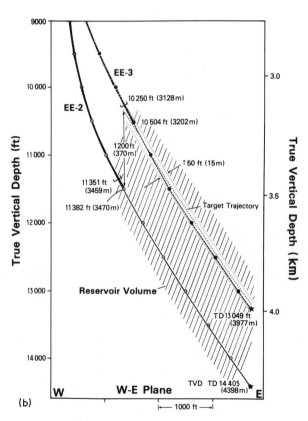

(b)

± 50 ft) directly above the EE-2 hole diameter reduction point. This would place the EE-3 wellbore in such a position and orientation that the tangent to the trajectory would be parallel to the tangent to the EE-2 wellbore trajectory 370 m (1200 ft) below.

(4) Drill a 25.1 cm ($9\frac{7}{8}$ in.) diameter transition hole about 6.1 m (20 ft) long.

(5) Drill reduced hole diameter with a 22.2 cm ($8\frac{3}{4}$ in.) diameter bit, and use packed-hole (stiff) inclination-angle-holding assemblies to maintain hole trajectory.

(6) Take single-shot magnetic surveys every 20 m (60 ft) and apply directional (azimuthal) corrections by use of high-temperature turbodrill runs as required to maintain the EE-3 borehole path parallel to EE-2 borehole to within the ± 30 m (± 100 ft) horizontal tolerances as listed above, and drill to total depth.

(7) The well would be cased with 24.4 cm ($9\frac{5}{8}$ in.) diameter production casing to the bottom of the 31.1 cm ($12\frac{1}{4}$ in.) hole section after reaching the planned total depth for each borehole. Production casing string would be run, cemented, and tensioned to prevent excessive thermal extension and stresses at the wellhead during normal production conditions.

The history of drilling EE-2 is summarized in Fig. 11.18 for the entire drilling period. Figure 11.19 shows the distribution of activities throughout the campaign. As can be seen, only 15% of the time was spent actually drilling while a major disaster involving a collapsed section of casing and other problems (twist-offs and fishing) occupied 34.5% of the total. The final casing schematic for EE-2 is given in Fig. 11.20. Readers interested in the actual day-to-day chronology of the drilling operation should consult Helmick *et al.* (1982) for details; we are only attempting to summarize their major points.

Many of the problems that appeared during the 410 days required to drill and complete EE-2 were not directly related either to the directional aspects of the hole or to the high temperatures involved. The collapsed casing was at the transition between the sediments and the Precambrian section and was more due to poor field procedures rather than to design error. Several of the drill string twist-offs and washouts were apparently just due to bad sections of pipe damaged before they reached Fenton Hill. Other causes include the possibility of stress corrosion initiated by dissolved H_2S found in the drilling fluid. These sulfides were not observed in the shallower drilling of EE-1 and

---

Fig. 11.17 (a) Plan view of EE-2/EE-3 well trajectories and tolerance band for EE-3 directional drilling. *Note:* depths are TVD. (b) EE-2 and EE-3 trajectories projected into east-west vertical plane showing EE-3 target inclination and vertical-separation target tolerance. (Rowley and Carden, 1982.)
* Borehole diameter reduction points.

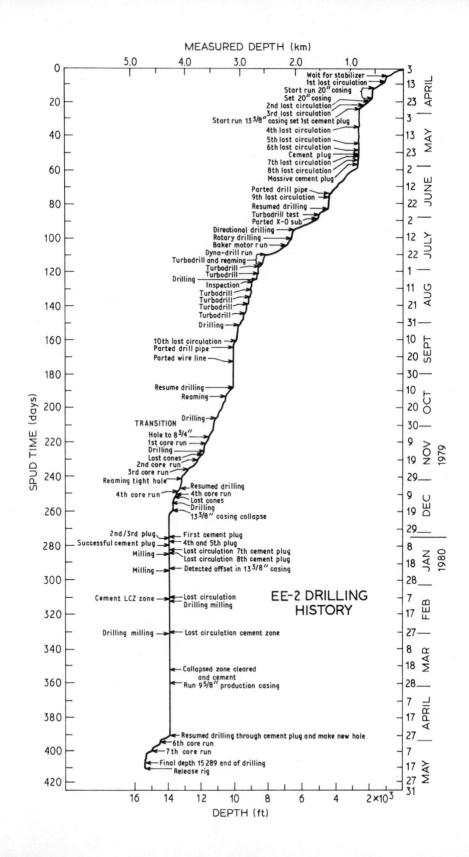

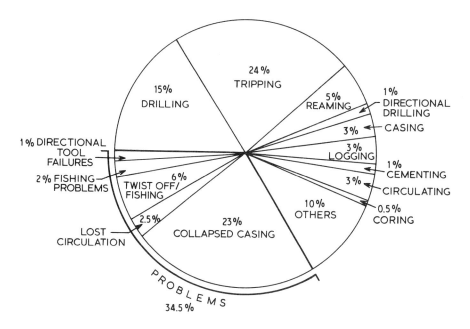

Fig. 11.19 Distribution of activities during the drilling EE-2. (Helmick *et al.*, 1982).

GT-2 and may have been connected with hydrothermal activity in the nearby Valles Caldera. EE-2 is deeper and presumably closer to the magma chamber. Helmick *et al.* (1982) in documenting the EE-2 drilling campaign, recommend stringent administrative controls to minimize risks in future drilling operations.

As reported by Rowley and Carden (1982), the EE-3 borehole target trajectory plan was based upon survey results from EE-2 using magnetic and gyroscopic measurements in the active reservoir region of EE-2 as shown in Fig. 11.17. The region consists of a 1000 m (3000 ft) length of borehole drilled at an inclination of 35° and was directionally oriented into the northeast quadrant. Alignment was chosen to place the EE-2 and EE-3 boreholes approximately normal to the predicted northwest–southeast direction of the plane of the hydraulic fractures of the earlier pair of HDR wells (GT-2/EE-1). The EE-2 borehole was drilled with a stiff or locked-in assembly that holds the inclination angle of the borehole but allows the azimuth to drift somewhat requiring downhole motor or turbodrill runs to make directional corrections (see Fig. 11.10(a)).

Fig. 11.18 EE-2 drilling history. (Helmick *et al.*, 1982).

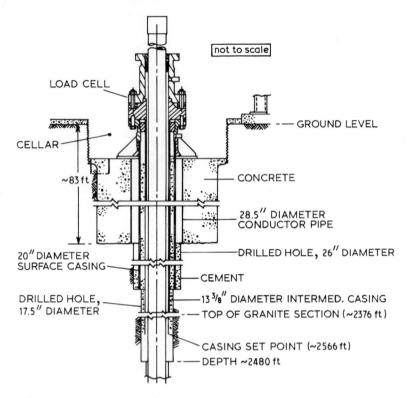

Fig. 11.20 EE-2 casing design schematic and pipe joint talley. (Helmick *et al.*, 1982).

The wellbore directional-drilling parameters set as targets for the reservoir portion of EE-3 were:

(1) 370 m ±15 m (1200 ±50 ft) vertically above EE-2,
(2) ±30 m (±100 ft) lateral deviations from the horizontal projection of the EE-2 trajectory,
(3) total depth (TD) at true vertical depth (TVD) of 4030 m (13 230 ft); directly above the TD of the EE-2 wellbore at a TVD of 4398 m (14 405 ft).

These objectives were met within practical drilling limits and financial constraints. The maximum deviations observed were a slightly greater vertical spacing of 400 m (1300 ft) over the upper few meters of the reservoir, decreasing to within 370 m ±15 m (1200 ±50 ft) below. The widest lateral deviation occurred at the bottom of EE-3 at 4247 m (13 933 ft) as measured

---

Fig. 11.21 EE-3 drilling history. (Rowley and Carden, 1982).

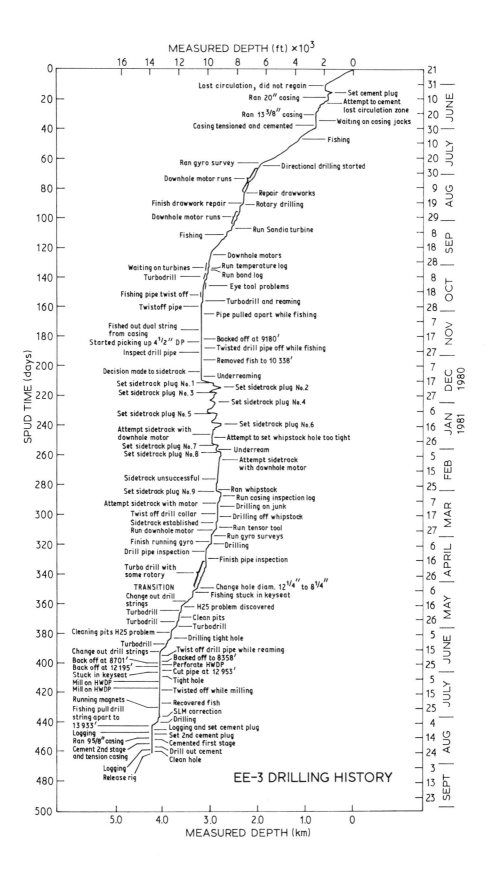

MEASURED DEPTH (ft) ×10³

Lost circulation, did not regain	Set cement plug
Ran 20" casing	Attempt to cement lost circulation zone
Ran 13³/₈" casing	Waiting on casing jacks
Casing tensioned and cemented	
	Fishing
Ran gyro survey	Directional drilling started
Downhole motor runs	
	Repair drawworks
Finish drawwork repair	Rotary drilling
Downhole motor runs	
	Run Sandia turbine
Fishing	
	Downhole motors
Waiting on turbines	Run temperature log
Turbodrill	Run bond log
	Eye tool problems
Fishing pipe twist off	Turbodrill and reaming
Twistoff pipe	Pipe pulled apart while fishing
Fished out dual string from casing	Backed off at 9180'
Started picking up 4¹/₂" DP	Twisted drill pipe off while fishing
Inspect drill pipe	Removed fish to 10 338'
Decision made to sidetrack	Underreaming
Set sidetrack plug No.1	Set sidetrack plug No.2
Set sidetrack plug No.3	Set sidetrack plug No.4
Set sidetrack plug No.5	Set sidetrack plug No.6
Attempt sidetrack with downhole motor	Attempt to set whipstock hole too tight
Set sidetrack plug No.7	Underream
Set sidetrack plug No.8	Attempt sidetrack with downhole motor
Sidetrack unsuccessful	
Set sidetrack plug No.9	Ran whipstock
	Run casing inspection log
Attempt sidetrack with motor	Drilling on junk
Twist off drill collar	Drilling off whipstock
Sidetrack established	Run tensor tool
Run downhole motor	Run gyro surveys
Finish running gyro	Drilling
Drill pipe inspection	
	Finish pipe inspection
Turbo drill with some rotary	
TRANSITION	Change hole diam. 12¹/₄" to 8¹/₄"
Change out drill strings	Fishing stuck in keyseat
Turbodrill	H25 problem discovered
Turbodrill	Clean pits
	Turbodrill
Cleaning pits H2S problem	Drilling tight hole
Turbodrill	
Change out drill strings	Twist off drill pipe while reaming
Back off at 8701'	Backed off to 8358'
Back off at 12 195'	Perforate HWDP
Stuck in keyseat	Cut pipe at 12 953'
Mill on HWDP	Tight hole
Mill on HWDP	Twisted off while milling
Running magnets	Recovered fish
Fishing pull drill string apart to 13 933'	SLM correction
	Drilling
Logging	Logging and set cement plug
Ran 95/8" casing	Set 2nd cement plug
Cement 2nd stage and tension casing	Cemented first stage
	Drill out cement
Logging	Clean hole
Release rig	

EE-3 DRILLING HISTORY

MEASURED DEPTH (km)

Dates (right axis):
21, 31 / 10 / 20 / 30 — JUNE
10 / 20 / 30 — JULY
9 / 19 / 29 — AUG
8 / 18 / 28 — SEP
8 / 18 / 28 — OCT
7 / 17 / 27 — NOV 1980
7 / 17 / 27 — DEC 1980
6 / 16 / 26 — JAN 1981
5 / 15 / 25 — FEB
7 / 17 / 27 — MAR
6 / 16 / 26 — APRIL
6 / 16 / 26 — MAY
5 / 15 / 25 — JUNE
5 / 15 / 25 — JULY
4 / 14 / 24 — AUG
3 / 13 / 23 — SEPT

along the well, or 3.977 km (13 048 ft) TVD, where EE-3 had drifted to the
north slightly and indicated a position displaced about 60 m (180 ft) north as
evaluated from the magnetic single-shot survey data. The final relation of EE-3
above EE-2 is depicted in Fig. 11.17, which illustrates the magnitudes of the
deviations from the targets that were achieved with the controlled-trajectory
drilling of EE-3.

   As indicated by the EE-3 drilling history in Fig. 11.21 and the distribution of
activities in Fig. 11.22, fate gave the EE-3 campaign even more trouble than
was encountered in EE-2. Two severe drill-pipe twist-offs and subsequent
fishing operations occupied 22.2% of a total of 461 days required to drill and
complete EE-3. Sidetracking problems were also encountered after a portion of
the hole had to be abandoned when a major (44 day) fishing operation to
recover a twisted-off BHA was unsuccessful. After numerous turbodrill,
downhole motor, and under reaming attempts, frequently with improperly set
cement plugs, the sidetracking eventually had to be done with a cemented-in-
place whipstock (Fig. 11.11) with the entire operation requiring 21.2% of the
total time. In this case only 12.1% of the time was spent drilling, and over 50%
spent on extraordinary problems. The final casing design for EE-3 is given in
Fig. 11.23.

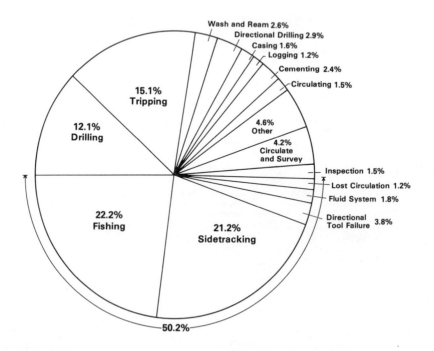

Fig. 11.22 Distribution of activities during the drilling EE-3. (Rowley and Carden,
1982).

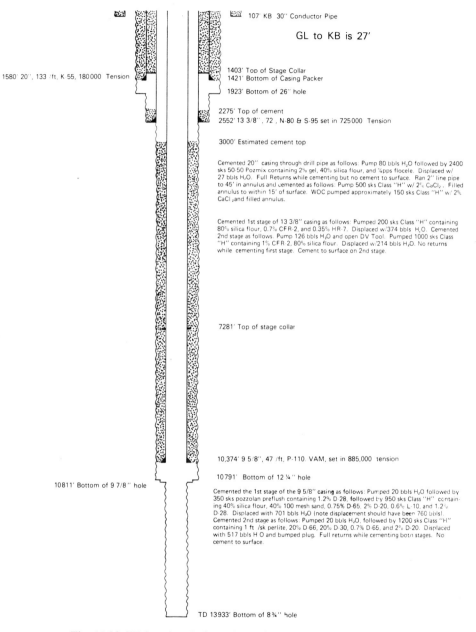

Fig. 11.23 EE-3 casing design schematic. (Rowley and Carden, 1982).

Even though these two holes were successfully completed, it might be concluded that the extended rig time and extra services required for fishing, sidetracking, pipe inspection, etc. were so costly that total costs for completing EE-2 and EE-3 would be excessively out of line for drilling at these depths. Amazingly, this turns out not to be the case as will be discussed in Section 11.3.

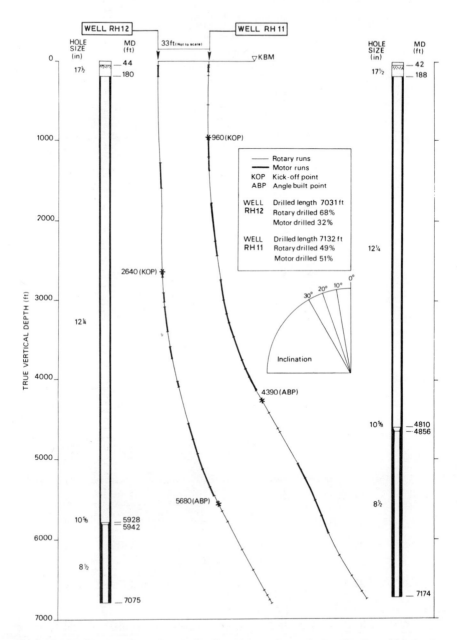

Fig. 11.24 Wellbore trajectories and casing design schematic for the Phase II system at Rosemanowes. (Batchelor and Beswick, 1982).

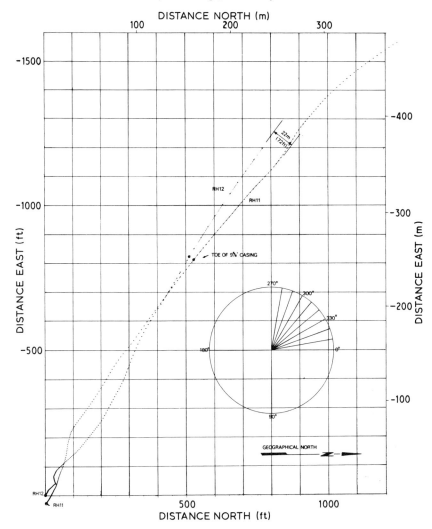

Fig. 11.25 Plan view of well trajectories at Rosemanowes. (Batchelor and Beswick, 1982).

11.2.4 Rosemanowes experience at lower temperature in granite

To parallel our discussion of drilling at Fenton Hill, the wellbore trajectories for RH11 and RH12 in the Phase II system at Rosemanowes are shown in Fig. 11. 24 along with a casing design schematic of the completed wells. A plan view is given in Fig. 11.25 while the distribution of activities are shown in Fig. 11.26. A chronological drilling history is summarized in Table 11.3.

The total drilling time required for both holes was short, only 74 days for RH12 from 18 August 1981 through 30 October 1981 and 50 more days for RH11 until 23 December 1981. From Fig. 11.26 it is obvious that drilling

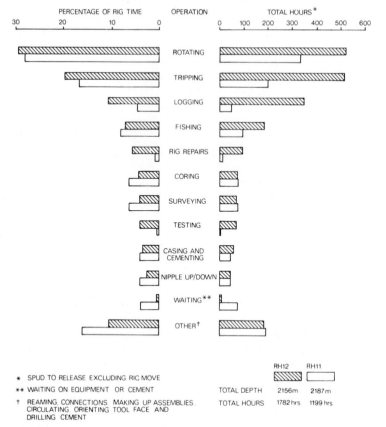

Fig. 11.26 Distribution of activities during the drilling of RH11 and RH12 at Rosemanowes. (Batchelor and Beswick, 1982).

RH11 and RH12 was enormously easier than either EE-2 or EE-3. Almost 30% for RH11 and RH12, as opposed to 12 or 15% for EE-2 and EE-3, of the total time was spent in actually drilling, while all problems occupied less than 20% of the total. Directional drilling was easier. In fact, as indicated by Fig. 11.27, estimates of 159 days for RH12 and 114 days for RH11 would be projected based on EE-2/EE-3 experience. These were 60% or more higher than the projected time estimates of 120 days for RH12 and 89 days for RH11 originally made by the Camborne team .

11.2.5 Summary of HDR drilling experience to date

One should not be left with the impression that the drilling successes at Rosemanowes were by mere chance. Careful planning, taking particular note of the experience at Fenton Hill, certainly played a major role in the successful drilling campaigns at Rosemanowes. There are, of course, other reasons why drilling was much more arduous at Fenton Hill.

Table 11.3. Drilling history for RH11 and RH12 at Rosemanowes; August-December 1981 (Batchelor and Beswick, 1982)

Date		Hole RH12	Depth (ft)	Day no.
August	10–12	Rig arrival		
	10–13	Rig erection		
	17	Derrick up		
	18	Spud in		1
	20	$17\frac{1}{2}$ in. section complete	188	3
	21	Reaming, casing, cementing wellhead assembly on		4
	22	Welding, heat treatment, etc. $12\frac{1}{4}$ in. section started		5
	23	Drawworks motor burnt out	351	6
	27	Resumed drilling		10
	31	Downhole motor in use		14
September	1	Downhole motor casing broke	1775	15
	1–5	Fishing		15–19
	5	Reaming run		19
	6	Resumed drilling		20
	10	Coring run	2640	24
	11	Resumed drilling, first deviation run		25
	15	Logging run	3035	29
	17	Deepest hole in Cornwall		31
	20	Inclination 11.3°, heading 315°	3740	34
	21	Logging—calliper, sonic, seisviewer		35
	23	Downhole TV, *in situ* stress	3753	37
	25	Resumed drilling rotary table		39
	30	Inclination 18°	4510	44
October	2	Temperature 60°C	4789	46
	5	One mile hole length	5280	49
	9	Angle build completed 30°	5835	53
	10	Ferranti–Eastman inertial navigation tool run and $12\frac{1}{4}$ in. section complete. $10\frac{5}{8}$ in. section	5924	54
	11	$8\frac{1}{2}$ in. section started	5940	55
	13	39 days ahead	6486	57
	14	Coring attempt—corebarrel stuck	6650	58
	15	Reaming run, core run, 6 ft recovered using diamond corebarrel	6926	59
	16	Planned depth reached (6912) (59 days from start—program projection 98 days)		60
	17	Coring run using tricone rock bit		61
	18	Coring run 7 ft recovered		62
	19	Coring run 9 ft 6 ins. recovered	7015	63
	20	Coring run 9 ft 6 ins. recovered. Well finished	7073	64
	21	Logging program 80°C temperature		65
	22	Geophysical logs		66
	24–25	*In situ* stress measurement		68–69
	26	Cement plug at bottom of $12\frac{1}{4}$ in.		70
	27	Casing installed		71
	28	Circulated well to cool slightly		72
	29	Cemented		73
	30	Cement plug drilled out and fish removed		74

Table 11.3—*continued*

Date		Hole RH11	Depth (ft)	Day no.
November	1	Rig skidded		76
	2	Second well started		77
	5	17½ in. section completed, reamed, cased, wellhead assembly welded	188	81
	8	Kick off point reached	1185	83
		Rates of 90 ft/hr achieved		
	11	Reaming well, inclination 5°	1542	86
	15	Fast rates up to 63 ft/hr	2824	90
	16	Gyroscope survey to check course	3050	91
	17	Motor correction run	3230	92
	20	Downhole motor broke, fishing started	3824	95
		FINDS tool run to 2400 ft		
	23	Fish recovered using an overshot assembly	3824	98
		Multishot survey run		
	25	Failure of two downhole motors	3824	100
		Replacement motor flown in from Aberdeen to Culdrose		
	27	Inclination 26°	4007	102
	28	Downhole motor failed due to sanding up	4190	103
December	1	12¼ in section complete, 10⅝ in. transition section drilled	4810	106
	2	Coring run recovered 7 ft 6 in. core	5070	108
		Rotary drilling at 22 ft/hr, second unsuccessful coring run using diamond core bit		
	5	Successful coring run using roller cone bit	5538	110
	7	8½ ft of core recovered	5840	112
	8	12 ft of core recovered	6054	113
		Directional correction assembly run		
	11	2 ft of core recovered using a diamond core bit	6696	116
	12	Final depth reached	7160	117
	13	Geophysical logs run	7160	118
	14	8 ft of core recovered	7172	119
		Cement plug set at 4900 ft		
		Casing run in		
	18	Cement delivered and cementing carried out		123
	19	Top section (300 m) cemented from the surface. Drilling out of cement plug started		124
	22	All cement drilled out of reservoir section		127
		Clear water circulated		
	23	Derrick lowered		128

Aside from the obvious differences in locale between Fenton Hill and Rosemanowes, one might have expected the drilling operations to be more similar. After all, they were both directionally drilled in crystalline granite with

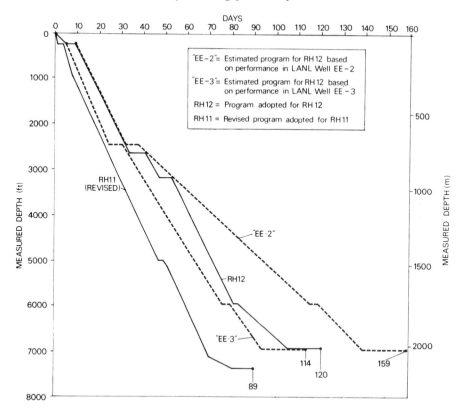

Fig. 11.27 Adopted drilling program for Rosemanowes Phase II compared with estimates based on Los Alamos performance at Fenton Hill.

about the same borehole diameters (see Table 11.4). But there are some real and subtle differences :

(1) The temperatures encountered in EE-2 and EE-3 were as high as 327°C whereas in RH11 and RH12 temperatures seldom rose above 90°C.
(2) The angle of inclination of the RH11 and RH12 bores in the reservoir region was actually only 30° versus 35° for EE-2 and EE-3.
(3) The length of the borehole drilled was roughly 2000 m for RH11 and RH12 versus 4500 m for EE-2 and EE-3.

Because of the lower temperatures, the superior performing Drilex downhole motors could be used for straight drilling as well as directional drilling. In RH12, 32% of the total drilled length of 2143 m was drilled with downhole Drilex motors, while in RH11, 51% of the total length of 2175 m was so drilled. At Fenton Hill, less than 15% of the total drilled length in either EE-2 or EE-3 was achieved with downhole motor drilling and turbodrilling combined.

Table 11.4. HDR well drilling summary

Well	Location	Dates drilled	Program phase	Total drilled length m (ft)	Undisturbed bottom hole temperature °C	Borehole diameter in active reservoir	Hole inclination angle in active reservoir	Time required for drilling and completion days
GT-1	Fenton Hill	1972	0	732 m (2400 ft)	100°C	10.1 cm (4 in.)	0°	40
GT-2	Fenton Hill	Feb.–Oct. 1974**	I	2932 m (9619 ft)	197°C	24.5 cm ($9\frac{5}{8}$ in.)	1–2°	180
GT-2A*	Fenton Hill	April 1977	I	~300m (~1000ft)	190°C	24.5 cm ($9\frac{5}{8}$ in.)	2–5°	60
GT-2B*	Fenton Hill	May 1977	I	~150m (~500ft)	190°C	24.5 cm ($9\frac{5}{8}$ in.)	5–6°	35
EE-1	Fenton Hill	May 1975–Oct. 1975	I	3064 m (10053 ft)	200°C	24.5 cm ($9\frac{5}{8}$ in.)	1–2°	180
EE-2	Fenton Hill	Apr. 1979–May 1980	II	4660 m (15289 ft)	327°C	22.2 cm ($8\frac{3}{4}$ in.)	35°	410
EE-3	Fenton Hill	May 1980–Aug. 1981	II	4250 m (13933 ft)	280°C	22.2 cm ($8\frac{3}{4}$ in.)	35°	461
RH11	Rosemanowes	Nov.–Dec. 1981	II	2175 m (7136 ft)	90°C	21.6 cm ($8\frac{1}{2}$ in.)	30°	50
RH12	Rosemanowes	Aug.–Oct. 1981	II	2143 m (7030 ft)	90°C	21.6 cm ($8\frac{1}{2}$ in.)	30°	74

* Sidetracked portion of GT-2.
** Conducted in two phases.

The large increase in drilled length and the more severe angle of 35° for EE-2 and EE-3 contributed much more difficulty than would be expected either from a linear extrapolation of previous experience in GT-2 and EE-1 at 3000 m or from what was observed in RH11 and RH12 at 2000 m. The temperatures encountered in EE-2 and EE-3 of greater than 300°C were too high to use most of the equipment and hardware currently available (see Fig. 11.8).

In retrospect, what is probably advisable for future HDR drilling using the inclined bore concept would be to reduce the inclination angle in proportion to reservoir depth to keep bending stresses within bounds. Perhaps 25° would have given much better results at 4000 m. Some reservoir volume could be lost but total rig time and costs should fall considerably. Also bottom hole temperatures should be kept within the current limitations of existing borehole logging and drilling equipment capabilities regardless of how attractive it may be to drill just a little deeper. Given that temperature gradients can change as a function of depth, a maximum bottomhole temperature should be the design target for drilling rather than depth.

The last point that needs to be made concerns costs. Total drilling costs should scale with required rig time since rigs are frequently rented on a daily fee basis. Thus, at first glance, one would expect the Cornish wells to be proportionally much more economic than the deep American wells at Fenton Hill. However, costs for both the Fenton Hill and Cornish wells are comparable when factoring in the effect of depth, as will be shown in the next section.

11.3   Drilling Costs

11.3.1   Economic motivation and tradeoffs

An important economic consideration is the fraction of the total costs for an HDR system that is represented by the drilling and well completion costs. Obviously this depends on a number of factors including the end use selected and its capital costs (electricity, process heat, or cogeneration) and well drilling and completion costs expressed as the product of the number of well pairs times the cost per well pair. The number of pairs will be completely determined by plant type, size, and performance. The only remaining factor then is individual well cost which is the topic of the next section.

Many economic studies of HDR and hydrothermal exploitation show that drilling-related costs typically represent somewhere between 50 to 70% of the total capital investment. This will be justified later in Chapter 14, but if for now we accept this observation then there is a clear incentive for reducing drilling costs. These drilling costs can be thought of as being comparable to the annualized 'fuel' costs of a fossil fuel- or nuclear-fired power plant integrated over its lifetime. The financial impact of drilling costs, like other capital investments for plant construction and equipment, however, is felt long before any revenues appear from selling electricity or heat because they must be incurred first, before the plant is built and operational. In this way, drilling costs are distinctly different from true fuel costs which appear as operating costs as they are incurred and the present worth of which is relatively modest at the outset. Nonetheless, on the positive side, geothermal developments are less prone to the hazards of inflation during the course of the plant's operating lifetime because the drilling costs represent a 'one-time' capital investment.

In addition to the motivation to keep individual well costs as low as possible, HDR systems, unlike hydrothermal reservoirs, require that the depth or initial rock temperatures be selected *by design* – not just by what nature has provided. This creates a need to optimize reservoir design temperatures or depths so as to minimize overall costs. A tradeoff exists between the cost reductions due to performance improvements that result from higher temperatures versus the higher costs associated with drilling deeper. As will soon be seen, drilling costs are strongly dependent on depth, approaching an exponential dependence, while performance improvements tend to scale linearly. Thus the economic optimum depth will occur far short of maximum performance. These specific points will be considered further in Chapter 14.

11.3.2 Individual well costs

Because only limited cost data are available for HDR wells, the more extensive cost data base for hydrothermal steam and hot water wells will be considered as a possible basis for interpolation and extrapolation of HDR well costs. Furthermore, any relationship between geothermal well costs and oil and gas wells to similar depths and of similar diameter that could be empirically established would be extremely helpful. Earlier work by the authors has dealt with developing suitable well cost models (see Armstead, 1983, Section 14.10; and Tester, 1982, p. 537ff). In any cost model the following facts need to be considered:

(1) Drilling costs are very sensitive to depth and rock properties;
(2) Many unforeseen difficulties arise in drilling operations even with experience at a site. Fenton Hill provides an excellent example of this. For instance, frequent directional drilling runs to make course corrections or twist-offs of pipe requiring fishing can consume massive amounts of rig time and greatly add to individual wellcosts. A large contingency provision is necessary and substantial scatter among costs for wells of similar depth and diameter in the same formation is to be expected.
(3) Early drilling at a site for exploration or engineering development purposes is more expensive, possibly by a factor of two or three over production holes in a commercially-mature HDR operation. This is partly due to coring costs, logging runs, and fracturing tests.
(4) Site location can be a major factor: the more remotely located the site, the higher costs are. This would include sites with difficult climates, high

altitude or being far removed from an experienced labor force and the necessary service operations for cementing, pipe inspection, fishing, geophysical logging, etc. The laws of supply and demand must also be closely followed. An extreme example would be drilling on the North slope in Alaska, where drilling costs can be as much as ten times higher than for the same well on the US mid-continent. Offshore drilling operations also have similar problems.

When we combine these factors with the fact that drilling and completion costs are the largest single cost component in HDR development and are largely incurred before any revenues from power generation appear, the risk attached to any HDR project would be perceived to be high. Therefore, any drilling cost estimate must be interpreted not only in the context of its impact on financing HDR development worldwide but also to allow for the fact that individual well costs are potentially subject to large variations from predicted values because of the contingencies involved.

Based on our previous work, we would suggest that a two-term empirical equation would represent drilling costs better than a single-term depth dependent approach. The general form for total completed well cost (C_{well}) might look like:

$$C_{well} = C_m + C_d \exp (A_z \cdot Z) \tag{11.1}$$

where

C_m = rig mobilization/demobilization costs
A_z, C_d = empirical constants to represent drilling and
  casing portion of costs.
Z = depth of the well

Equation (11.1) can be regarded as quasi-exponential to represent the strong depth dependence cited earlier and its form is consistent with earlier studies of HDR drilling costs (Milora and Tester, 1976; and Garnish, 1976). It would also be true that mobilization or setting-up costs would depend on depth, since drilling deep wells requires a larger rig which would cost more to transport and set up. The constants A_z and C_d would also depend on rock type, especially rock hardness, and the presence of unstable formations as they would influence bit performance and overall penetration rates. All three constants may depend on site location to some degree. The choice of an exponential term is somewhat arbitrary but mathematically convenient. Other forms of polynominals in depth could be substituted, but in view of the uncertainties involved would offer no advantage. Figures 11.28 and 11.29 compare drilling costs for US commercial and US DOE funded hydrothermal wells from Carsen and Lin (1981) with values estimated for four wells at Fenton Hill (GT-2, EE-1, EE-2, and EE-3), one exploratory well in Barley Canyon about 2.5 km from Fenton Hill (GT-1) and the three deep wells at Rosemanowes (RH11, RH12 and RH15). All costs shown have been corrected to 1984 US dollars using Carsen and Lin's (1981) recommendation for US-based wells of 17% yearly inflation for the period 1970–1982, and our estimate of 10% from 1982 to 1984. Garde-Hansen (1981) in a separate study indicated an inflation rate of 15%, per year from 1970 to 1979 for oil and gas well drilling costs. This average rate of 17 to 15% includes three years of modest inflation before the 1973 Middle-east oil embargo. For the UK-based wells, adjustments were made following recommendations by A.S. Batchelor (1985). Although a direct conversion of British pounds sterling to US dollars is questionable given the variable exchange rate, he has assumed an appropriate conversion rate of approximately two US dollars to one pound sterling which was the exchange rate in 1981.

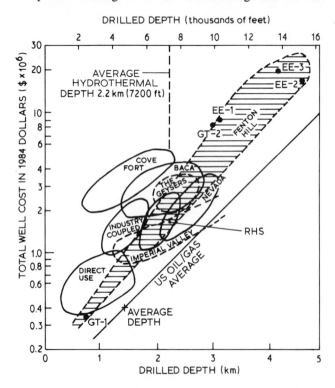

Fig. 11.28 Comparison of US commercial hydrothermal well and Fenton Hill well drilling costs. (Adapted from Rowley and Carden, 1982). Costs updated to 1984 US dollars.
RHS=Roosevelt Hot Springs.

Several things are apparent from Figs. 11.28 and 11.29. First of all, costs are subject to large uncertainties and the fact that they are correlated by a straight line on a semi-logarithmic plot is not unequivocal proof that an exponential depth dependence is correct. Nevertheless, it certainly seems reasonable given the quasi-exponential trend of the massive number of oil and gas well statistics

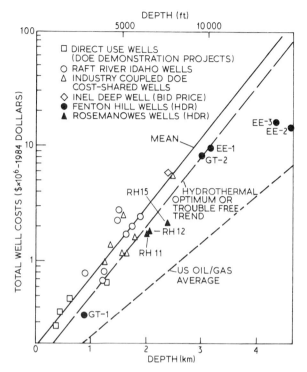

Fig. 11.29 Comparison of US DOE-sponsored hydrothermal well costs to HDR well costs at Fenton Hill and Rosemanowes. (Data from Rowley and Carden, 1982; Batchelor and Beswick, 1982, and Murphy *et al.*, 1982). Costs updated to 1984 US dollars.

that comprise the average oil and gas well line. By using ten years or so of data provided by the Joint Association Surveys (1974–1984), literally thousands of well costs over a wide range of depths are included in the correlation. Hydrothermal drilling costs for the commercial US wells (Fig. 11.28) and the DOE funded wells (Fig. 11.29) range from two to four times those for an average oil or gas well to the same depth. These higher costs for hydrothermal wells are due to high temperatures, corrosive and erosive fluids, lost circulation problems and generally more difficult drilling in harder, more fractured rock. The trend of the Imperial Valley costs to approach the oil and gas average line results from the similarity of the sedimentary rock formations encountered there to those found in many oil and gas wells. The large spread of hydrothermal well costs reflect the uncertainties and contingencies cited earlier.

HDR well costs based on Fenton Hill and Rosemanowes experience fall below the mean cost line for the other USDOE funded hydrothermal system wells shown in Fig. 11.29.

The cost data for HDR wells GT-2, EE-1, EE-2, and EE-3 are summarized in Table 11.5. Both the costs at the time of completion, as well as normalized 1981 costs (inflated at 17% again), are shown. A striking effect of 17% annual

Table 11.5. Drilling and completion costs for Fenton Hills wells (from Murphy et al., 1982)

Well	Drilling time (months)	Completion date	Total depth along the wellbore		Actual cost, millions of dollars		Oil/gas average cost,* millions of 1981 dollars	Ratio, 1981 actual cost to oil/gas average	Learning and disaster free cost, millions of 1981 dollars	Ratio, learning and disaster free cost to oil/gas average	Major disaster events
			km	feet	At comp. time	Escalated to 1981†					
GT-2	8	10/74	2.93	9620	1.9	5.7	0.94	6.1	3.3	3.5	Stuck drill pipe, washover required.
EE-1	5	10/75	3.06	10050	2.3	5.9	1.1	5.4	3.2	2.9	Experiments at 6500 ft, surveying experiments.
EE-2	13	5/80	4.66	15290	7.3	8.5	3.6	2.3	6.3	1.8	Collapsed casing.
EE-3	15	8/81	4.25	13930	11.5	11.5	2.8	4.1	6.9	2.5	Major fishing job, and sidetracking.
							Average, all wells=4.5		Average, all wells=2.7		
							Average, EE-2+EE-3=3.2		Average EE-2+EE-3=2.2		

* Based on Joint Association Survey data (1974–1981).
† Drilling cost escalation taken as 17% per year.

inflation is seen with the costs of GT-2 and EE-1 at the times of their completions nearly tripling at today's drilling costs. For each well we also present the cost, in 1981 dollars, for the average oil and gas well drilled to the same depth based on the Joint Association Survey (1974–1984) tabulations. A consequence of the nearly exponential cost-depth relationship with 17% inflation is that EE-2 and EE-3 were actually less expensive than GT-2 or EE-1 when compared on equivalent time and depth bases. To see this more clearly, refer to the heading, 'Ratio of 1981 actual cost to oil/gas average' in Table 11.5. Wells GT-2 and EE-1 cost about five times the oil/gas average, whereas EE-3 costs four times the average, and EE-2 costs only two times the average. Thus, drilling has improved significantly at Fenton Hill, in the sense that HDR well costs are approaching those of oil/gas average costs. This fact is even more apparent when one recalls that GT-2 was drilled nearly vertically, with minimal directional control, and that EE-1 was directionally drilled only for the bottom 150 m (500 ft), at a maximum deviation of 4° from the vertical, and rather inaccurately at that. In contrast, EE-2 and EE-3 were directionally drilled with closer tolerances to an angle of 35° from the vertical for the bottom 2.3 to 2.7 km (7500 to 8800 ft). This convergence of HDR and oil and gas well costs was predicted in a Republic Geothermal, Inc. (RGI) study by Nicholson *et al.* (1979). Their study suggests several aspects of deep HDR well drilling that may result in economic advantages over oil and gas wells to the same depth. Well costs from their model, shown in Fig. 11.30, vary linearly with depth from 3 to 6 km and show slightly higher costs than the exponential model of Fig. 11.29 for depths less than 3 km. Above 4.5 km the model indicates a lower well cost.

The actual reasons for this lower predicted cost behavior of the RGI model center around expected bit penetration rate and lifetime and the absence of drilling mud and hole stability problems critical to very deep oil and gas wells in sedimentary formations. The RGI study predicted 'normal' bit performance in deep environments (2–5 m/h, 70–100 m lifetime with rotary drilling) for HDR wells, whereas oil and gas well drilling in deep systems typically show a strong decline in penetration rate. Because HDR wells in crystalline rock will be drilled with water rather than mud, several million dollars are saved by avoiding elaborate mud programs. Furthermore, the six deep wells (3–5 km) drilled so far have been very stable, requiring a much more modest casing program than that for sedimentary wells of similar depth.

Improvements of bit performance could have a very large impact on reducing geothermal development costs (Newson *et al.*, 1977). For directional as well as conventional drilling, the use of high-temperature downhole turbodrills may result in reduced costs (Maurer, 1979). Further field testing will be required to demonstrate this, however.

In addition to drilling, methods used to produce the fluid may require significant capital expenditure. For example, in the hot dry rock case, artificial stimulation may be costly, involving complex fracturing techniques to produce

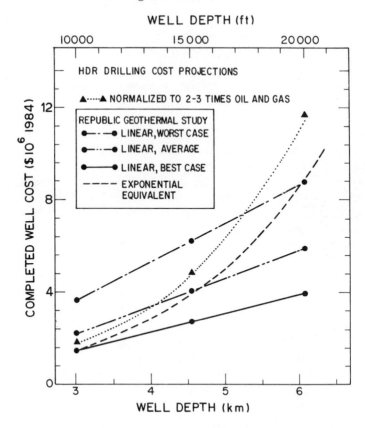

Fig. 11.30 Comparison of the Republic Geothermal, Inc. drilling cost model. (Nicholson *et al.*, 1979) and the exponential, oil and gas based, geothermal well cost model for HDR wells in crystalline basement formations of low permeability.

multiple-fracture systems with proper flow control. This will be required to provide uniform flow distribution through the fractured reservoir. Costs associated with stimulation will be discussed in Chapter 14.

HDR drilling experience to date has had comparable costs to hydrothermal drilling despite harder formations and stricter directional drilling requirements. HDR drilling is, however, far from being commercially mature, implying that future improvements must be vigorously sought. Refer again to Table 11.5 to the column headed 'Learning and disaster free costs'. These costs are the extrapolated 1981 costs, from which have been subtracted those costs due to delays for experiments and 'disasters'. According to Murphy *et al.* (1982) these were *not* the same as 'trouble free' costs. Drilling wells will always have the usual, unavoidable troubles, but in estimating HDR well costs, they suggest subtracting costs due to disasters as well as to experiments that one might reasonably expect to avoid as drilling matures and the number of wells at the site increases. The costs of these disasters and experiments that need not be repeated have been identified with the help of Don Brown and John Rowley, both of Los Alamos National Laboratory. As examples, for GT-2, costs have been subtracted for the continuous coring experiments and for the stuck pipe and subsequent washover effort. For EE-1, the cost of 26 days of experiments at 2 km and the cost of excessive rig time lost in locating the bottom of the hole in relationship to GT-2 – an art now mastered with EE-2 and EE-3 – have been subtracted. For EE-2, the costs due to the casing collapse have been deleted from the total. For EE-3, the cost due to the prolonged fishing job and subsequent sidetracking has been removed. Murphy *et al.* (1982) did *not* subtract the costs of more typical troubles such as losses of circulation, twistoffs and the more usual fishing jobs, breached casings, and directional drill motor and tool failures. Nor, of course, did they subtract costs of reaming, cementing, circulating, inspection, logging, and casing as they were considered normal required operations.

The situation in Cornwall at Rosemanowes is somewhat different from that at Fenton Hill. First of all, the Rosemanowes site would be considered very remote by US standards since in many cases services had to come from Aberdeen, Scotland, some 700 miles (1100 km) away, or from Europe. The rig mobilization and daily rates were much higher than at Fenton Hill (see Table 11.6 for a cost breakdown). Even with this, normalized costs for the 2000 m holes RH11 and RH12 in US dollars correlate with Fenton Hill well costs on the optimum or trouble free trend line for hydrothermal wells shown in Fig. 11.29. This agreement is encouraging as it strengthens our predictive capability, but it may be fortuitous. As Batchelor and Beswick (1982) point out, the more massive scale of drilling activities in the continental US coupled to the high costs of North Sea offshore drilling makes a direct comparison of US and UK costs difficult.

At this point, we can summarize by saying that HDR well costs using existing conventional and directional rotary drilling technology are likely on the average to be two to four times more expensive than comparable oil and gas wells drilled to depths over 3 km. This provides a clear economic incentive to explore possible drilling alternatives to rotary methods.

It is important to recognize that the quest for deeper penetration per dollar of expenditure is not to be regarded as 'scraping the bottom of the barrel' for the sake of some minimal economic gain. The potential benefits from quite a small incremental improvement in the depth penetration per dollar are disproportionately greater than might at first be expected. For example, a gain of 100 metres of additional penetration at 3 km depth without incremental cost would be a mere 3.3% penetration gain, but about 17% gain in restricted crustal heat content above 85°C in a typical thermal area as shown in Fig. 4.8. Moreover, this gained heat would be at a slightly higher grade (by about 3.8°C maximum) than any of that of the restricted crustal heat to 3 km depth.

Table 11.6. Overall cost summary for the RH11 and RH12 2000 m doublet at Rosemanowes in 1981 cost figures*† (from Batchelor and Beswick, 1982)

Cost group	Allocated to drilling (£K)	Allocated* elsewhere (£K)	Total (£K)	%
1 Civil engineering	—	108	108	3.6
2 Rig moves	241		241	8.1
3 Payments to contractor	642		642	21.6
4 Directional drilling	149		149	5.0
5 Surveying	41		41	1.4
6 Downhole motors	126		126	4.3
7 Tool rental	87		87	2.9
8 CSM equipment purchases	22		22	0.7
9 Stabilization and reamers	92		92	3.1
10 Rock bits	188		188	6.3
11 Drilling muds	24		24	0.8
12 Casing	283		283	9.5
13 Wellheads	19		19	0.7
14 Cementing	74		74	2.5
15 Christmas trees	—	106	106	3.6
16 Logging	16	96	112	3.8
17 Coring	106		106	3.6
18 Testing	—	63	63	2.1
19 Fuel and lubricants	68		68	2.3
20 Water supply	20		20	0.7
21 Engineering	69		69	2.3
22 Supervision	83		83	2.8
23 Environmental	26		26	0.9
24 CSM support	—	94	94	3.2
25 Transport	31		31	1.0
26 Cranage	9		9	0.3
27 Abnormal drillstring wear	31		31	1.0
28 Inspection	35		35	1.2
29 Miscellaneous	19		19	0.7
Total	£2501K	£467K	£2968K	100%

* Allocated to other provisions in the overall research budget.
† The US $ conversion at that time was 2.0 US $ per 1 £.

11.4 Advanced drilling concepts

11.4.1 Physical constraints to rotary drilling

Up to now we have only considered the application of mechanical forces to drill rock by a disintegration mechanism that induces stresses of sufficient magnitude to cause failure. Fracturing, crushing, and grinding are the main components of this disintegration or comminution process. Although the rotary tricone bit with tungsten carbide inserts has greatly improved the penetration rate and bit lifetime in hard rock drilling, it is basically an energy inefficient device that is prone to wear.

If major strides are to be made in lowering drilling costs in deep (>2 km) HDR wells, the basic problem of bit wear and energy inefficiency has to be solved. There are two approaches: either to develop a mechanical type bit that will take longer to wear out or to derive an alternative drilling method that will disintegrate the rock by a non-mechanical mechanism.

As discussed by Maurer (1975, 1980) and Armstead (1983), many possibilities exist for the first alternative. Sandia National Laboratories have been developing concepts for a renewable bit that could be changed downhole to reduce the number of trips required. In another analogous approach, they have suggested a continuous chain drill bit that would expose a new set of cutting surfaces for five or ten complete changes. Unfortunately, even though the concept is a good one, only limited bench-scale testing has been conducted. One chain element drilled twice as far and required a lower bit loading than a conventional diamond bit on Sierra White granite. With multiple chain elements, the potential time and cost savings could be substantial. According to Maurer's (1980) commentary on the development of the replaceable cutterhead bits, emphasis has shifted to the use of diamond impregnated Stratapax cutting elements manufactured from synthetic diamonds using a process developed by the General Electric Company. When the Stratapax elements are incorporated into the drill bit assembly by rigidly mounting them adjacent to the roller elements, penetration rates are better than for conventional tricone roller bits. Thus, both a cutting and crushing mechanism is used to drill.

Research and development is continuing on these prototypes, but their eventual commercial viability for HDR drilling still awaits the results of a field testing program.

The second alternative may eventually have more impact in the long term as it would avoid the inherent mechanical limitations of the comminution process, and with many of the novel concepts proposed would avoid the problems arising from drill string rotation, metal to rock contact, and the need for frequent replacements at the cutting interface. Maurer (1975, 1980) reviews over 25 of these concepts and is very objective in evaluating their potential for successful commercialization. Four of the most promising involve different methods of rock disintegration. They include:

(1) Thermal spallation
(2) Fusion or rock melting
(3) Fluid or jet erosion
(4) Chemical attack.

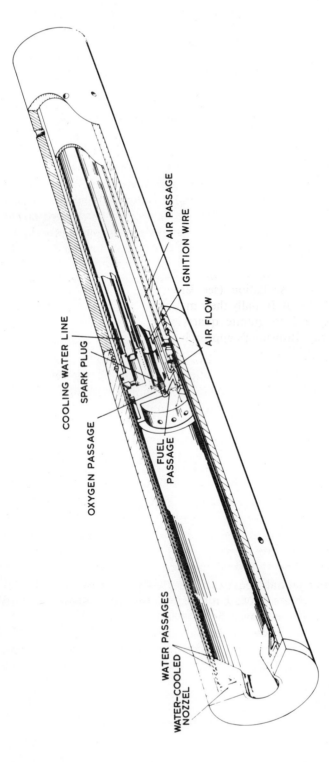

Fig. 11.31 Browning Engineering flame-jet drill schematic. (Browning, 1969). Dowhnhole ignition using spark plug with power supplied from the surface.

Each of these is discussed in the sections that follow.

11.4.2 Thermal spallation techniques

It has been known for over fifty years that the application of heat from a flame or torch can cut certain materials, especially granite and many other silicate-containing rocks. Rapid heating of the rock surface induces sufficient differential thermal stresses within the rock to propagate fractures which form spalls after the surface temperature reaches about 400°C. The actual spallation mechanism has been under investigation for some years with many controversial theories appearing in the literature (Soles and Geller, 1964; and Yagupov, 1963). Experiments are being conducted at the Massachusetts Institute of Technology and at Los Alamos National Laboratory which hopefully will determine the proper mechanism (Rauenzahn, 1986; and Rauenzahn and Tester, 1985). Browning Engineering, Inc. and the Linde Division of the Union Carbide Corporation have developed commercial techniques for cutting and drilling rock using spallation (see Browning, 1963, 1969; Rubow, 1956; Calaman *et al.*, 1960). Initially their methods were applied to taconite mining but quickly spread to granite quarrying operations where they are used extensively today. Browning's approach is to use the heat of combustion from oxidizing fuel oil in air or oxygen-enriched air to generate a hot combustion flame that is expanded through a converging-diverging nozzle to yield a supersonic 'flame-jet'. The jet impinges directly on the rock surface imparting a large heat flux to spall the rock; the combustion gases then sweep the surface clean of rock chips and spalls and continue the cutting process.

A schematic of Browning's rock spallation burner shows how the concept might be adapted to drilling with downhole ignition (see Fig. 11.31). The Union Carbide approach, which actually preceeded Browning's work, involved the use of pure oxygen and fuel oil to produce what they called a jet-piercing drill. The physical principle is the same as Browning's tool except that the oxygen-rich jet-piercing drill produces a much higher temperature and flame velocity (see Fig. 11.32). As cited by Maurer (1980):

Flame jet (air + fuel oil 2)   1650°C at 760 m/s
Jet piercing (pure O_2 + fuel oil 2)   2980°C at 1220 m/s

Since the flame jet only requires compressed air and not pure oxygen, it will be less costly to operate, because it apparently operates without any significant reduction in performance over the jet piercing drill.

In the early 1970s, Browning began to explore the possibilities of flame-jet spallation for deep drilling as opposed to surface rock channelling, rock cutting, and shallow blast hole drilling which were by then already proven commercial techniques. By 1980, he had combined forces with R. Potter at Los Alamos and others to look at the feasibility for deep HDR well drilling using flame-jet techniques in spallable granite formations. A conceptual rig design is given in Fig. 11.33.

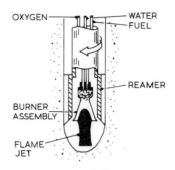

Fig. 11.32 Union Carbide jet-piercing drill schematic. (Rubow, 1956).

Two major field tests to demonstrate the technical feasibility of flame-jet drilling were conducted under US government sponsorship in 1981 and 1982 at Conway, New Hampshire, and at Barre, Vermont, in well-characterized surface-exposed granite formations (Browning, 1981, 1982). The drilling apparatus used was similar to that shown in Fig. 11.33 where a downhole flame-jet burner using compressed air at 14 to 20 bar (200 to 300 psi) is mixed with no. 2 fuel oil in a water-cooled combustion chamber nozzle that is supported and fed through umbilical hoses. The supersonic flame-jet applied to the rock drilling face creates intense thermal stresses resulting in rapid spallation and penetration. The spalls are lifted to the surface through the annulus by the high velocity of the combustion products. The hoses are fiber-reinforced and protected from the upcoming combustion gases by quenching the gases with a radial outflow of cooling water just above the nozzle as shown in the photograph of Fig. 11.34. This remarkably simple system has produced large diameter, stable holes at rapid penetration rates in high compressive strength competent quarry granite (see Fig. 11.35). The main results from the two field tests are:

Location/rock type	Hole depth m (ft.)	Average hole diameter cm (in.)	Penetration rate m/hr (ft/hr)	
			Average	Maximum
Conway, New Hampshire Granite	335(1100)	22.9(9)	15.8(52)	>30.5(>100)
Barre, Vermont Granite	131(430)	38.1(15)	7.6(25)	9.1(30)

If these penetration rates could be maintained at present levels, which are about three to ten times faster than normal rotary drilling, a large cost

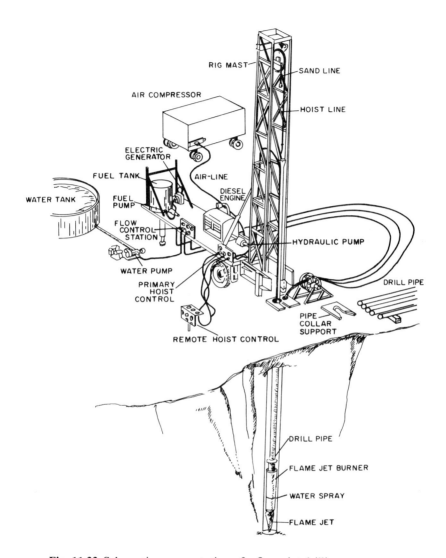

Fig. 11.33 Schematic representation of a flame-jet drilling apparatus.

reduction in drilling may be possible. Furthermore, trip time and 'bit' wear may be almost eliminated.

In addition to the fuel oil and air or oxygen systems developed by Browning and Union Carbide, more exotic systems have been proposed by the Soviets using rocket fuels in an expanding rocket nozzle-type drill and by Flame Jet Partners Ltd. using hydrazine as the fuel and either nitrogen tetraoxide or nitric acid as the oxidizer (Anon., 1983). Both of these methods can generate higher

Fig. 11.34 Browning flame-jet tool under subsonic flow condition before drilling in Conway granite in New Hampshire. Radial cooling water flow also visible. (Courtesy of R. Potter, Los Alamos National Laboratory.)

temperatures, gas velocities, and heat fluxes than are possible with other combustion jets. This should enhance the spallation process. In one design, periodic application of the flame is followed by a pulse of water to quench the rock creating an even higher rate of spallation. Flame Jet Partners claim that they will have a commercial unit operational by 1988 using some form of drill string hardware to carry the reactants to the drilling face (Maurer, 1980; Anon., 1983). They hope to substitute kerosene as the primary fuel.

Even though the Browning field tests cited above have produced impressive results, they also have indicated some serious limitations of using flame jets for deep borehole drilling in HDR. There is considerable doubt whether an umbilical hose concept can be developed to carry the large volumes of air needed at great depths without requiring pressures in excess of attainable hose strength. The capability of downhole ignition will also have to be assured. Another important concern is how to deal with a zone of non-spallable rock if it is encountered. To continue drilling, conventional methods would be required and a stand-by rotary rig may be needed. This could destroy any potential economic advantages of spallation techniques. If the flame-jet process could be

Fig. 11.35 Flame-jet drilled hole in Conway Granite.

made to be compatable with rotary rig hardware, a solution to many of the limitations may be possible.

Current studies at Los Alamos (Rauenzahn and Tester, 1985) are underway in the field to modify Browning's flame-jet tool so that it could be used on a conventional rotary rig. The jet nozzle assembly will be supported on $2\frac{7}{8}$ in. (7.3 cm) standard drill pipe which will be tied back onto the drillstring. An electrical lead will provide high voltage to a spark plug to permit downhole ignition of the fuel-oxygen mixture. Following ignition, the oxygen flow will be cut off and only compressed air will then be used. Drilling operations started during the summer of 1984 at a site in the Pedernal Hills in central New Mexico, where a number of well-characterized granites are exposed at the surface.

11.4.3 Fusion/rock melting techniques (Subterrenes)

Another method of rock penetration involves the application of sufficient heat to fuse or melt the rock and the displacement of the molten rock outwards in a consolidating penetrator (Fig. 11.36(a)). An alternative to this would be to force

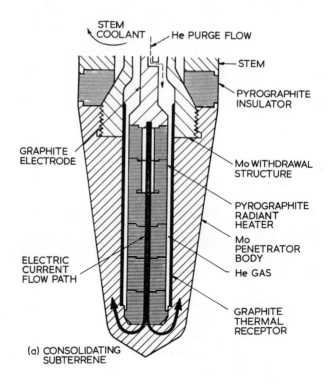

(a) CONSOLIDATING SUBTERRENE

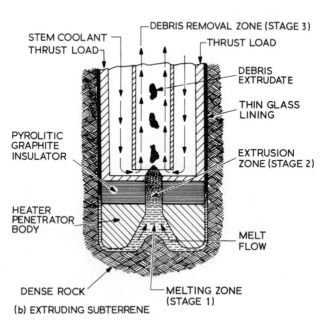

(b) EXTRUDING SUBTERRENE

Fig. 11.36 Consolidating and extruding subterrene penetrators developed by and tested at Los Alamos National Laboratory from 1971–1976. (Neudecker, 1973; Hanold *et al.*, 1977).

the molten rock inwards and up the bore as shown in the extruding penetrator schematic of Fig. 11.36(b). Most igneous rocks melt at temperatures of 1200 to 1500°C, so heating devices capable of 2000 to 3500°C may be possible choices with refractory metals such as tungsten or molybdenum as a primary material of construction. If such a penetrator could be perfected, geothermal drilling even into magma chambers or lava lakes would become feasible.

In the late 1960s, scientists at Los Alamos began to develop prototype designs which they called Subterrenes. As summarized by Hanold *et al.* (1977), the initial concept was to use electrically-heated elements which could be ultimately replaced by a nuclear-heated penetrator that could be used for deep drilling or tunnelling.

For porous rocks such as volcanic tuffs or sandstones, consolidating penetrators as shown in Fig. 11.36(a) could be used to melt the rock forcing it to the outside of the penetrator as a melt which eventually forms a glass-like liner on the interior bore hole wall, effectively forming a casing to stabilize the formation. In dense rock, as will be encountered in many deep HDR reservoirs, extruding Subterrenes would be used. As shown in Fig. 11.36(b), molten rock flows inward and passes up through the central section of the penetrator where it is solidified into small glass-like chips which can be brought to the surface by using a suitable drilling fluid.

Both of these devices have been tested in bench-scale and shallow-hole experiments at Los Alamos. As Hanold *et al.* (1977) report, the testing program was encouraging but tests did establish several inherent physical limitations of the method itself.

Current subterrene designs are limited to penetration rates of 0.25 mm/s or about 3 ft/hr which is considerably slower than current rotary methods, even if trip time is included. This low penetation rate is caused by limitations in the rate of heat transfer from the penetrator surface to the rock. Because of the high viscosity and relatively low thermal conductivity of molten and solid rock, heat transfer is primarily by conduction which keeps the heat flux low. Unfortunately, before an alternative design could be proposed and tested to improve penetration rates, the US DOE decided to terminate funding because of other priorities. Several entrepreneurs have attempted to develop the concept with private funds but so far have not reported any breakthroughs that would lead to commercialization.

11.4.4 Erosive systems

Maurer (1975, 1980) describes a number of concepts that involve using pressurized jets of gas, water or drilling mud with or without actual bits to fracture and spall rock. Some years ago EXXON Production Research Company and eight other major oil companies combined forces to develop an erosion drilling system using high pressure diamond, tricone roller, or drag bits as shown in Fig. 11.37. In preliminary testing pressures up to 1000 bar (15 000 psi)

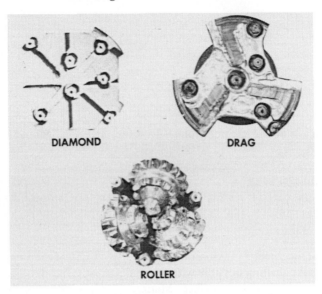

Fig. 11.37 High pressure water-jet erosional assisted rotary bit designs by Exxon Corporation. (Maurer, 1980).

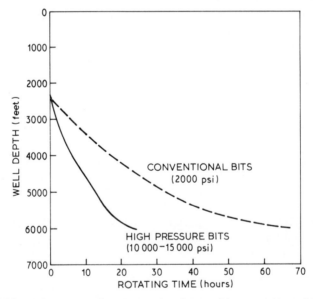

Fig. 11.38 Field test data comparing conventional rotary bit penetration with that of the Exxon Corporation designed erosion bit. (Maurer, 1980).

provided by positive displacement pumps were used to force drilling mud and water through nozzles in the bits. Results were encouraging as shown in Fig. 11.38 where the erosion-assisted bits drilled two to three times faster than conventional rotary designs over a depth interval from 700 to 1830 m (2300 to 6000 ft) in an oil bearing formation. High pressure drilling systems of this type should be adaptable to geothermal HDR drilling and may provide a method of greatly reducing costs. It seems at this time that erosion-assisted rotary methods have shown superior performance in field and bench-scale tests over the continuous and explosively pulsed jets that do not use any mechanical cutters. However, as technology improves, a pure jet-erosion system might prove to be the better choice for deep HDR wells because of the elimination of drill bit wear and subsequent reduction in rig trip time to replace worn bits.

11.4.5 Chemical attack techniques

As mentioned in Chapter 10, it is possible to dissolve rapidly and selectively certain minerals that are contained in granitic rock by using corrosive fluids such as hydrofluoric (HF), hydrochloric (HCl), and organic-based acids (acetic or formic). This selective dissolution mechanism could weaken the rock fabric enough to make drilling possible with a jet of corrosive liquid or gas to remove debris from the advancing surface. Maurer (1975, 1980) describes chemical drill systems and their possible application to geothermal drilling. A conceptual design is given in Fig. 11.39. The chemical would react vigorously with specific minerals to produce non-corrosive by-products which would be circulated out of the hole with unreacted chips in the normal manner used for rotary drilling. Safety problems, chemical costs, and the development of a suitable downhole delivery system for chemical reagents have hampered development of this concept.

11.5 Status of drilling practices for HDR: a summary

Even though none of the advanced drilling techniques for HDR have been brought into commercial practice yet, it is important to reemphasize that HDR wells have been drilled to required directional specifications with conventional drilling techniques at costs that scale reasonably with oil and gas wells to similar depths. Experience at Fenton Hill and Rosemanowes vividly points out that directional drilling in hard rock requires a well-planned, well-managed effort to keep costs in line. The undisputable fact remains that operational HDR wells have been successfully drilled to almost 5000 m (15 000 ft) in hard crystalline rock to temperatures in excess of 300°C. This achievement alone opens up a significant portion of the HDR resource base to potential exploitation. Eventually, assuming that large workable reservoirs can be created, only the economics will influence the rate of development of HDR at these depths. The economic picture, in turn, is controlled by reservoir output and lifetime from a given well pair relative to the cost of drilling and completing the wells. This could be expressed as the mass flow rate or net enthalpy of the fluid per dollar of investment. As the mass flowrate per well and

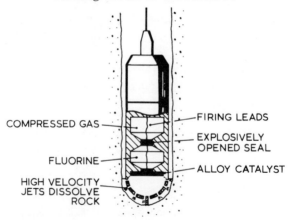

COMPRESSED GAS

FLUORINE

HIGH VELOCITY
JETS DISSOLVE
ROCK

FIRING LEADS

EXPLOSIVELY
OPENED SEAL

ALLOY CATALYST

Fig. 11.39 Chemical drill concept. (Maurer, 1980).

temperature of the fluid increase, higher individual well costs can be tolerated and drilling deeper becomes more economically attractive.

In all situations, because individual well cost is such a large component of the total investment, lower drilling costs as a result of experience gained in the field and from technological improvements will certainly have a positive impact on HDR development.

Chapter 12

Direct recovery of magmatic energy

Up to this point we have examined only the possibilities of recovering heat from the earth's crustal rocks; but there is an even more tempting prize to be considered. Embedded within the crust in certain volcanically active parts of the world are to be found intrusions or chambers of magma – very hot material exuded from the mantle, that has penetrated upwards towards the surface. These magma reservoirs represent immense local concentrations of energy, and some of them lie within range of our present drilling capabilities. Attention has naturally been devoted to the possibilities of recovering some of this energy, and a large amount of research has been undertaken in the USA, the USSR and Japan with this goal in view. The Sandia National Laboratories of Albuquerque, New Mexico, USA, have been particularly active in this direction, and much of what follows in this chapter is supported by two Sandia reports (Colp, 1982; and Gerlach, 1982).

12.1 The nature of magma chambers

The presence within the crust of pockets or intrusions of magma is undoubtedly associated with the process whereby volcanoes come into being – a process that is itself mainly caused by the relative movements of the great tectonic plates of which the earth's crust is composed and sometimes by the presence of mantle plumes that partially penetrate the crust from below or places far from their boundaries, e.g. in Hawaii and Yellowstone. It is at these tectonically active places that volcanoes are formed by a process which, though imperfectly understood, could be somewhat as shown (greatly simplified) in Fig. 12.1. Owing to intense tectonic activity in these regions temperatures just beneath the crust become so high as to impart a degree of plasticity to the underside of the local crust and relative buoyancy to the local overheated mantle material. As a result, the mantle material intrudes into the crust and is impelled upwards by lithostatic pressure so as to form 'gulps' of magma which rise slowly towards the surface of the crust, rather in the manner of bubbles rising in a pool of water. If these gulps are sufficiently hot and large, they will ultimately break the

381

surface to form a volcano; if less hot or smaller, they may cool off and solidify beneath the surface – perhaps giving rise to a hydrothermal field where the crustal formation is favorable, or to a zone of high temperature gradient eminently suitable for HDR exploitation.

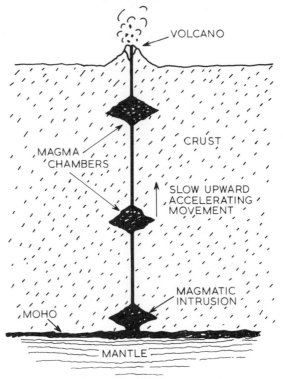

Fig. 12.1 The formation of a volcano.

Regardless of the mechanism of their formation, large magma chambers are known to exist in several parts of the world, notably in the western USA, Alaska, Kamchatka, Iceland, Sicily, the Azores and elsewhere. It could scarcely be coincidental that all these places are close to tectonic plate boundaries. Between 300 and 400 active volcanoes are known to exist in the world, plus many more that are dormant and potentially capable of renewed activity. (The alternation between activity and dormancy could be caused by a sequence of 'gulps' of magma rising from the same weak point in the crust, as shown in Fig. l2.1.) There are probably also many rising magma chambers that have not yet broken the surface for the first time, for new volcanoes are not infrequently being born. The existence of magma chambers need not therefore be confined to locations directly beneath known active or dormant volcanoes.

No one has yet penetrated an underground magma chamber, so not very much is known with certainty as to their size, shape and physical properties.

Much, however, can be deduced from the indirect evidence of lavas at the surface and from geophysical observations. Underground magma differs from surface lavas mainly in that it is rich in dissolved gases, whereas lavas have lost most of their gas content as a result of the removal of the intense pressures to which the underground magma is subjected. One of the dissolved gases present in virgin magma is water vapor, the source of 'magmatic waters' that are richer in deuterium content than meteoric waters. Chemically, both magma and lava consist mostly of silicates of calcium and magnesium, with lesser amounts of compounds of iron, aluminium, titanium, sodium and potassium. Apart from water vapor, the dissolved gases within the underground magma chambers consist mainly of CO_2 and SO_2, which volatilize as the surface is approached. It is expected that the presence of these gases will give rise to a highly corrosive environment when deep magma chambers are penetrated.

The temperatures of magma chambers range from about 600°C to 1300°C, depending on their origin, size and age since they first penetrated the lower boundary of the crust. Magmas originating in rift zones are generally hotter than those in subduction zones, and are basaltic. As to size, magma chambers measure up to several tens of cubic kilometres. In shape they are probably rather in amorphous, though of vaguely lenticular form. Little is known of their boundary configuration, whether there is continuity from core to boundary, or whether they are honeycombed at the edges or dispersed among the surrounding crust in the form of dyke swarms. Within the heart of the chambers the high temperatures would usually ensure sufficient liquidity which results in pronounced convection currents, though heat losses would induce high viscosity in the outer layers, especially at the upper sides, and probably a solid crust at the edges. Even surface lava, after losing much of its primordial heat, can flow under the influence of its own gravity – sometimes at speeds of 3–5 km/h.

The prevailing pressures are enormous – probably of the order of 250 bar per kilometer of depth – and it is the release of these huge pressures as the chambers rise towards the surface that causes the emission of the dissolved gases and vapors that are ultimately discharged from the craters of active volcanoes. Before an escape vent has been formed, the gradual evolution of gases would cause the chambers to swell somewhat as they rise through the crust. This would give added buoyancy that would tend to accelerate their upward movement, as with bubbles rising in a liquid. Evidence of such acceleration is provided by the greater scarcity of magma chambers in the upper crust by comparison with those that are believed to exist at greater depths. For although many magma chambers lie within the upper 5 km of the crust, it is thought that far more exist within the next lower 5 km layer.

12.2 Size of the resource

In thermal energy a single cubic kilometre of magma, in cooling from, say, 1200°C to 50°C would release something like 2.5 quads, or 82 000 MWyr of energy. A magma chamber at 1200°C and of about 30 km³ in size (equivalent to a sphere of only about 3.9 km diameter) would theoretically be capable of supplying as much energy as was used by the USA in 1986. Even though such a degree of cooling would undoubtedly be impracticable, these figures serve to show the potential importance of the resource. Smith and Shaw (1975, 1979) have estimated that within the upper 10 km of the crust in the USA alone there is at least 50 000 and probably as much as 500 000 quads of energy in the form of molten, or partially molten, magma. It is believed that something between 3 to 15 times as much energy again is contained in magma chambers at the tectonic rifting zones of the world at depths of only 6 or 7 km below the land surface and at temperatures of 1000 to 1200°C. On a world scale, therefore, the resource has been estimated at 200 000 quads *minimum*, and probably a lot more. Although tentative, this figure is very impressive. Moreover the resource is of very high grade, so that the recovered energy could be used to generate electricity at high efficiency. In 1982 the world consumed 8436 millions of MWh of electrical energy, which is equivalent to 28.6 quads – a mere 7000th part of the estimated world total of magma energy. It will be seen from Fig. 4.9 that with a source temperature of 600°C it should be possible to generate electricity at about 39% efficiency. Taking a mean magma temperature of 1100°C, it would be possible to remove approximately (1100 – 600)/(1100 – 15), or 46% of the gross thermal potential to generate electricity at 39% efficiency; so the electrical potential of the magma heat could be reckoned at $0.46 \times 0.39 \times 200 000$, or 35 880 quads (electric). This is approximately 1250 times the world electricity consumption in 1982. Considering that this estimate is derived from the product of two minima – 50 000 of the (50 000–500 000) range, and 3 of the (3–15) range and that the first minimum (50 000 quads) relates to the USA alone – the whole resource is clearly immense. Moreover, the residual heat below 600°C has been ignored.

Furthermore, the thermal component of the resource is not all; for magma has also a significant chemical potential. Northrup *et al.* (1978) has estimated that a cubic kilometre of basaltic magma contains sufficient chemical energy to produce 2×10^9 kg of hydrogen gas which, at a net calorific value of 1.142×10^{-10} quads/kg, is equivalent to 0.2284 quads/km³. This is about 1/12th of the thermal potential of the magma. Other chemical reactions are possible besides the production of hydrogen, as will be shown in Section 12.5, but this simple calculation will serve to show the order of magnitude of the chemical energy of magma pockets.

12.3   Detection of magma chambers

As will be seen from Fig. 12.1, many magma chambers will lie more or less directly beneath active and dormant volcanoes, the locations of which are self-evident and do not have to be 'discovered'. Nevertheless, their concealed underground structure, size, temperature and other properties would have to be divined by exploration methods. Other magma chambers generated from intrusions that have not yet broken surface would, however, have to be identified as to location also.

The most obvious indication of the presence of an underground magma chamber would be an abnormally high temperature gradient and local surface heat flow. Other conventional exploration methods can be applied to the detection of the locations and properties of concealed magma reservoirs – e.g. gravimetric, resistivity, seismicity and other standard geophysical aids. Geochemistry also could be helpful in detecting some of the gases emitted from the magma as they come out of solution during its slow upward movement and seep through the surrounding crust.

With active volcanoes there will usually be a magma reservoir more or less directly beneath the crater, at quite moderate depth. With dormant volcanoes it may be necessary to explore much more deeply.

12.4 Drilling into magma reservoirs

If magma reservoirs are to be exploited, it will be necessary not only to approach them by drilling, but also to penetrate into them. Doubts have been felt as to how this can be accomplished. If conventional drilling methods are to be employed, will it be possible for the drilling equipment to withstand the aggressive environment that is likely to be encountered? How will it be possible to drill into a substance that is not merely plastic, but liquid? It is natural that misgivings should have been felt as to whether such problems could ever be overcome.

Fortunately nature has placed at our disposal a convenient and gratuitous laboratory at which extensive tests can be made before risks are taken in deep underground operations. In Hawaii, which lies over a hot spot in the mantle, the Kilauea Volcano erupted in 1959, leaving its crater half filled with molten magma, over which a crust has formed. Here is an ideal place where experiments may be performed and experience gained before the exploitation of an underground magma chamber is ever attempted. It is true that conditions in the Kilauea Lava lake will differ in many respects from those that will be encountered in deep magma reservoirs – particularly in pressure and in gas content – but the temperatures are realistic and much preliminary and rewarding research has already been undertaken there. To supplement the work in Hawaii, extensive laboratory tests have also been performed under simulated conditions of pressure and temperature resembling as closely as possible those that are expected to exist in actual magma chambers.

The conclusions of all this research are very optimistic. By using high velocity water jets in advance of the drill bits it is possible to maintain a stable borehole not merely in crustal rocks but also in molten magma at confining pressures equivalent to those prevailing at depths of up to 10 km. When penetrating molten magma the water jet produces a protuberance of chilled crust extending into the molten mass and effectively preventing any flowback of magma into the drillhole for as long as an adequate supply of coolant is maintained. If the coolant is curtailed or interrupted, the chilled magma may be softened and even re-melted. Molten rock at 1100°C has been successfully penetrated to a depth of 33 m in Hawaii. To retain the stability of holes to depths of up to 10 km it is necessary that they be filled with a fluid of density equal to or greater than that of water. If the bore walls can be maintained at 700°C or less, even an unfilled bore will remain stable at depths of about 10 km. High temperature, high-strength and corrosion resistant materials, primarily nickel- and cobalt-based, have proved successful for drill bits and stems in withstanding the arduous corrosive and physical conditions that are expected to be encountered when penetrating magma chambers at depths of up to 10 km. The crusty walls at the boundaries of the magma chambers are brittle, and should offer no special problems in drilling.

12.5   Exploitation: heat extraction

The most obvious way of extracting heat from a molten mass of magma would be to bury a heat exchanger into the magma and circulate a primary fluid through it to remove the heat and supply it for direct or indirect use for power generation or other purposes. Experiments performed with heat exchangers in the Kilauea lake and in the laboratory show that an exchanger having a certain melting temperature can survive when immersed in magma of a much higher temperature so long as heat is removed by the primary working fluid at an adequate rate. What happens is that a crust of chilled magma forms on the walls of the heat exchanger and builds up to a thickness at which a balance is attained between the rate of heat removal and the rate of heat transfer from the molten magma through the crust. The heat is of course transmitted through the solid crust by conduction. In stagnant molten magma the effect of continuous heat loss would be to thicken the crust until no molten magma remains, after which the temperature would gradually fall as further heat is removed. However, in a large pool of molten magma there will be natural convection currents due to temperature and liquid density differences that will ultimately supply heat to the exchanger at exactly the same rate as that of removal by the primary working fluid. When this steady state is attained, no further growth of the solid crust will occur. With very hot magma the natural convection rate will be high and the crust thin, but at lower temperatures the crust will be thicker and the heat replenishment from convection will be less. The mere presence of a heat exchanger in a molten pool of magma will enhance natural convection currents by creating a temperature differential and density gradient that increases the thermosyphonic action or buoyancy effect over whatever levels are naturally present.

As the crust forms on the surface of the heat exchanger the transferred heat is augmented by latent heat surrendered by the magma in its change-of-state transition from liquid to solid. Moreover, as the crust grows in size its increased surface area will to some extent offset the increase in conduction resistance resulting from the additional thickness of the crust.

Because the thermal conductivity itself does not change, the heat transfer resistance scales with the ratio of thickness to thermal conductivity. But for a cylindrical pipe in a liquid it does not scale linearly with thickness but more like a logarithmic dependence on $(1/\lambda_r) \ln (r_o/r_i)$ where λ_r = thermal conductivity, r_o = outer radius of crust, r_i = inner radius of crust and of course $r_o - r_i$ = crust thickness (t) thus:

$$(1/\lambda_r)\ln(r_o/r_i) = (1/\lambda_r)\ln((r_i+t)/r_i) \qquad (12.1)$$

which will only be linear for small values of t. So what really needs to be examined is how the total heat transferred to the pipe varies as t increases. This total heat transferred per unit length (q/L) is given at steady state by:

$$q/L = \frac{2\pi[T_{\text{melt}} - T]}{t \ln[(r_i+t)/r_i]/\lambda_r(r_i+t) + 1/h(r_i+t)} = \frac{2\pi(r_i+t)[T_{\text{melt}} - T]}{t \ln[(r_i+t)/r_i]/\lambda_r + 1/h} \qquad (12.2)$$

where h is the outside heat transfer coefficient in the molten magma, T = working fluid temperature, and T_{melt} = bulk magma temperature. As t gets very large, the second term in the denominator becomes negligible and q/L becomes proportional to $1/\ln t$ and goes to zero as $t \to \infty$. Therefore the compensation is not quite complete.

Even with these complications, these effects combine to give overall heat transfer rates of approximately 16 kW/m² with a magma temperature of 1210°C to 40 kW/m² with magma at 1310°C. It is of course essential that a continuous flow of primary fluid is maintained; for if it were interrupted, the heat exchanger would be destroyed by excess temperature. Emergency cooling arrangements are therefore necessary.

An alternative to a closed heat exchanger would be an open design in which water would be directly injected into the molten magma so as to form a large fractured crust of highly permeable solidified rock around the end of the borehole. The permeability of this crust should offer a large surface area for heat transfer. The injected water would be instantaneously converted into steam, which could be withdrawn through a pipe concentric with the bore and used as a working fluid. Experiments have shown that heat transfer rates could probably be attained up to 10 or even 100 times as high as possible with surface heat exchangers. This would enormously improve the economics of heat extraction. This concept is shown in simplified form in Fig. 12.2.

12.6 Exploitation: fuel production

Although magma chambers can probably be exploited for the sake of their heat content alone by methods that have been briefly described above, there are

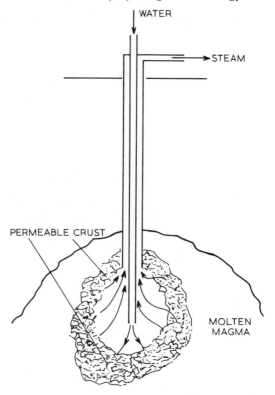

Fig. 12.2 Concept of an open heat exchanger.

alternative chemical methods of exploitation whereby it may be possible to produce synthetic combustible fuels, which possess the great merits of transportability and value as industrial feedstocks (Colp, 1982; Gerlach, 1982). These alternative methods may either make use of the ferrous oxide that occurs in the basaltic content of magma, or alternatively of the heat content of the magma in combination with the chemical content of organic matter extraneously introduced.

By injecting water directly into hot magma containing ferrous oxide, it is possible to produce hydrogen from the following chemical reaction:

$$2FeO + H_2O \rightarrow Fe_2O_3 + H_2 \qquad (12.3)$$

Not all magmas are rich in ferrous oxide. For those that are not, a more useful reaction could be obtained by mixing the water with biomass, the cellulose content of which ($C_6H_{10}O_5$) can combine with water vapor under certain conditions to form methane, so:

$$H_2O + C_6H_{10}O_5 \rightarrow 3CH_4 + 3CO_2 \qquad (12.4)$$

At higher temperatures a further reaction can produce other combustible gases – carbon monoxide and hydrogen, e.g.:

$$CH_4 + CO_2 \rightarrow 2H_2 + 2CO \tag{12.5}$$

The temperature at which this second reaction occurs is sensitive to the pressure. These last two reactions use the heat only of the magma, the chemical energy being introduced extraneously; and since biomass may be regarded more or less as a product of solar energy, the reactions offer a partnership between geothermal and solar energy. Cellulose accounts for more than half of the organic matter in the bio-sphere; and although it is itself combustible, its conversion to combustible and liquefiable gases renders it far more convenient as a commercial fuel. The energy balance for the biomass reaction is very favorable, offering a six to twelve-fold advantage in terms of the energy input, even after allowing for the energy consumed in the production, harvesting, collection, transport and refinement of the biomass.

The most favorable chemical reaction, as well as the fuel productivity of the process, is very dependent upon the type of magma exploited. Basaltic magmas are rich in ferrous oxide and thus favor the straight production of hydrogen from water, whereas andesitic and rhyolitic magmas would be more suited to methane production from cellulose. There would seem to be no reason why both reactions should not take place simultaneously in different proportions according to the magma type exploited.

Commercial production of gaseous fuels from magma would require the availability of large quantitites of magma and of water in close proximity. The rift zones where tectonic plates are separating would seem to offer the best conditions, for it is at these zones that the magma is basaltic, temperatures generally higher and depths more moderate. Moreover, basaltic magmas can support fuel generation rates of two to three times by comparison with andesitic, and five to six times by comparison with rhyolitic magmas. Also, being generally hotter, basaltic magmas would have low viscosities, and therefore more rapid convection for sustaining heat extraction operations for very long times. All this, however, does not imply that magma could not be successfully exploited in subduction zones either for methane production or simply for heat extraction. It is not yet certain whether sea water could be used instead of fresh water for either process.

12.7 Exploitation: ocean-based magma

Yet another method of extracting magmatic heat has been considered recently. Although at present it is a purely conceptual notion, it could perhaps one day become a practical reality.

In several places beneath the oceans (for example off the Western coasts of the Americas and perhaps in the Gulf of Mexico) active submarine volcanoes are to be found along the lines of tectonic rifts. Underwater photographs have been obtained which show the ejection of luminously hot rock into the lower depths of the ocean. The hydraulic pressures at these depths (up to about 8 km) are so great as to inhibit boiling because they are above the critical pressure of water, so that layers of supercritical hot brine occur near the points of eruption. Fluid buoyancy or natural convection causes these zones of very hot water to be dissipated over wide regions; but if the superheated water could somehow be 'funnelled' to the surface through insulated ducts, huge quantities of thermal energy could become available for exploitation. The practical difficulties of achieving this are obviously great – particularly as the rift zones occur at considerable distances from the shores. Nevertheless, it has been estimated by the USGS that the available energy from this source is equivalent to about 2000 times the present energy requirements of the entire world or about 500 000 quads. Sandia National Laboratories have estimated the total energy from submarine rifts to be from 150 000 to 7 500 000 quads. Thus, it would seem safe to use a range of 150 000 to 500 000 quads for a conservative estimate of the resource. It has been suggested by Rumsthaler (1984) that a practical way of using this energy would be by means of a reductive formulation process, whereby biomass would be converted into petroleum in a matter of days, rather than in the millions of years normally required by natural processes. A valuable by-product of this form of magmatic energy would be the rich source of minerals exuded from these submarine rifts, including copper, iron, silver, zinc, cadmium, and molybdenum.

12.8 Summary

Magma exploitation must at present be regarded as a long range, high risk, project. It would essentially be a large scale activity involving very great capital expenditure; so it must be cautiously approached on a step-by-step basis. The results of research to date, however, are encouraging and confirm that there is nothing implausible about the concept and that it is without doubt scientifically feasible – either for direct heat extraction or for the production of gaseous fuels. This chapter claims to be no more than a very elementary introduction to the subject, as the main emphasis of this book is upon heat extraction from HDR – an activity that would now seem to be rather nearer to practical realization, though, given the possibility of technical breakthroughs occuring, circum-stances could quite possibly change in favor of magma exploitation. Magma energy has one particular potential usefulness that HDR lacks (except indirectly by way of electricity generation): it could possibly provide synthetic fuels suitable for transportation. It is emphasized that the magma energy resource has been excluded from Table 3.1, Figure 3.1(a) and Figure 4.10.

Chapter 13

Environmental aspects

13.1 Local and remote environmental impacts

When an industry or power plant is set up on some particular site it would be wrong to judge its environmental impact solely on the local pollutive effects, or their absence, at that site. For a proper environmental assessment it is necessary to think beyond the confines of the installation and to consider both the supply lines for the materials needed to sustain it and also the means of delivering the end product to the consumers or customers. The establishment of almost any industrial activity is liable to have consequential ramifications that extend far beyond the activity itself.

For example, a fossil-fired thermal power plant may be equipped with the most sophisticated means of suppressing the emission of toxic combustion products into the atmosphere; but however immaculate the power plant itself may be it must be fed with fuel, and the generated electricity must be delivered to the consumers. The supply of fuel might, for example, involve open-cast mining operations many miles away, perhaps with the consequent destruction of valuable agricultural or timber-growing lands. Or, if the power plant is oil-fired, it is necessary to consider the transportation of petroleum, maybe over thousands of miles, with the associated risks of marine pollution from spillage. The environmental aspects of delivering electricity from a power plant to the consumers are mainly aesthetic, concerning the unsightliness of overhead high voltage transmission lines, but questions of aircraft and wildlife hazards may sometimes also arise.

13.2 Positive and negative aspects of environmental impact

Whilst the environmental aspects of some particular industrial activity may be assessed in positive terms of 'guilt' or 'innocence' of causing pollution in one way or another, a more comprehensive judgment can be made by looking further afield and in weighing up alternatives; for there can also be *negative* advantages in an innocent process if it can replace a more guilty alternative.

Thus heat mining should be viewed not only in terms of its own extent of guilt or innocence in giving rise to pollution, but also in terms of how its non-adoption would affect the environment. Since the world must have an abundance of energy, and since that energy must now be supplied predominently from combustion – a pollutive process – then if heat mining can be shown to be almost or totally non-pollutive it can be credited with the *avoidance* of such pollution as would be caused by the combustion that it can replace.

13.3 Broad assessment of the environmental impact of heat mining

With heat mining, everything up to the production stage would be locally self-contained. The energy would be collected locally and converted locally. As there would be nothing equivalent to fuel supply lines, there would be no corresponding adverse remote side effects. Only in the distribution of the end product, be it heat or electricity, could remote environmental ill effects ever conceivably arise – and then to no greater extent than if the industry or other activity were dependent upon any other source of energy. Thus the environmental aspects of heat mining may be viewed almost entirely from a local standpoint. Perhaps the simplest way of assessing these aspects for heat mining is to consider the relevance of all potential sources of pollution that could arise from the exploitation of a *hydrothermal* field, some of which can be dismissed because of a totally different mode of operations. These are as follows:

(1) Hydrogen sulfide emission
(2) Carbon dioxide emission
(3) Water-borne poisons
(4) Air-borne poisons
(5) Silica deposition and disposal
(6) Heat pollution
(7) Land erosion
(8) Noise
(9) Escaping steam
(10) Scenery spoilation (esthetic aspects)
(11) Ecological disturbance
(12) Subsidence
(13) Seismicity

The first six of these potential sources of pollution can arise from natural hot fluids when brought from belowground to the surface in large quantities. Such fluids are not directly used in the heat mining process: only a closed loop of circulated water is artificially introduced into the reservoir, and although small quantities of natural pore fluids can migrate into the closed circulation loop (see Section 10.4) their influence would be minimal because the working fluid is primarily confined to a system of pipes and rock fractures, and need never 'see the light of day'. None of these six potential sources of pollution will therefore constitute an environmental hazard for heat mining.

Land erosion has caused some problems in the Geysers field in California, but this is strictly a matter that is peculiar to the local terrain and would seldom arise elsewhere; in any case it can be prevented by the exercise of care. Land erosion need not be considered as an inherent side effect of heat mining.

Noise is mostly caused by the emission at high pressure of huge quantities of unwanted fluids to waste; there would be no such fluids in heat mining activities. The only other sources of noise would be the whirr of power station machinery which could be controlled by the use of sound-proof buildings, and the noise of drilling operations as a prelude to heat mining activities. Although temporary, drilling noise could sometimes present a serious problem at sites in, or close to, populated areas; for even if it is of less intensity than those associated with road repairs, it would sometimes endure for a year or more without respite. There could be occasions where the construction of elaborate sound-insulating barriers as at Rosemanowes, or perhaps the payment of substantial compensation would be involved.

Escaping steam should cause no problem as there would be no need to discharge natural steam at any time. On rare occasions, if corrosion inhibition additives or other hazardous chemicals are present, plumes of vapor rising from cooling towers may cause concern, for example, if Cr^{+3} ions are present the drift of water droplets could be troublesome. It is also true that where the extracted heat is used for low efficiency power generation, cooling towers would have to dispose of greater quantities of unwanted heat than in the case of conventional generation of the same amount of power. That could cause some aesthetic offence.

Scenery spoliation should be modest with heat mining except for the cooling towers required in the case of low efficiency power generation, for the mining operations themselves would be underground and the utilization plants would have to handle only piped hot liquids and to accommodate 'clean' machinery. It should be possible to ensure that heat mining installations are neither unsightly nor obtrusive.

Ecological disturbance and subsidence have already been discussed in Chapter 5 and effectively dismissed as potential environmental hazards. The only local factor that could perhaps, though improbably, cause adverse side effects is the last of those listed above – seismicity, induced by either reservoir formation and/or operation; but, as explained in Chapter 5 and as a result of the extensive monitoring at Fenton Hill and Rosemanowes, it is unlikely to be a source of trouble. Problems should be easy to avoid as experience is gained circumspectly in regard to seismic source mechanisms and their dependence on geologic factors.

Thus from local considerations it is advisable to give further thought to noise problems and to the adverse aesthetics. If heat mining in non-thermal areas should ever become a commercial reality it will usually be possible to recover the energy so close to the market as to be virtually *at* the market, thus obviating the need for energy transmission over more than very short distances. There could, however, be instances where very high population densities would necessitate heat mining operations rather more remotely from the market, thus requiring energy transmission over moderate, but not great, distances. This question will now be examined more closely.

13.4   The problem of energy delivery in densely populated areas

To take an example, it may be noted that the island of Manhattan has an area of 13 km^2 and supports a resident population of about 5 million. The present energy consumption per capita in the USA is about 12 tce p.a. Thus the total energy requirements of Manhattan, assuming the national mean energy density to apply, are about $5 \times 10^6 \times 12/13$, or 4.6 million tce p.a. per km^2 of area. It was shown in Chapter 4, Section 2, that heat mining could produce about 773 000 tce per km^2 per °C of average cooling of the uppermost 10 km of the earth's crust. Thus even now *local* crustal heat could supply Manhattan with all its energy needs for no more than $773\,000/(4.6 \times 10^6) \times 12$, or about 2 months per °C of average cooling to 10 km depth. If energy consumption per capita continues to rise, local resources of crustal heat will become even less adequate. Under present conditions, and if only 1°C average crustal cooling to 10 km depth were permissible, it would be necessary to mine heat from an area of about 600 times that of Manhattan – e.g. a circular area of about 50 km radius – to supply that island with its total energy needs for a century. As Manhattan borders the ocean, heat and/or electricity would in this case have to be transmitted over distances appreciably greater than 50 km.

This feature of densely populated areas need not be regarded as an argument against heat mining, but it should nevertheless be noted. It will have some economic impact where relevant, but environmentally it would render heat mining no more blameworthy than any other source of energy: in fact usually it would be less so. In any case the problem will not arise until it becomes economic to mine heat in non-thermal areas.

13.5   Environmental aspects: summary

It may reasonably be claimed that heat mining would be environmentally more acceptable than any other form of energy production. Not only would it intrinsically be virtually pollution-free itself, but to the extent that it could partially replace combustion it would have an additional positive environmental effect. By obviating the need for fuel transportation it would help to free the world from one of the most intractable causes of pollution, namely the fouling of the oceans with oil spillage. By reducing the emission of toxic combustion products and CO_2 into the atmosphere, or at least by moderating the rate at which this emission is growing, the health not only of humans but also of the entire animal and vegetable kingdoms would be conserved. It is alleged by many, though disputed by others, that one of the most insidious effects of combustion is the consequent rise in the carbon dioxide content of the atmosphere, a process that is being accelerated by the wholesale destruction of forests. For many years the CO_2 content of the air has been rising at a compound rate of 0.7% per annum, which implies a doubling in a century. Many warnings have been uttered of the dire consequences if this process is allowed to continue unchecked, for CO_2 induces a 'greenhouse effect' whereby the average temperature of the atmosphere must steadily rise. This is because the gas admits the entry of ultraviolet waves of solar radiation but obstructs the outward flow of infra-red waves into which the ultraviolet waves are degraded on contact with the earth's surface. As a result there is a steady accumulation of heat that cannot be lost into outer space. The slow warming of the earth might have short term beneficial effects on the growth of crops, but ultimately there could be a gradual melting of the polar ice caps and a consequent gradual rise in the level of the oceans, resulting in land inundation. If carried too far, life could ultimately become insupportable long before conditions on the earth approached those of the planet Venus! In so far as heat mining can replace combustion, these dire events can be avoided, or at least postponed. Even though these alarming predictions are challenged by some as being exaggerated, or even groundless, evidence of a gradual rise in atmospheric temperatures is not entirely lacking. Heat mining would provide a measure of insurance against the possible hazards of rising CO_2 content in the atmosphere.

Finally it may be claimed that heat mining would require minimum land use, as a very large volume of hot rock could be exploited by means of a single pair of bores accommodated within a few square metres. Only the conversion equipment, if electricity production is the aim, or the distribution equipment if direct heat utilization is employed, would occupy very moderate areas of ground space.

For further reading see Kestin *et al.* (1980), Chapter 9.

Chapter 14

Economic aspects of heat mining

14.1 Overall economic considerations

Because heat mining has not yet been practiced commercially, economic estimates are inherently clouded with uncertainty. Nevertheless, some broad conception of the probable cost range of mined heat for specified purposes is essential if public or private enterprise is to be stimulated into activity. As with any other new form of engineering development a fairly large element of risk must inevitably be attached to early heat mining operations, but as experience is gained, the risk element will recede. Although the only direct experience of heat extraction from artificial HDR reservoirs has been confined to a few experimental activities financed on a governmental or intergovernmental scale, there is a wealth of practical and economic experience on which estimates can be made of many of the component costs of an HDR installation. These include costs for site exploration, drilling, fluid pumping, electric power generation equipment, and heat exchange equipment needed for district or process heating. Thus, only part of the total costs for heat mining will be at high risk.

The presentation in this chapter only summarizes the economic aspects of HDR development. Readers interested in details should consult earlier publications by the authors (Tester, 1982; Tester *et al.*, 1979; Milora and Tester, 1976; Cummings and Morris, 1979; Murphy *et al.*, 1982) and by others on the economics of hydrothermal development (Armstead, 1983; Kestin *et al.*, 1980).

In many cases, heat mining will be a large-scale activity – in terms of hundreds rather than tens of MWt at a single site: to that extent costs will be beneficially aided by the 'economy of scale effect'. But it is also true that construction times are likely to be long, especially in the creation and testing of reservoirs; this means that whereas much of the capital investment will be needed early, the revenues ultimately to be earned will be adversely affected by their discounted 'present worth' at the time when the project is launched. Future expected earnings will then have to be discounted accordingly, and their present worth will be correspondingly lowered.

Another important economic consideration in heat mining will be the expected 'life' of a reservoir. A given mass of HDR will contain a certain finite

396

quantity of recoverable heat that can be mined in a manner between two extremes. If the rock mass is only coarsely fractured it is possible to extract heat at a moderate rate over a relatively long life; but if it is finely fractured – fully 'rubblized' ideally – it would be possible to extract heat at a far higher rate over a much shorter life. As the second possibility would give quicker returns on the investment, it has obvious attractions over the first; but it may be much costlier in terms of the capital costs for reservoir stimulation and production. An optimum between the two extremes must be sought.

As explained in the earlier chapters of this book, the potential economic rewards of heat mining are much greater in places having high temperature gradients than in normal non-thermal regions, because of the substantially lower drilling costs associated with the shallower depths required to reach a desired rock temperature. Early commercial heat mining operations will therefore undoubtedly be sited at places having abnormally high temperature gradients and situated close to potential heat markets, whether for electric power generation or for direct application. Only when extended experience has been gained, both practical and commercial, are places of lower temperature gradients likely to be exploited for heat mining purposes. This approach is commonly termed 'high-grading'; it applies equally to other non-renewable energy sources such as oil and gas.

To improve our objectivity in the light of the many technological uncertainties that exist, we have sought to parameterize, when possible, certain key engineering factors to assess their relative effects on the economics. Fortunately, as mentioned earlier, several components of an HDR enterprise are common to other commercially operating systems, so their costs are less subject to controversy. In addition, several institutional and regulatory factors (taxes, ownership, etc.) applying to hydrothermal installations are also likely to be similar for HDR systems. However, no matter what is done to reduce the *perceived* or *real* risks, there will always be some residual uncertainty which, for certain engineering cost components, particularly those related to drilling and reservoir stimulation, may be very large. Interest rates and rates of return on invested capital will therefore be well in excess of prevailing market levels as an inducement to the investor for assuming these risks. This will be particularly true for early heat mining developments where risks will be perceived to be greatest.

14.2 Adaptability of heat mining to market needs

14.2.1 Temperature

HDR differs from fossil fuel resources in that the source temperature is largely a matter of choice rather than of combustion chemistry or of material limitations. Subject to over-riding economic or practical considerations, we can choose any desired source temperature by fixing the depth to which we drill into a field of known temperature gradient. Solar energy also to some extent has this quality of adaptability in so far as high temperatures (say greater than 300°C) obtainable by concentrators are conducive to high conversion efficiencies. For HDR, high temperatures can generally be obtained only by drilling to great depths, and as drilling costs rise very rapidly with increasing depth, temperatures will be limited, resulting in some sacrifice of efficiency. Fortunately, direct applications for industrial, farming or space heating purposes for the most part need heat at only moderate temperatures – seldom above 150°C. It seems probable, therefore, that until or unless the cost pattern of deep drilling changes for the better, applications for heat mining will be limited to low efficiency electric power generation and direct use for process and space heating purposes. As pointed out earlier in Sections 5.1 and 6.3, utilization efficiency could be improved by mining at multiple depths, or by employing cogeneration systems if economically justifiable.

14.2.2 Demand pattern and availability

The adaptability of heat mining to market needs, as covered in Chapter 5, showed that a close match between demand and source can be achieved with only a moderate degree of heat degradation and with little wastage. Later in Chapter 7, Section 8, we described how HDR energy supply (unlike hydrothermal energy) could easily be regulated to meet wide short-term variations in demand without wastage. Technically, therefore, HDR can easily be adapted to any pattern of demand, though obviously high load factors are commercially preferable.

The issue of load or availability factor is common to any form of utilization, but obviously has different implications for each. HDR systems could be used for base load electrical applications and for seasonal space heating or base load process heating needs in the same way as natural hydrothermal resources. Typical load or availability factors would be expected to range from 80 to 85% for electricity production, 90–95% for industrial process heat needs such as in a paper mill, and considerably less for space heating depending on seasonal and diurnal use patterns. Depending on location, annual space heating load factors may be only 20 to 30% for places with climates like New York and as high as 50 to 60% for those similar to Iceland (see Chapter 11 in Armstead (1983) for details).

14.2.3 Efficiency

A finite form of thermal waste in the heat mining process itself is the irrecoverable permeation of the circulating fluid away from the confines of the reservoir into the surrounding rock. If the fluid throughput varies, it is possible that a certain amount of seepage may be recoverable at times of reduced fluid input, but there will always be some that is permanently lost unless downhole pumping is used.

Fortunately, for most practical cases, the fraction of energy lost by fluid permeation is small. Therefore, the thermal efficiency of the whole heat mining process will mainly be governed by the nature of the utilization process itself, and in this respect heat mining is as suitable as alternative sources of heat, except that it is available at lower temperatures than provided, for example, by fossil fuels. Thermodynamically it is of course far less wasteful to use the mined heat directly at high efficiency, rather than to convert it into electricity and incur the inevitable losses inherent in the 2nd law of thermodynamics. Nevertheless, since electricity possesses the immeasurable attractions of convenience and versatility there will always be a demand for it, even at some considerable loss in efficiency.

14.2.4 Fluid quality

In district space heating applications, as practiced in Iceland, the USSR, Italy, Hungary, France, Japan, Romania and the USA, hydrothermal fields have been extensively and economically used. Hydrothermal fluids, however, are often chemically ill-suited for this purpose, so that expensive heat exchangers fabricated with corrosion resistant materials may be required. The fluid circulated through an HDR reservoir is much less likely to be chemically aggressive, so that in this respect heat mining may be more adaptable to space heating requirements.

Except for high efficiency power generation – and this may be a limitation that will pass as deep drilling technology improves – heat mining would provide a very adaptable source of energy particularly well suited to base load power generation at moderate efficiency, large scale district heating systems, and to multi-purpose power and heat applications. However, as temperatures increase above 300°C, the reactivity of the rock with the circulating fluid may increase to a point where high levels of dissolved minerals create a chemical problem for the system.

14.2.5 Economy of scale

The costs of drilling and of reservoir stimulation are at present so great that even a single pair of bores could never be commercially justified for any application involving heat extraction rates at less than perhaps 30 or 40 MWt; and once the necessary equipment for drilling and fracturing has been assembled at any one site there would be every incentive to form a reservoir large enough to embody a multi-bore extraction system. Whilst the economies of scale would have a markedly favorable influence on the cost of the reservoir formation, they would confer even greater cost benefits on the electric power plant up to a capacity of about 50 to 100 MWe. Beyond that point, well piping distances, multiple units for turbogenerators, heat exchangers and condensers, and the process control problems associated with such complexities will limit any further cost savings on a cost per installed kWe basis. This fact has been borne out by worldwide development of electric power from natural hydrothermal reservoirs. For example, by 1987 at The Geysers field in northern California approximately 2000 MWe of total generating capacity are achieved with 24 self-contained plants of around 50 to 100 MWe each. For individual plants of this size, total capital investments will typically be limited to 50 to 100 million dollars. This is quite different from a large coal-fired or nuclear installation having a generating capacity of 1000 to 2000 MWe and requiring capital financing of 1 to 3 billion dollars per plant. In fact, recently, this high total investment cost has motivated construction of smaller 200 to 300 MWe capacity coal-fired units in a number of cases in the USA, even though the cost in dollars per kilowatt would be somewhat higher.

In many non-electric or cogeneration applications, much smaller unit capacities may be the best economic choice. For example, consider a smaller industrial user who operates a high load factor plant that is remotely located requiring only 3 to 4 MWe and 20 to 30 MWt. For this application, an HDR reservoir with only two wellpairs would be sufficient to meet the entire energy demand.

14.3 End use options and economic tradeoffs

14.3.1 Electric power generation

Several design options are available for converting the heat content of any hot geothermal fluid into electricity. As described in considerable detail by Kestin *et al.* (1980), Milora and Tester (1976), Tester (1982), and Armstead (1983), most of the realistic possibilities involve open- or closed-loop Rankine cycles as depicted in Fig. 14.1. Heat from the geothermal fluid can be utilized to produce steam by flashing the geothermal fluid itself to a lower pressure and expanding the steam through a condensing steam turbine, with the unflashed liquid fraction either flashed again and expanded or reinjected. An alternative to these single or multiple stage flashing cycles, which are in commercial operation throughout the world, is the so-called binary-fluid cycle. It too is a Rankine cycle, but one that has only recently been put into commercial-scale operation. Heat contained in the produced geothermal fluid is transferred in a primary heat exchanger to vaporize a secondary working fluid which then expands through a turbine, and is condensed with heat finally rejected to the environment in a cooling tower or similar device. A closed cycle is completed on the secondary side by pumping the working fluid up to its maximum operating pressure.

Compounds utilized as secondary working fluids in binary cycles include low

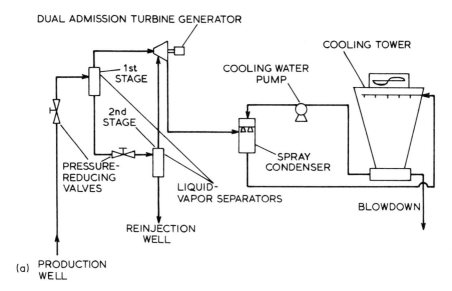

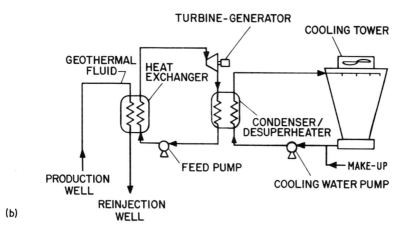

Fig. 14.1 Electric power cycle configurations possible for use with an HDR system using circulating pressurized water to extract heat. (a) 2-stage flashing cycle. (b) Binary-fluid Rankine cycle.

molecular weight hydrocarbons such as isobutane (C_4H_{10}) and propane (C_3H_8), their halogenated derivatives such as $CHClF_2$ (R-22), $C_2Cl_2F_4$ (R-114), and ammonia (NH_3 or R-717). Because these non-aqueous fluids have reasonably high vapor densities relative to steam at condensing temperatures of 15 to 30°C, much smaller and less expensive turbomachinery would be required for the same power output. On the other hand, flashing systems do not require a primary heat exchanger or secondary fluid feed pump. Economic tradeoffs for each system exist: for flashing plants, large, expensive, low-pressure steam turbines dominate the capital costs while heat exchanger and condenser costs are the major items for a binary-fluid plant.

It is important to emphasize that the technology associated with electric power production from low temperature fluids will not limit the development of HDR. Existing, 'off-the-shelf' equipment is available now, and improvements using binary fluids, direct two-phase expansion turbines, hybrid cycles using mixtures and/or combinations of flashing and heat exchange steps are currently under development for hydrothermal and solar applications (see Chapter 10, Armstead, 1983, for details).

An HDR reservoir is distinctly different from a natural hydrothermal system in one important respect: the HDR consists solely of a rock formation with a defined temperature gradient (see Fig. 6.9). Unlike a hydrothermal reservoir, it does not depend on the presence of hot fluids at a particular temperature. In fact, reservoir sizes and temperatures and fluid circulation or heat removal rates can be selected by design rather than by what nature has provided. All we depend on is the presence of hot rock with certain properties. This provides us with an important choice that does not exist for hydrothermal reservoirs. We can drill deeper to produce higher reservoir temperatures and produce more efficient electric power or heat per unit of fluid mass circulated through the reservoir. Thus if a constant fluid circulation rate per pair of wells in an HDR system is assumed, fewer wells will be required for a plant of specified MWe or MWt output as one drills deeper.

In order to explore the economic impact of this design choice, we must first quantify the effect of fluid temperature on efficiency. Figure 14.2 illustrates the improvement in electric conversion efficiency (η_c) that results in a binary cycle as the wellhead fluid temperature rises. The various possible conversion cycle options show similar trends. In all cases, the cycle efficiency would be equivalent to the ratio of net power produced to the rate of heat removal from the geothermal fluid. The utilization efficiency (η_u) is defined as the ratio of the net power produced to the maximum power that could be produced in an idealized, fully reversible process only limited by the 2nd law of thermodynamics. It is a measure of how efficiently the fluid is being used:

$$\eta_u = \frac{\dot{W}_{net}}{\dot{W}_{max}} = \frac{P_{net}}{P_{max}} \tag{14.1}$$

where the maximum power level ($\dot{W}_{max}$ or P_{max}) can be defined in terms of the so-called availability of the fluid from its state at the wellhead (T_{gf}, P_{gf}) to the ambient or 'dead-state' condition at (T_o, P_o) where heat is ultimately rejected to the environment. Thus,

$$\dot{W}_{max} = P_{max} = \dot{m}(\Delta H - T_o \Delta S) \left.\right]_{T_0,P_0}^{T_{gf},P_{gf}} \tag{14.2}$$

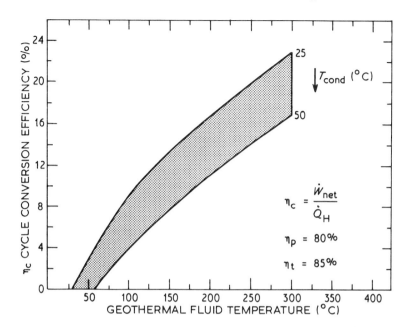

Fig. 14.2 Typical Rankine power cycle conversion efficiencies for geothermal power plants employing non-aqueous fluids as a function of fluid wellhead temperature with full allowance for parasitic auxiliaries. (From Tester, 1982).

As shown in Fig. 14.3, for a given conversion cycle configuration η_u generally increases as fluid temperature increases. This implies that cycle efficiency is increasing with rising temperature T_{gf} faster than the corresponding increase in $\dot{W}_{max}$.

Having established these thermodynamic performance criteria, we can proceed to analyze the economics. Individual well costs are the major cost component and are controlled by rock-type, well diameter, depth, and a number of other location-related factors. By specifying a certain HDR reservoir of fixed fluid production capacity for a given resource, rock-type and well diameter are defined. That leaves depth, on which well costs are strongly dependent. As was fully justified in Section 11.3.2, this depth dependence can be expressed empirically as a power or exponential function of depth:

$$\text{well cost} \sim (\text{depth})^{n^\circ} \qquad n^\circ = 1.4 \text{ to } 3.0 \qquad (14.3)$$

or

$$\text{well cost} \sim \exp(\text{depth}) \qquad (14.4)$$

Economic tradeoff possibilities exist between increased plant efficiency and lower equipment costs at higher temperature versus higher individual well

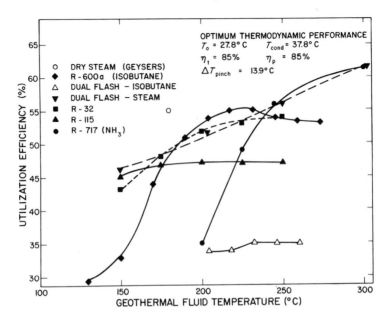

Fig. 14.3 Power utilization efficiency as a function of geothermal fluid temperature for seven different Rankine cycles involving direct expansion of dry steam, flashing, and organic binary cycles. Refer to the glossary of terms for chemical formulae of working fluids R-600a, R-32, etc. (Adapted from Tester, 1982.)

costs. The choice is either to drill deeper to produce hotter fluids from fewer, more expensive wells which results in a more efficient, less expensive power plant on the one hand, or to produce cooler fluids from shallower depths resulting in more, but less costly wells and a less efficient plant on the other hand. Milora and Tester (1976) created a simple model for HDR systems to explore this tradeoff quantitatively. In effect, thermodynamic performance was parameterized using a correlation as shown in Fig. 14.4, where the geothermal fluid flow rate per MWe is plotted against fluid temperature for several binary Rankine cycle configurations using different working fluids. An exponential drilling cost model and a linear plant cost model were used. Over a temperature range from 100 to 300°C, installed plant costs can be represented as an approximately linear decreasing function of wellhead fluid temperature (see Fig. 14.8).

Figure 14.5 shows the results of one particular case study where busbar generating costs in cents/kwh are plotted as a function of wellhead fluid temperature and mean geothermal temperature gradient at flow rates of 45, 113 and 227 kg/s per pair of wells. Although the actual generating cost figures shown depend on the assumed component costs, the relative effects would be expected to be generally true. Although the minima are fairly shallow, there is

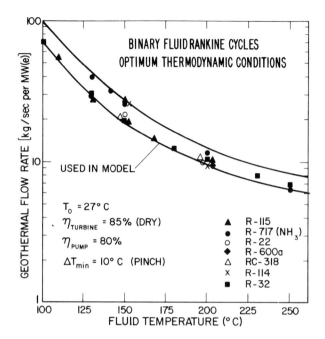

Fig. 14.4 Geothermal fluid flow rate requirement per megawatt as a function of wellhead fluid temperarture. Refer to the glossary of terms for chemical formulae of working fluids R-115, R-22, etc. (From Tester, 1982.)

an optimum temperature corresponding to a certain reservoir depth given a ∇T, where the optimal busbar costs occur. Figure 14.6 plots this optimum reservoir design temperature as a function of gradient at a well pair flow of 75 kg/s. As individual capital costs or other economic assumptions change, the actual position of this optimum temperature or depth will vary but the general trends should be similar.

14.3.2 Space and process heating

Direct application for space heating raises certain practical concerns regarding fluid transmission and retrofitting. As the HDR plant site becomes more remote from the place where the heat is utilized, its economic attractiveness is less with the need to establish pipeline right-of-ways to transmit the heated working fluid to the point of use and possibly to return it to the HDR plant.

Except in a few countries where district space heating is commonly used, regulatory and institutional barriers may have to be overcome before widespread implementation of HDR space heating systems. Industrial process heat applications are a different story. In cases where individual companies will utilize the heat internally to produce process steam, they may be willing not

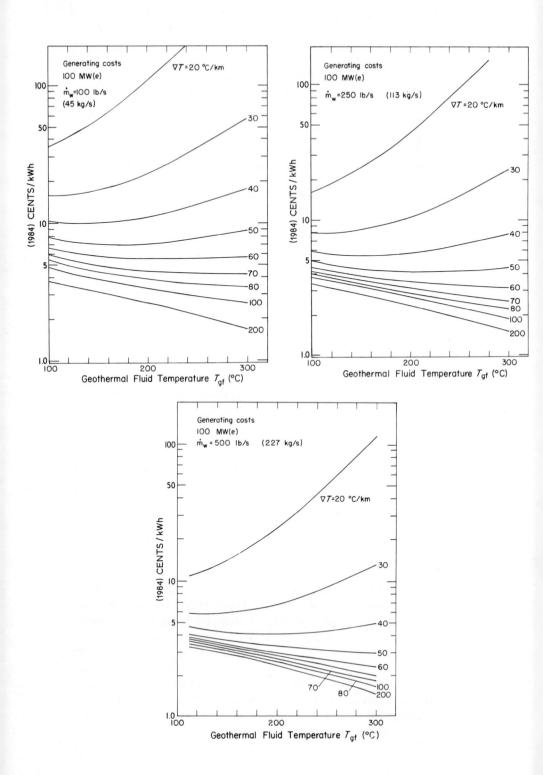

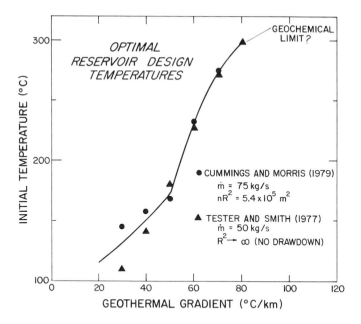

Fig. 14.6 Effect of resource quality (geothermal temperature gradient) on the optimal design temperature for HDR systems. (Adpated from Tester, 1982).

only to arrange for fluid transmission but also may financially support the development of the HDR reservoir itself.

There are three major factors that influence the cost of supplying process or space heat from an HDR system: *the magnitude and location of the demand* (or the size and proximity of the market), *the grade* (or required fluid temperatures), and *the capacity or composite load factor* (net percent utilization of the available HDR supplied heat on an annual basis).

One would expect that the total cost of providing geothermal space heat in the range of 100° to 150°C would depend linearly on the distance between the plant site and the user. As the size of the market grows, larger diameter pipes would be used, causing the unit costs to decrease in proportion to economy of scale factors. Einarsson (1975) has provided a very detailed analysis of geothermal heating costs in Iceland to show the quantitative aspects of these effects. Other published cost data for geothermal space heating costs are scarce, but once the conditions regarding a particular resource are known, then

---

Fig. 14.5 Estimated busbar generating costs in 1984 dollars for an HDR reservoir of specified performance (10% drawdown in 20 years) as a function of resource quality (geothermal gradient) and wellhead fluid temperatures (well depth) for fluid flowrates of 45, 113 and 227 kg/s per pair of wells ($\dot{m}_w$).

accurate cost estimates could be made easily since very little sophisticated machinery or equipment is normally required for direct heat applications. Capital costs for the surface system would be largely determined by heat exchanger costs (if required) and by installed fluid transmission piping costs. One would anticipate that heat exchanger unit costs would scale linearly with heat exchange area and would be affected by the presence of HDR reservoir fluids that are more corrosive than normal (Tester, 1982). The remaining costs for a direct use application are related to well drilling, completion, and stimulation costs which are discussed later in Sections 14.4 and 14.5.

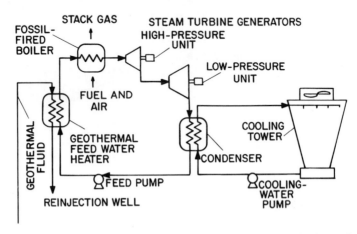

Fig. 14.7 Hybrid steam power cycle (500 to 1000 MWe typical) with a fossil-fired boiler and HDR preheater. (Adapted from Tester, 1982.)

14.3.3   Cogeneration

As depicted in Fig. 5.2, a cascaded system may sometimes be possible where the high temperature fraction of the fluid's thermal energy is used to generate electricity or to produce high grade process heat while the lower temperature portions are utilized as direct heat at those temperature levels. Cogeneration in this manner has obvious advantages from a thermodynamic standpoint. Because the cost of producing the hot fluid is substantial, this thermodynamic advantage translates into a real economical advantage as well.

Other forms of cogeneration using HDR are possible involving dual and topping-bottoming cycles with summer and winter configurations. In some cases, when space heating needs are nil during the summer, peak electrical demand often occurs, and a second electric power cycle using a low-boiling point organic fluid could be used to generate power from the fluid's heat content exhausted from the primary heat exchanger in the dual cycle or from the heat rejected in the condenser of the topping cycle. During the winter, with space heating load factors high, these heat sources would be used directly for district heating purposes.

Figure 14.7 shows one other hybrid concept that has been proposed for generating electricity where coal and geothermal resources are nearby (Anno *et al.*, 1977 and Kingston *et al.*, 1978). This cycle configuration can significantly improve overall conversion efficiencies by reducing both coal and geothermal fluid consumption on a per MWe basis. Basically, heat is being utilized in the cycle closer to its source temperature reducing the thermodynamic losses associated with larger temperature differences. The major disadvantage of this type of system lies in the increased complexity of building and controlling the plant operation because two different 'fuels' are being used.

The capital costs associated with a cogeneration system will be a composite of the electric and non-electric components. Obviously, each case will be different, depending not only on the HDR resource characteristics but also on the split between electric and non-electric needs.

14.4 Component costs of heat mining

The risk elements of HDR exploitation are concentrated in the creation of artificial reservoirs and in the extraction of adequate quantities of heat from them. The costs of utilization plant and equipment are not difficult to estimate as they involve no new technology and as all the necessary components are available in well established markets. Other cost components, especially those that involve high risk elements such as drilling and stimulation, are to some extent speculative; but sufficient evidence is available for such costs to be broadly estimated to show whether heat mining is but an impracticable dream or whether it looks as though it stands a good chance of being commercially competitive with other forms of energy. Each cost component will now be examined in turn.

14.4.1 Plant equipment costs

For conceptual designs, the *total installed cost* of an HDR power plant can be related to the total *purchased cost* of the major equipment items in the plant using a factored estimate approach (Tester, 1982; Milora and Tester, 1976). The major equipment items used will vary depending on end-use and the particular plant design selected. For example, for electric power generation, primary equipment would include heat exchangers, pumps, condensers, cooling towers, and turbines and electric generators. For non-electric use, only costs for pumps and heat exchangers would be included in the major equipment. Both direct and indirect cost factors are used to account for such things as instrumentation, buildings and structures, piping, insulation, environmental controls (under direct costs); and engineering and legal fees, overhead, contingency, and environmental and regulatory fees (under indirect costs). The combined effect of these factors typically causes the installed plant cost to be about 2.7 times the total purchased equipment cost.

Based on the analysis of a large number of conceptual designs for organic binary power plants at various resource temperatures, estimated total plant costs decrease more or less linearly with wellhead fluid temperature over a range of 100 to 300°C as shown in Fig. 14.8 (Milora and Tester, 1976; Tester, 1982). The major cause of this decrease is the smaller equipment sizes, particularly heat exchangers and condensers, that result as the conversion efficiency improves at higher resource temperatures. Even though the resulting correlation is obviously oversimplified, it serves to quantitatively account for a real economic effect important to modeling and eventually optimizing the design of a commerical heat mining development.

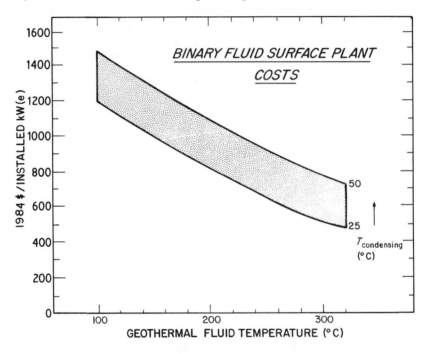

Fig. 14.8 Estimated geothermal power plant costs in dollars 1984 as a function of wellhead fluid temperature. A binary fluid Rankine cycle utilizing an organic working fluid is assumed.

14.4.2 Exploration and land acquisition

The location of suitable heat mining sites should be less troublesome than in the case of hydrothermal or fossil fuel resources. For a long time to come, interest will be focused on regions of high thermal gradient; and to locate these, to identify rock types, to measure thermal gradients, to study the local geochemistry, etc., it will be necessary to undertake hydrologic, geologic and geophysical surveys. Shallow exploratory drilling for heat flow and gradient measurements may often be appropriate, but at least one deep exploration hole would be needed to substantiate earlier estimates of rock structural and fracturing characteristics and temperature gradients.

Land lease or purchase costs, which can vary quite widely, must be determined; and local water usage rights must be secured to provide for the make-up losses in reservoir permeation and in cooling towers, as well as for drilling activities and for general usage.

Table 14.1 provides a reasonably realistic range of the relevant costs for low and for high perceived risk categories. Although these costs will form a small fraction of the total project costs, their impact can still be significant because they may be incurred several years before any revenues are earned from exploitation.

Table 14.1. Estimated exploration and land acquisition costs for a 200 MWe HDR development (updated from Tester, 1982, Table 10.5)

Cost category	*Cost in US $1000 (1984)*	
	Low perceived risk	*High perceived risk*
(1) Power plant site purchase	500	220
(2) Leased sites for well drilling	800	220
(3) Geophysical, hydrologic, and geological surveys for site reconnaissance and selection	220	400
(4) Shallow heat flow/temperature holes (100–200 m)	470	470
(5) Deep exploration hole* to reservoir depth of		
3 km (9800 ft)	8000	8000
4 km (13 100 ft)	12 000	12 000
5 km (16 400 ft)	19 000	19 000
6 km (19 700 ft)	30 000	30 000

* Costs based on data presented in Fig. 14.9. (Present HDR costs.)

14.4.3 Reservoir proof-of-concept costs

Before investment capital can be guaranteed, it may be necessary in the early stages of the technology development to undertake a proof-of-concept or demonstration on a less than full commercial scale. The costs of such a 'pilot' scheme may be quite substantial, partly because they are incurred long before any revenue appears, and partly because fairly long test periods may be required with a complete plant and prototype reservoir. An engineering staff will be needed to operate the system and analyze the results. All of this can be quite costly. The US and UK HDR research projects have already made a combined investment of over $150 million US dollars. This should be regarded as a probable upper cost limit for two such projects, as they are both pioneering, first-of-a-kind HDR developments. With many of the key questions answered by these two projects, substantial cost reductions would be expected in future demonstrations of HDR feasibility.

14.4.4   Well drilling and completion costs

This question has been treated in considerable detail in Chapter 11, Section 3.2, and will here only be summarized. Up to about 3 km depth HDR wells are expected to be broadly comparable in cost with hydrothermal wells; that is about two to four times the cost of an 'average' oil or gas well to the same depth. At depths exceeding 3 km it is expected that the cost trend of HDR wells will be somewhat less than suggested by the quasi-exponential form of equation (14.4) as:

$$C_{well} = C_m + C_d \, \exp[A_z \cdot Z] \qquad (14.5)$$

where the first term on the righthand side is due to rig mobilization/demobilization charges and the second term accounts for the depth dependence in Z with A_z, C_d and C_m empirical constants.

Comparisons between costs for HDR wells drilled at Fenton Hill and Rosemanowes and those for hydrothermal and oil and gas drilling to similar depths were shown earlier in Figs. 11.28 and 11.29. Although these figures verified this form of the well cost equation, the uncertainty of estimating actual well costs at any depth is apparent from the scatter in the data. Readers are cautioned not to extract more than general trends from the costs shown in Figs. 11.28 and 11.29 because of the inherent uncertainty and risks of drilling in general. However, in order to make economic prognostications for HDR, a predictive well cost model is required. Figure 14.9 gives a composite prediction for HDR wells in 1984 US dollars, assuming two levels of commercial development: (1) *present level* with actual costs for EE-1, EE-2, EE-3, and GT-2 wells at Fenton Hill and RH-11 and RH-12 at Rosemanowes forming the basis of the correlation, and (2) *commercially-mature level* for rotary drilling with disaster-free costs, a favorable rig availability environment, and some advances in rotary bit technology assumed. Of course, it may be possible to lower costs even further than the commercially-mature levels for rotary drilling by using advanced drilling methods as discussed earlier in Section 11.4. But for some time to come, it would be reasonable to expect the cost of HDR wells to lie within the shaded area of Fig. 14.9.

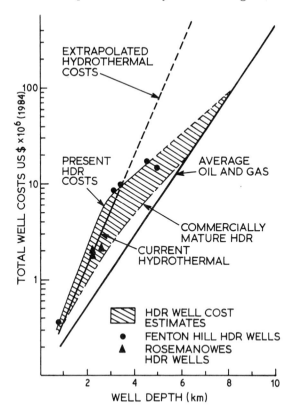

Fig. 14.9 Estimated rotary drilling and completion costs for an individual HDR well in hard crystalline rock as a function of well depth. Current and commercially-mature projections for HDR wells are given. Completion in this case refers to the placement and cementing of all steel casing liners and wellhead valving assemblies. No account is made of the potential costs of a downhole valving or production tubing assembly nor of stimulation costs for hydraulic fracturing and the like.

14.4.5 Reservoir stimulation costs

Murphy *et al.* (1982) were the first to consider fracturing costs as a separate item in estimating HDR heat production costs for Fenton Hill conditions. Previous work by Tester *et al.* (1979) and by Cummings and Morris (1979) assumed that these costs were minor, compared with the well-drilling costs themselves, so they became incorporated into the direct costs of the completed wells. Recent data at Fenton Hill and Rosemanowes suggest that fracturing costs will be large enough to be considered separately.

What makes the estimation of stimulation costs difficult at this time is that the procedures for forming commercial-sized HDR reservoirs have not yet been fully determined. High pressure pump rental charges and fracture fluid and proppant costs can be quickly obtained from vendor cost schedules, but in order to calculate the full stimulation costs it is necessary to know the total volumes to be pumped, the pressures, flow rates, etc. If, however, we use the experience at Fenton Hill and Rosemanowes as a guide, crude estimates can be prepared. For example, in an economic analysis presented by Murphy *et al.* (1982) for a 75 MWe power plant installation for Fenton Hill, they assumed:

(1) water used as a fracture fluid;
(2) self-propped fractures, each 360 m diameter;
(3) reservoir configured as a staggered 5-spot system with nine inclined wells and twelve sub-reservoirs (see Fig. 9.41(b));
(4) 192 fractures required to generate total necessary reservoir heat transfer area;
(5) injection volumes and pressures per fracture comparable with what had been observed during the enlargement of the Phase I system to a 300 m fracture (see Chapter 9, Section 9.4.1)

The resulting estimated costs were $530 000 (US 1982) per well, or $4 800 000 for nine wells, which is somewhat lower than for comparable oil and gas field hydraulic fractures. This is primarily because of assumptions (1) and (2). The hundreds of tons of solid proppants, the expensive fracture fluids with exotic properties, the blending trucks and tanks, and the pad and clean-up fluids used for typical oil and gas field stimulation are assumed not to be required for HDR. However, Murphy *et al.* (1982) are quick to point out that expensive downhole flow isolation hardware to balance flows into multiply fractured systems and repeated packer runs for fracturing could raise the costs to perhaps $14 million for the nine wells. Furthermore, more recent work at Rosemanowes suggests that high viscosity fluids may sometimes be desirable; this would raise the costs by an amount proportional to the total fluid volume used. At this point it is only possible to make a rough approximation that, given a reasonable extrapolation of current field results, the total stimulation costs will be of the order of $3 ± 1 million (US 1984) per *pair* of wells.

14.4.6 Fluid gathering and transmission costs

The pipework for gathering the hot water from the production bores, delivering it to the point of use, and returning the cooled water to the injection bores would have to be thermally insulated and provided with expansion facilities (see Armstead, 1983, Chapter 9). Costs will depend on well spacing and on flow rates, which fix the pipe diameters. By comparison with hydrothermal fields with more scattered well placement, an HDR system would be more compact, so that route lengths should be considerably shorter in the latter case, but of course return piping is always necessary for HDR systems. Hot water transmission is very much cheaper than steam transmission in terms of cost per unit of energy transmitted, for despite the lower permissible water velocities, the fluid density is so very much greater. These costs are best dealt with at this stage by providing a reasonable allowance per kW(th) – say $20 (US, 1984) or even more for long transmission distances.

14.4.7 Fluid pumping

The power required to pump the fluid through the reservoir will depend on the degree of buoyancy and on the flow impedance through the system. These factors have been considered in Chapter 7, and have later been compared with field data at Fenton Hill and Rosemanowes in Chapter 10, Section 2. The presence of high water losses may create a need for downhole pumping to reduce the overall pressure driving force for fluid permeation into the rock formation surrounding the active reservoir. Actual pump costs will depend on the volumetric flow rate as well as the type of service – downhole or surface-mounted. As shown by Tester (1982), pump prices can be correlated to drive shaft horse-power or MWe of electric power, as shown in Fig. 14.10. One particular advantage of HDR systems over hydrothermal is that fluid quality should be better and easier to maintain, so that pump service life will be extended. Experience at Fenton Hill and Rosemanowes suggests that dissolved and suspended solids levels should be low in HDR reservoirs created in low permeability crystalline rock at temperatures below 250°C.

Because of potential applications to hydrothermal reservoir production enhancement, several manufacturers have been developing hardware for downhole pumping (see Tester, 1982 and Armstead, 1983, for more details). Present materials limitations have restricted the upper operating temperatures of conventional shaft or electrically driven submersible pumps to about 200°C. However, novel designs have been proposed to operate at higher temperatures, perhaps up to 300°C. For example, Sperry Research (1977) has developed a downhole steam turbine-driven submersible pump for geothermal applications.

14.4.8 Institutional and regulatory costs

In addition to the actual capital costs for plant equipment and for drilling, the method of financing the development project as well as how the government taxes, regulates, and/or provides incentives for HDR development will play a vital role in determining the economic future for HDR.

(a) LOANS AND LOAN GUARANTEES

Just how bankers and others who might invest in HDR projects will perceive the risks of investing their capital is unclear. On the one hand, if they follow the 'oil and gas investment mentality', risks will be assumed high, causing an escalation of interest rates over what might be expected for other more conventional power development projects. If, however, they regard HDR as

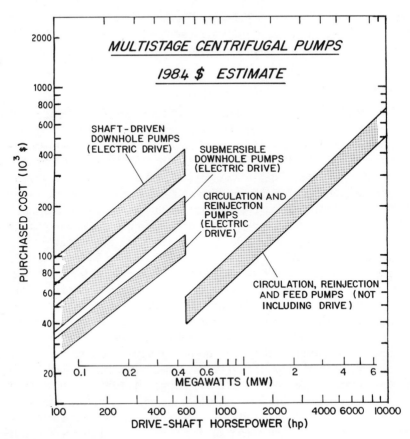

Fig. 14.10 Estimated costs for pumps for HDR service. The high cost figures correspond to high pressure operation at pressures up to 200 bar (3000 psia) whereas the lower cost figures refer to lower pressures up to 70 bar (1000 psia). (Updated from Tester, 1982.)

less risky than searching for oil or gas or for natural hydrothermal fluids, which, in our opinion, is a much more realistic assessment, then interest and debt equity rates will be lower.

In the USA, the Federal Government, and in other countries, the World Bank, have been underwriting low-interest loans for new commercial geothermal projects. Although these loan guarantees have dealt with only natural hydrothermal resources so far, the groundwork is in place to support HDR projects.

(b) TAXES, TAX INCENTIVES, AND CREDITS

HDR developments, at least in the US, would be subject to state revenue, *ad valorem* or property, and income taxes. In other countries, such as the UK or

Japan where all utilities are public government-run authorities, no taxes would be involved.

In the USA, where large taxes are imposed on private developments, tax incentives and investment tax credits are frequently used to promote development. For example, in 1977, the National Energy Act was passed by Congress to allow intangible geothermal drilling costs as a tax deduction using the same criteria applied to oil and gas drilling. Expensing of intangible costs, which may represent up to 70% of the total drilling costs, in the year that they are incurred increases cash flow by effectively creating a tax deferment. In another move to promote alternate energy development, the US government recently enacted legislation to force large utilities to purchase power from smaller (<75 MWe) suppliers at the so-called 'avoided' cost which is the cost that the large utility avoids by not having to add additional power generation units to meet its demand.

(c)   OWNERSHIP

Traditionally, in the USA, the resource- and reservoir-related development aspects of a field and its costs are borne by a 'producer' (typically an oil company). The hot water or steam is then sold to a second company or 'operator' who converts this to electricity or distributes it as heat to its customers. Although this separation between producers and operators has historical roots and may, in fact, be the most practical arrangement, it has generated an economic infrastructure that places different risks on the field development versus the power generator. In this case, a higher risk premium is placed on the producer's investment, which has led to higher interest rates, higher anticipated rates of return from equity capital, and higher equity-to-debt ratios. If the venture were singly-owned, more flexibility would exist in financing, resulting not only in lower effective interest rates but also lower required power prices to give a more favorable investment structure for the developer. Although it is hard to say just how HDR systems will evolve in this respect, one might expect more wholly-owned systems during the early phases of development.

14.5   Estimated total costs for heat mining

Obviously, from all that has been said so far, reliable prediction of generating costs for electricity or of producing heat from HDR is difficult. In fact, it would be ludicrous for us to propose that accurate HDR cost figures could be developed before a commercial-sized plant has been built and tested. Nevertheless a key question that needs to be addressed at this stage is: just what is required to make HDR economically competitive? This includes not only cost factors such as plant capital costs and well drilling costs but also performance factors such as conversion efficiency and reservoir capacity and lifetime. In order to answer these questions, a basis or model of HDR economics is needed.

14.5.1 The expected pattern of heat mining costs

If a certain volume of HDR is to be fractured and its heat exploited for some practical application, it is necessary to estimate the optimum 'life' that should be assigned to the resource, or, to put it in another way, to estimate the optimum rate at which heat should be extracted. As a general business principle it is usually preferable, when an investment has been made, to secure revenues from that investment as rapidly as possible; and when the enterprise ceases to be profitable, to sell off the residual assets and to move on to some new venture. Although this principle may be broadly desirable, it would be necessary to modify it for various practical reasons in the case of heat mining.

In order to study this problem it is helpful at first to neglect the effects of inflation and to assume that the sale price of heat or electricity remains fixed throughout the life of the exploited resource. On this assumption, the *total* revenue that can be earned from the resource would be fixed, while the *annual* revenue would vary more or less inversely with the number of years (i.e. the life) during which the resource is fully exploited. The investment is partly, but not wholly, dependent upon the reciprocal of the resource life. It is instructive to examine a very much simplified example, as follows:

Assume that a hyperthermal region of HDR is to be exploited by cooling the crust beneath it through an average temperature drop of 1°C; and that a horizontal layer of rock, 500 m thick, has been adequately fractured to form a reservoir of average depth of 3 km and at an average temperature of 200°C. This implies a thermal gradient of $(200-15)/3$, or 61.7°C/km. According to Chapter 4, Section 2, the total recoverable heat would be about 2500 MWy(th) per square kilometre of exploited area. An effective idealized way of perforating the exploited area would be to arrange the bores in a hexagonal array as shown in Fig. 14.11, with alternative bores serving as injection and recovery points. Assume further that 37 bores in all are sunk, and that an approximately spherical zone of fractured rock be formed around the base of each bore – each spherical zone having a radius of about 300 m. These fractured spherical zones would overlap with one another in the manner shown in the lower part of Fig. 14.11, thus providing good fluid communication between adjacent bores. Figure 14.11 shows that the reservoir so formed would have an effective thickness of about 500 m if the bores were spaced at 450 m apart. In practice, of course, an actual fractured reservoir would differ considerably from this simplified and idealized model, but the assumptions will suffice to serve the purpose here required.

The total heat capture zone would be bounded approximately by the broken line hexagon shown in the upper part of Fig. 14.11: its area would be six times that of the triangle ABC, namely $6 \times 1639^2 \times \sin 60°/2$, or

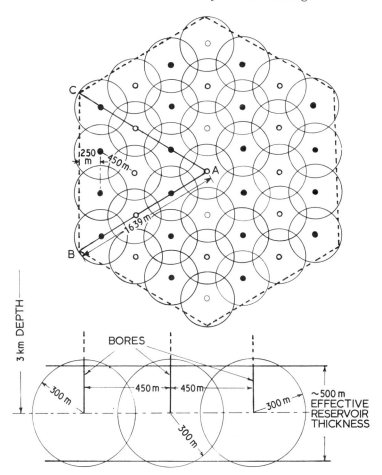

Fig. 14.11 Hexagonal array of alternating injection and production bores at 450 m spacing, each communicating with an approximately spherical zone of fractured rock at the base, thus forming an artificial reservoir approximately 500 m thick and 7 km² in horizontal area.

6 979 267 m², 6.98 km² or about 7.0 km². The total recoverable heat would therefore be 2500 × 6.98, or 17 450 MWy(th).

The cooling of a 35 km thick crustal column through 1°C (average) would be effected by cooling the 0.5 km thick reservoir through 70°C during its expected lifetime – i.e. from 200°C to 130°C. According to Fig. 4.9, these rock temperatures should enable electricity to be generated at efficiencies ranging from about 17.6% initially to about 9.2% finally – average 13.4%. However, it would be impracticable to make continuous adjustment for a gradual and steadily falling admission temperature so as to obtain optimum efficiency throughout the full temperature range. Stepped adjustments could be made by bypassing turbine stages or by periodically reblading, but at some cost in efficiency. Assume that a mean efficiency of only 12% is attainable over the life of the reservoir, instead of the theoretical figure of 13.4%.

The electrical energy that could be generated during the life of the reservoir would be 17 450 MWy(th) × 1000 (kW/MW) × 0.12 (efficiency), or 2 094 000 kWye. If the life of the reservoir is N years, and if the generating plant operates at an average annual load factor of 90%, then the capacity of the power plant should be

$$P = \frac{2\,094\,000}{0.9 \times N}, \text{ or } \frac{2\,326\,667}{N} \text{kWe}$$

An estimate of the total capital investment may now be prepared:

Cost of power plant at \$900/kW (Fig. 11.31) $\quad = \quad \dfrac{900 \times 2\,326\,667}{N \times 10^6}$

$\qquad\qquad\qquad\qquad\qquad\qquad\qquad\qquad = \quad$ 2094 US \$ millions/N

Cost of pipework at \$20/kW(th) (Section 14.4.6) $\quad = \quad \dfrac{20 \times 2\,326\,667}{N \times 0.12 \times 10^6}$

$\qquad\qquad\qquad\qquad\qquad\qquad\qquad\qquad = \quad$ 387.8 US \$ millions/N

		US \$ millions
Site acquisition and exploration, (Table 14.1), say	=	9.5
Proof of concept costs, (Section 14.4.3), say	=	40.0
Drilling 37 bores at \$3 million each (Fig. 14.9)	=	111.0
Stimulating 37 rock zones at \$1.5 million each (Section 14.4.5)	=	55.5
Pumping equipment (Fig. 14.10), say	=	4.0
Total investment		US \$(220 + 2482/N) millions

The annual recurring costs and charges would consist of the following elements:
(1) *Interest* on the investment, here assumed to be 10% p.a.
(2) *Depreciation or amortization*, which may be taken at the 'sinking fund' rate for 10% interest according to the reservoir life, N years.
(3) *Operation, repairs and maintenance*, for which an allowance of $1\frac{1}{2}$% p.a. on the investment will be assumed.
(4) *Contingencies and unspecified charges*, for which an allowance of 10% of the sum of items (1), (2), and (3) above will here be made.

The quantity of electricity generated during the life of the reservoir has been shown to be 2 094 000 kWye, or 1.8343×10^{10} kWhe. At an average price of c cents per kWhe the annual revenue would be ((\$183.43 c)/N) millions.

The break-even price will therefore be the value of c when the annual expenditure equals the annual revenue (on average).

EXAMPLE

Assume $N=20$ years. Then the annual cost will be at the following percentage of the total investment:

Interest	10.00%
Amortization at sinking fund rate for 10% interest and for 20 years	1.746%
Operation, repairs and maintenance	1.5%
	13.246%
Contingencies and unspecified charges at 10% of this	1.325%
Total	14.571%

Capital investment $=\$(220+2482/20)$ millions, or \$344.1 millions

Annual expenditure $=14.571\%$ of \$344.1 millions $=\$50.139$ millions

Annual revenue $=\$183.43/20 \times c$ millions $=\$9.1715$ c millions

The break-even price will then be c when 9.1715 $c=50.139$, or about, $c=5.5$ cents/kWhe

For other values of N the average break-even prices may be calculated in a similar way. The results are shown plotted in Fig. 14.12, which shows an apparent optimum reservoir life of 12 years and a break-even price of 5.0 cents/kWhe.

A similar exercise may be performed to ascertain the pattern of *heat costs* in terms of reservoir life. In this case the investment for providing heat in the form of hot water at a single point of delivery would be $\$(220 + 387.8/N)$ millions, and the heat supplied annually would average $(17\,450/N)$ MWy(th), or $8760 \times 17450/N \times 1000$ kWh(th), i.e. $((1.5286 \times 10^{11})/N)$ kWh(th). This is $(1.5286 \times 10^{11} \times 3412.5)/N$, or $5.2163 \times 10^{14}/N$ BTU p.a.

$$\text{or} \quad \frac{5.2163 \times 10^8}{N} \text{millions of BTU p.a.}$$

If p is the average price of heat per million BTU in cents, the average annual revenue will be

$$\frac{\$5.2163 \times 10^8}{N} \times \frac{p}{100}, \text{ or } \frac{\$5.2163}{N} \times p \text{ millions}$$

The average annual expenditure may be calculated at the same proportions on the investment as for electricity, and the break-even prices thus determined. The results are as shown also in Fig. 14.12, from which it will be seen that they follow a similar pattern to that for electricity costs, but with an apparent optimum at a much shorter life of five years, at which the break-even price would be about 87.5 cents/10^6 BTU.

As would have been expected, both curves of Fig. 14.12 show that a short life is generally more profitable than a long life, but that too short a life involves excessive depreciation charges. In both cases the apparent optimum reservoir life is inconveniently short for such costly investments; but for various reasons

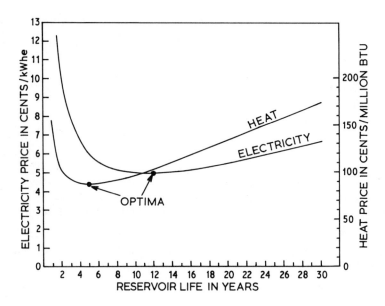

Fig. 14.12 Theoretical pattern of break-even prices for electricity in terms of the life of the reservoir.

much longer reservoir lives would almost certainly be adopted even though Fig. 14.12 would suggest that the break-even prices would thereby be raised.

In the first place it would be impacticable to extract heat so rapidly as to lower the reservoir temperature by 70°C in five, or even twelve, years however finely the rock is fragmented. Consider first the twelve-year case, in which the rate of heat extraction would average 17450/12, or 1454 MW(th). Assuming the working fluid has a constant heat capacity of 4187 J/kg K and is cooled through 110°C on average after leaving the production bores and before reentering the injection bores, the total circulation rate would have to be 1.454 × 10⁹/(110)(4187) or 3157 kg/s. If 18 of the 37 bores were production bores, the flow per bore would be 3157/18, or about 175 kg/s. At around 150°C this flow would be equivalent to about 191 liters/s, or 0.191 m³/s. The corresponding water velocity in a 10 inch (25.4 cm) bore would be 3.77 m/s, which is on the high side. Certainly a five-year life would be unthinkable, as the water velocity in a 10-inch bore would have to be about 9 m/s (30 ft/s). Thus the apparent optimum reservoir lives suggested by Fig. 14.12 would be impracticable. Reservoir lives of at least 15 years would probably have to be adopted for hydraulic reasons alone.

Another reason for favoring longer reservoir lives than the apparent optima suggested by Fig. 14.12 is that a power plant is usually expected to have an economic life of at least 20 to 25 years – often longer. Similar considerations would apply to any heat-consuming industrial plant or district heating installation. Thus if a reservoir were allowed to become obsolete in 5 or 12 years, the developers would be left with a serviceable plant on their hands, with years of unexpended life available and deprived of their source of motivating energy. There would be much to be said for equating the life of the reservoir to that of the utilization plant, so that all physical assets would become obsolete simultaneously. Figure 14.12 therefore suggests that more probable break-even prices would be somewhat as follows:

Heat   135 to 175 cents per million BTU
Electricity   5.5 to 6.5 cents per kWhe

However, there are other factors that make it most advisable to treat Fig. 14.12 with the utmost caution and to regard the above cost ranges as no more than order-of-magnitude figures (quite apart from the uncertainties of the various assumptions on which they were based).

One adverse factor is that no allowance has yet been made for interest during construction – i.e. for the deferment of revenue as compared with the incidence of expenditure. Even though the capital expenditure may be spread over several years before any revenues start to flow in, the significant period for which interest during construction would be incurred is the lapse of time reckoned from the 'center-of-gravity' of capital payments and the date from which revenues start to come in regularly. If, for example, this significant period were three years, the effect would be to raise the costs by 30% if the interest rate were 10% p.a.

This adverse factor is mitigated by a favorable factor, in that Fig. 14.12 has been computed on the implicit assumption that the whole of the capital expenditure is incurred at the outset, whereas in practice it would be possible to develop the reservoir more economically in stages so as to keep pace with the gradual depletion of the resource, thus deferring much of the capital expenditure. By preparing a scheduled development chart covering the expected program of construction and exploitation, and by adopting the method of discounted cash flow costing (present worth) over the whole time span of the project, a more rational estimate of the break-even prices could be made than by the simplified method used in deducing Fig. 14.12. The result would probably raise the break-even prices by a moderate percentage by comparison with those shown in that figure. However, this adverse effect would almost certainly be more than offset by yet another favorable factor, which arises from the fact that after the reservoir has been cooled through 70°C there would remain a huge quantity of accessible heat in the ground in the form of hot rock at 130°C, which is 45°C above the minimum useful temperature of 85°C that was provisionally laid down for Category I countries in Chapter 6. Thus on completion of the reservoir 'life', as limited by the assumption of 1°C average local crustal cooling, there would remain $45/(70 + 45)$, or about 40% of the useful available heat, together with all the necessary heat extraction facilities that have already been written off according to the accounting method on which Fig. 14.12 has been based. Although the permissible extent of crustal cooling has yet to be established as applicable to large tracts of land, there is little doubt that an isolated project such as has here been postulated could tolerate an average degree of local crustal cooling well in excess of 1°C. If full use could be made of the lower temperatures down to 85°C, in the example being considered, the average extent of crustal cooling would be (200−85)/70, or 1.64°C, which would almost certainly be acceptable. It is not possible to assign a quantitative value to this unused heat, as everything would depend on the availability or otherwise of suitable local heat markets. Being of lower grade than that which was first recoverable, it could perhaps in part be used for low efficiency power generation by means of a binary fluid cycle; but preferably a market for industrial or farming purposes or for district heating could be of greater economic value.

The outcome of all these considerations would seem to be that in the assumed circumstances the break-even prices would probably be substantially below the ranges suggested earlier. Precise figures cannot be given owing to the market uncertainties for the lower grade heat, but something of the following order would seem to be not unreasonable:

Heat   about 1.30 dollars per million BTU
Electricity   about 5 cents per kWhe

An entirely independent estimate of electricity production costs from HDR has been made by Murphy *et al.* (1982) in a study made of the economics of a 75 MWe HDR power plant based on the design of the Phase II reservoir at Fenton Hill. The authors deduced a break-even busbar cost of 4.4 cents (US)/kWhe in 1981 dollars – say about 5.3 cents/kWhe in 1984 dollars. This estimate was based on a ten-year reservoir life and a somewhat similar temperature gradient of 58°C/km to that assumed in our example, a mean well depth of about 4200 m and a mean temperature of 230°C at the point of use. Despite the differences in assumed conditions, this result is in reasonable accord with our own rough estimate, falling well within the suggested range.

14.5.2 Structure and assumptions of HDR cost models

Because of several unique characteristics of HDR, particularly the choice of design temperature, more complex cost models can be appropriately used to estimate required parameters for economic feasibility. The basic idea of any model is to combine capital and operating costs, reservoir and power plant operating parameters related to performance, and regulatory constraints and financial variables to produce an estimate of *minimum economic feasibility conditions*. This will set the required price for the product (HDR electricity or heat) which must be met in the market-place to cover all costs. This break-even price includes the cost of maintaining and borrowing the money; for example, principal and interest payments on bank loans, return on equity to share holders, *ad valorem* and income taxes, and plant maintenance costs. The selling price of electric power produced that meets these conditions is often termed a *levelized busbar cost*.

In our modeling of HDR economics, two basic financial models have been used to estimate power generating costs for a given set of resource and reservoir performance conditions.

(1) *Fixed charge rate (FCR) model*: Annual cost expressed as a specified fraction of the present value of the total capital investment superimposed onto annual operating and maintenance costs. Capital costs parameterized as a function of critical system variables (see Milora and Tester, 1976).

(2) *Intertemporal discounted cash flow model*: Estimates the optimal management path to maximize aftertax net benefits of an HDR system. Allows for various powerplant and reservoir design and operating choices to maximize net profits. Revenues and costs incurred in each time period are discounted to the beginning of production or plant operation (Cummings and Morris, 1979; Tester, *et al.*, 1979).

Before either of these models can be used, the total capital investment must be estimated. Usually, individual component costs are multiplied by an appropriate factor to account for indirect and direct cost factors relating to installation and contingency effects.

As stated earlier, the total installed cost of an HDR plant can be approximated to be 2.7 times the cost of the major pieces of equipment. Drilling and stimulation costs are treated with only an indirect cost factor of 1.38 applied to actual costs. For reservoir fluid downhole pumping equipment and fluid distribution costs a combined indirect cost factor of 1.51 was used, while for exploration costs it was 1.31. Given these assumptions, the total capital investment required for an HDR system can be expressed as:

$$\Phi = 2.7\Phi_e + 1.38[\Phi_w + \Phi_{stim}] + 1.51[\Phi_{dwp} + \Phi_f] + 1.31\Phi_s \qquad (14.6)$$

where the Φs refer to actual costs for purchasing equipment or services:

Φ_e = combined plant equipment cost (Fig. 14.8)
Φ_w = total drilling and completion cost (Section 14.4.4 and Fig. 14.9)
Φ_{stim} = stimulation costs (fracturing, downhole valving, etc.) (Section 14.4.5)
Φ_{dwp} = downhole pump equipment cost (Fig. 14.10)
Φ_f = fluid transmission costs (Section 14.4.6)
Φ_s = site development and exploration costs (Section 14.4.2–14.4.3)

If a proposed HDR plant was actually to be built, capital costs for all equipment would be obtained by bids from vendors or manufacturers, and drilling and stimulation costs would be obtained from actual field data for wells in that area or from bids from drilling contractors. However, at this stage of technology development, this kind of data or bidding capability just does not exist for HDR, so the approximations proposed earlier were used to specify the individual costs needed to estimate the total capital investment for each model.

(a) FIXED CHARGE RATE (FCR) MODEL

The FCR model produces a levelized busbar cost estimate by simplifying the rigorous discounting of cash flow. The annual cost of invested capital is obtained by multiplying the total capital investment by the FCR. The total annual cost is then obtained by adding this levelized capital cost to operating and maintenance costs. The FCR is a single number which, in today's financial climate, is about 15 to 17% for power generation facilities. The FCR incorporates the effects of several complex financial parameters including rates of return on equity capital, debt interest rates, income and *ad valorem* taxes, insurance, and depreciation (Tester, 1982):

$$FCR = CRF + f(CRF, \text{tax credits}) \tag{14.7}$$

where the *CRF* is the capital recovery factor, which is the annual payment as a fraction of the principal required to fully repay a loan (principal plus interest) over a specified period of n years at an annual interest rate i

$$CRF = \frac{i}{1-(1+i)^{-n}} \tag{14.8}$$

In general, as the interest and rate of return on equity capital increase, the *CRF* and *FCR* increase.

(b) INTERTEMPORAL DISCOUNTED CASH FLOW MODEL

Cummings and Morris (1979) recognized the need to incorporate the economic effects produced by the depletable nature of HDR. Because finite drawdown will occur and redrilling, restimulation, or additional drilling may be required to restore the power output of an HDR system, proper management of the reservoir and plant is essential to optimize return on investment. They developed a model for optimizing the after-tax net benefits that could be obtained from an HDR commercial venture with specified performance characteristics, capital costs, and energy selling prices. The most significant contribution of their model resulted from the inherent flexibility provided by its structure to explore the sensitivity of costs, not only to the key resource and reservoir performance parameters but also to consider the best way to manage the system. Optimal drilling, reservoir stimulation, and plant operating strategies could be determined to maximize profits. Specific aspects of their model included:

(1) *Reservoir performance.* Thermal drawdown rates could be specified by setting the effective heat transfer area contained between a pair of wells (see equation (10.31)).
(2) *Plant conditions.* Design conditions for the plant, including inlet fluid and ambient temperatures, plant size, and load factors, are specified to match

the resource characteristics. Penalties for performance decline as inlet fluid temperatures drop below design levels are included.

(3) *Financial factors.* Capital costs for the surface plant and well/reservoir system are specified during the time period where they are incurred. Taxes, tax credits, depletion allowance, and interest and equity rates of return are used to calculate net revenues for each time period. All costs and revenues are discounted to the present value in year 0.

(4) *Drilling/redrilling and stimulation strategies.* When and how deep (to what temperature) HDR wells should be drilled for a given temperature gradient. Initial rock/reservoir temperatures could be specified at levels equal to or greater than the plant design temperature.

(5) *Reservoir flow rate.* The thermal drawdown rate can be altered for a specified size reservoir by changing the mass flow rate $\dot{m}$ in each decision period.

The intertemporal model determines the optimal time path of management decisions including drilling activity and flow rate changes to maximize the present value of after-tax profit, which is expressed in a mathematical objective function. The model considers all possible options to maximize the objective function in a deterministic manner. Because of the mathematical structure of the intertemporal model, break-even selling prices are determined indirectly rather than directly as they are for the FCR model. The selling price is specified as a variable parameter in the revenue term, until the objective function becomes equal to zero. At this point, a break-even condition is reached, where the revenues generated from selling power are just enough to cover costs and debts including interest and principal payments, return on equity, and taxes.

Both the FCR and intertemporal models have been used under various assumed financial conditions to determine what was required in terms of reservoir performance for economic feasibility. In addition, the models have been used to explore the economic sensitivity caused by changing parameters such as drilling costs, interest rates, or plant design temperatures – see Fig. 14.16.

14.5.3   Important economic parameters for heat mining

The first parameter that has a strong influence on economic feasibility is end use. Assuming everything else equal, the costs for a non-electric system would normally be less than an electric plant of similar capacity. This may seem like comparing apples with oranges but two major reasons lead to this conclusion. The first is the inherent inefficiency of converting low-grade heat into electricity and the second is the simpler, less costly, plant equipment required for non-electric applications. In a typical case of heat mining where the wellhead fluid temperatures are 200°C, the overall transfer of heat would be about 80% efficient, while the conversion of heat to electricty would be about 14%

efficient, (see Fig. 14.2). Thus if we were to compare just the cost of producing the fluid at the wellhead for electric and non-electric applications, the ratio of supply costs for non-electric to those for electric applications would be a 0.8/0.14 or a factor of 5.7. Superimposed on this would be the cost of the plant itself which for a typical geothermal electric plant would be approximately five times more expensive than for a non-electric plant. Thus for wellhead plant conditions, heat mining would be about five times less costly for direct use than for generating electricity on an equivalent heat rate basis. These benefits could be completely offset, of course, if the fluid transmission distance was long.

With this rationale, we may proceed with a summary of modeling results obtained in earlier studies of the economics of electric power production from HDR done by Cummings and Morris (1979), Tester *et al.* (1979), and Murphy *et al.* (1982). The key economic factors affecting HDR can be divided into four major categories:

(a) Resource characteristics [average geothermal gradient ∇T].
(b) Reservoir performance [mass flow rate, thermal drawdown rate].
(c) Plant performance [fluid temperature].
(d) Financial aspects [capital costs, equity discount and interest rates, tax treatment].

Within these categories are embedded a relatively small number of parameters that effectively control costs.

(a)   RESOURCE CHARACTERISTICS

Lithology (rock type), depth to the reservoir, and temperature and ambient conditions, will determine the 'economic quality' of an HDR resource at a particular site. Geothermal temperature gradient is the key parameter as it interrelates depth and temperature. Earlier studies by Milora and Tester (1976), Garnish (1976), Tester and Smith (1977), and Cummings and Morris (1979), determined the optimum reservoir design temperature or reservoir depth corresponding to minimum generating costs. Earlier in Section 14.3, the study by Milora and Tester (1976) showed the dependence of busbar generating costs on produced fluid temperature T_{gf}, mass flow rate or reservoir production capacity $\dot{m}$, and geothermal gradient ∇T (see Fig. 14.5). These effects were then translated into a plot of optimum reservoir temperature or depth as a function of gradient as given in Fig. 14.6. Although the Milora and Tester approach was admittedly grossly simplified, later more comprehensive studies by Cummings and Morris (1979) showed that the predicted trends were qualitatively correct.

Figure 14.13 generalizes the effects by showing relative changes to busbar generating costs as a function of T_{gf}, ∇T, and $\dot{m}$. With high gradients ($>60°C/km$), the economics are controlled by surface plant capital costs, while

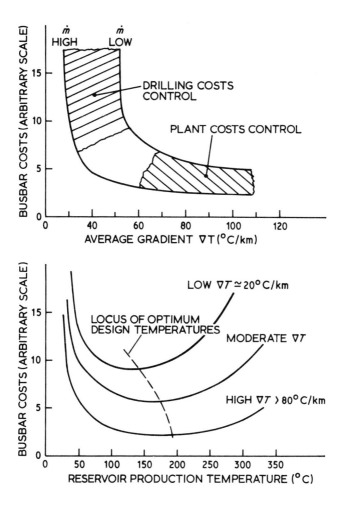

Fig. 14.13 Generalized effects of resource quality and reservoir performance on busbar generating costs for HDR-produced electricity.

with lower gradients, drilling costs dominate. Also, as the well pair production capacity $\dot{m}$ increases with no corresponding increase in thermal drawdown, busbar costs would decline because fewer wells would be required to provide the total mass flow of fluid to the plant. Superimposed on these effects is the fact that, for each set of resource and reservoir parameters, (∇T and $\dot{m}$), an optimum reservoir temperature (or depth) exists as shown by the dotted line in Fig. 14.13(b). A more quantitative representation of these effects is shown in Fig. 14.14 for the base case conditions given in Table 14.2. The effect of thermal drawdown on costs could be removed by increasing or decreasing the number of fractures in proportion to $\dot{m}$ in order to keep $\dot{m}/nR^2 \geqslant 6.67 \times 10^{-5}$ kg/m²s which would result in negligible drawdown as shown by Fig. 10.25. At base case conditions for a 40°C/km resource, and a 75 kg/s flow rate, the break-even price is about 6.2 cents/kWh.

Table 14.2. Base case reservoir performance and economic assumptions for a commercial-sized power plant utilizing an HDR resource at 40°C/km in the US (adapted from Cummings and Morris, 1979)

RESERVOIR PERFORMANCE	
Design well flow rate per pair of wells $\dot{m}$	75 kg/s
Effective fracture radius R	300 m
Number of fractures per pair of wells n	6
$\dot{m}/nR^2$ (20% drawdown in 10 years)	1.39×10^{-4} kg/m²s
Power plant design temperature T_d	160°C
ECONOMIC ASSUMPTIONS	
Exponential drilling costs (see Fig. 11.29)	$\Phi_w \sim \exp[Z]$
Surface plant costs decrease with temperature (see Fig. 14.8)	$\Phi_e \sim -T_{gf}$
Capacity factor	0.85
Plant size	50 MWe
Operating and maintenance costs	.13 cents/kWh
Contingency	13%
Real debt interest rate	9%
Real rate of return on equity	12%
Working capital	10% of Φ_e
Taxes as a % of taxable income	51%

(b) RESERVOIR PERFORMANCE

From the power producer's viewpoint, a constant fluid inlet temperature and chemical composition to the plant is desirable. Once the plant design has been specified, required mass flow rates at the design temperature can be determined to provide a certain power output level (see, for example, Fig. 14.4). However, finite drawdown is inevitable in most practical situations. The effect that finite thermal drawdown has on plant performance and on the economics must be considered for HDR as it is for a hydrothermal system. As the reservoir's production temperature declines, plant efficiency will decrease. Practically speaking, the effect of lowering the power producing potential of the fluid may actually be smaller than the effect of operating the plant at off-design conditions. The combined result is that plant output would drop off faster than might be expected from just the reduction in cycle efficiency at lower temperature. For conceptual design modeling, a parametric equation was utilized to lower the plant's output and reduce revenues as fluid temperature declined. The effective drawdown for this situation could be defined as:

$$(T_d - T(t))/T_d \tag{14.9}$$

where T_d = plant design temperature, °C
     $T(t)$ = wellhead fluid production temperature at time t, °C

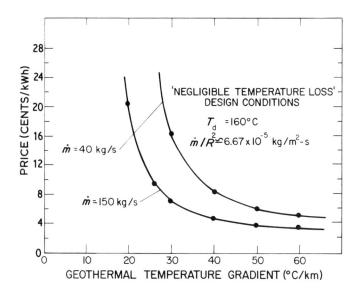

Fig. 14.14 Effect of geothermal gradient (∇T) and reservoir fluid production rate ($\dot{m}$) on breakeven busbar generating costs. ∇T and $\dot{m}$ conditions varied with other base case parameters given in Table 14.2 (Costs updated from Tester (1982) using a factor of 1.56 for inflation from 1979 to 1984 and a factor of 1.5 to account for higher relative drilling costs based on Fig. 14.9.)

Cummings and Morris (1979) examined the effect of finite thermal drawdown quantitatively. They used the idealized reservoir performance model introduced in Chapter 10 (equation (10.31)) and the intertemporal discounted cash flow model to develop optimum strategies for operating and managing an HDR system by changing mass flow rate and by redrilling and restimulating to alter the reservoir's drawdown characteristics. The reservoir consists of a set of equally-sized circular fractures contained between pairs of inclined boreholes and spaced horizontally by 50 m to avoid thermal interference effects between them over a 20 to 30 year period. Consequently, the total fracture area is proportional to the number of fractures times the area of one fracture or $\pi n R^2$ so the reservoir mass loading parameter becomes $\dot{m}/nR^2$. Although this non-interacting, multiple fracture model is admittedly only an approximation of real reservoir geometries, it clearly introduces the concept of finite drawdown in a systematic manner so that its economic impact can be studied.

At initial stages of operation or after redrilling and restimulation, $T(t)$ could be larger than T_d for a while as shown in Fig. 14.15. No further improvement beyond the design output was used, however, even if $T(t)$ was greater than T_d. In this system, finite drawdown occurs because $\dot{m}/nR^2$ is 1.39×10^{-4} kg/m²s, which, according to Fig. 10.25, would result in about 20% drawdown of the reservoir in ten years of operation.

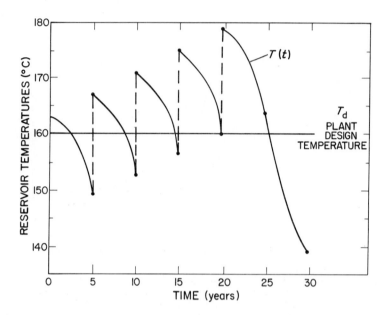

Fig. 14.15 Reservoir fluid production temperature as a function of time showing drawdown during periods of production and rejuvenation by redrilling at five year intervals. (From Cummings and Morris, 1979.)

The Cummings and Morris (1979) study showed that for the base case conditions cited in Table 14.2, redrilling about every five years led to maximum net benefits. Figure 14.15 shows the reservoir's outlet production temperature as a function of time over the 30 year operating period for the plant. Redrilling to a new, deeper region of hot (undepleted) rock is warranted for the first, five five-year periods. For this management strategy, a break-even selling price of electricity of 4.3 cents/kWh (1979 US dollars) results, which, when inflated to 1984 levels, would be about 6.7 cents/kWh.

Readers are cautioned not to take the predicted costs literally at this point, but rather to concentrate on the relative effects of these parameters controlling reservoir performance. In attempting to generalize these results, we would clearly state that for a given relationship between well cost and depth, reservoir performance is the single most important factor controlling costs. Assuming that well diameters will be large enough to accommodate all reasonable mass flow rates, then the key engineering objective should be to maximize the flow rate per well pair while holding drawdown within acceptable bounds. This has the effect of reducing drilling costs on a dollar/kg of fluid basis and can only be done by increasing the size of the reservoir's heat capture volume in proportion to the fluid circulation rate. For idealized monolithic fractures, areal heat extraction requires larger effective fracture surface areas. For fully rubblized volumetric labyrinths, more volume is required which might need larger well bore separations or longer production lengths. What these modeling studies tell us for drilling costs given by Fig. 14.9 is that effective heat transfer areas of order 1 to 2 million square metres ($1-2 \times 10^6$ m^2) per pair of wells are needed for economic viability, with well production flows of 45 to 100 kg/s. This will result in a thermal production capacity of 40 to 80 MWth per well pair at acceptable low levels of thermal drawdown. As expected, these figures are comparable to those for operating hydrothermal reservoirs such as those in Wairakei, New Zealand and Cerro Prieto, Mexico.

(c) PLANT PERFORMANCE

Clearly, there is an economic incentive for having an efficient HDR power plant, because of the high relative drilling and stimulation costs required to produce the fluid. In fact, it was just exactly that sort of motivation that has led to such interest in organic binary cycles for geothermal power production (Milora and Tester, 1976; Kestin *et al.*, 1980). Binary cycles have a distinct advantage over steam Rankine cycles because their high vapor densities at low temperature permit turbine expansions to exhaust at conditions closer to ambient. However, even with this advantage, there are practical factors which would keep the economic optimum binary plant design performance well below its thermodynamic Second Law limit. These factors include: (1) finite turbine, pump, and generator efficiencies, (2) practical temperature differences between streams of 10 to 20°C in the primary heat exchanger and condenser and (3) finite pressure drops through piping. Studies described by Tester (1982) and Kestin *et al.* (1980) point out that economic optimum values for utilization efficiency (η_u) will normally lie in the vicinity of 40 to 55% for hydrothermal systems, when drilling costs represent roughly 50% of the capital investment. We would expect similar conditions to apply to HDR electric power plants.

In non-electric applications, heat losses from equipment and pipeworks control performance to limit practical operating efficiencies to 80 to 85% of the reservoir's thermal power output being finally realized by the user.

(d) FINANCIAL ASPECTS

Levels of assumed risk, as expressed in the equity rate of return r_d and the debt interest rate i and individual well drilling costs are the most important financial factors in determining overall HDR commercial feasibility. Cummings and Morris (1979) studied the relative effects of these and other financial parameters by varying conditions from the base case cited in Table 14.2. The results given in Fig. 14.16 are cited as a + or − % deviation from the break-even busbar price $p*$ of 6.7 cents/kWh (1984 US dollars) predicted by the intertemporal discounted cash flow model.

14.5.4 Summary of requirements for a commercial HDR system

Readers are cautioned to recognize that the estimated breakeven busbar price of 6.7 cents/kWh for generating electricity is subject to all the conditions cited in Table 14.2 as well as the assumptions of the intertemporal model. Using these same conditions and assumptions, a selling price of $1.80/$10^6$BTU for hot water at the wellhead of a direct-use plant would result, which on an equivalent energy-use scale is a factor of five net reduction in cost of generating electricity from the same resource.

The results from the intertemporal model with discounting and optimum drilling strategies employed compares reasonably well with the break-even price results of 5 cents/kWh or $1.30/$10^6$ BTU obtained from the simpler model shown in Fig. 14.12. This model used a different set of assumptions regarding the resource and reservoir performance with an annual fixed charge rate approach. In addition, an entirely independent estimate of electricity production from HDR has been made by Murphy *et al.* (1982) in a study of the economics of a 75 MWe HDR power plant based on the design of the proposed Phase II reservoir at Fenton Hill. The results of their study also compare favorably with those shown in Fig. 14.12 or from the intertemporal model, for the authors deduced a break-even busbar cost of 4.4 cents/kWh in 1981 US dollars – say about 5.3 cents/kWh in 1984 dollars compared to 5 and 6.7 cents/kWh. Their figure was based on a ten-year reservoir life, an average thermal gradient of about 58°C/km, a mean well depth of about 4200 m, a mean temperature of 230°C at the point of use, and a staggered 5-spot reservoir design as shown in Fig. 9.41(b).

Although the specific assumptions for all three of these approaches differ widely, the break-even cost estimates from them are comparable, which places bounds on the degree of confidence in estimates of this type. The results regarding reservoir performance (mass flow rate and lifetime) and resource characteristics (such as average gradient) on costs have been interrelated empirically to map out a range of break-even prices for comparison with competing energy sources. Even though these estimates are admittedly very preliminary in nature, it is concluded that the economic feasibility requirements for the reservoir are not unreasonable and are not, in fact, that far-removed from demonstrated field performance. These required parameters are summarized in Table 14.3 for modest 40 to 60°C/km gradient HDR resources. They are presently regarded as general goals for the Phase II field demonstration programs at Fenton Hill and Rosemanowes.

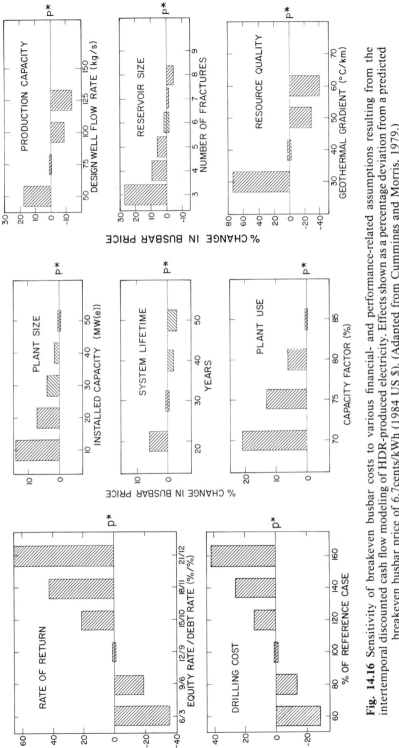

Fig. 14.16 Sensitivity of breakeven busbar costs to various financial- and performance-related assumptions resulting from the intertemporal discounted cash flow modeling of HDR-produced electricity. Effects shown as a percentage deviation from a predicted breakeven busbar price of 6.7cents/kWh (1984 US $). (Adapted from Cummings and Morris, 1979.)

Table 14.3. Minimum reservoir performance parameters required for commercial feasibility for HDR resources ranging from 40°C/km to 60°C/km

Parameter	Performance bounds
(1) Mass flow rate per well pair	$\geqslant 45$ kg/s
(2) Effective heat transfer area per well pair	$\geqslant 1 \times 10^6$ m^2
(3) Thermal drawdown rate	$\leqslant 20\%$ in 10 yrs
(4) Fluid loss rate	$<5\%$ of injected flow

The first three parameters of mass flow rate, heat transfer area, and thermal drawdown rate relate directly to the economics of how much power can be produced from a well pair. The last parameter listed is water loss rate at less than 5% of the injected flow. This figure is somewhat arbitrary as it relates more to the potential environmental hazards as well as to the availability of water at the site. If the reservoir is pressurized above its normal ambient level of pore pressure, fluid will be lost by permeation with the rate of loss determined by (1) formation permeability, (2) fluid viscosity, (3) permeating area, and (4) pressure gradient. Since the first three are fixed by *in situ* conditions pertaining to the reservoir, only the pressure gradient can be adjusted to control water losses. As mentioned earlier in Chapter 10, downhole pumping has been used at Rosemanowes to reduce losses to negligible levels. Providing that reservoir impedance is stable to pressure reductions, for example with a labyrinth of sufficiently self-propped fractures, downhole pumping should provide an effective means for managing water losses.

14.6 Dealing with the competition

The major factor that distinguishes both hydrothermal- and HDR-produced energy from fossil or nuclear alternatives is that the 'fuel' cost is absent from consideration as a separate operating cost. All geothermal 'fuel' costs are embedded in the capital costs of drilling, redrilling, and stimulation. Therefore, they are less susceptible to changes in the marketplace that might be caused by normal supply and demand effects or by political upheaval. A second positive factor for HDR, as a part of the general geothermal development worldwide, is that plants producing electricity and/or heat from low-grade hydrothermal resources have developed a reliable and cost-effective record which is quite different from energy produced by such new alternatives as solar, fusion, or wind power. Table 14.4 summarizes estimates for base load electricity and space or process heat supplied by oil, coal, or nuclear plants. In addition, cost estimates for vapor-dominated and liquid-dominated hydrothermal resources are also cited for comparison purposes.

Table 14.4. Busbar generating and heat supply cost estimates for new base load capacity in the USA (adapted and updated from Tester, 1979; Murphy *et al.*, 1982; Armstead, 1983)

Resource type	Busbar base load electricity costs (1984 US $)				
	Installed power plant capital costs $/kW	Annual power plant costs** cents/kWh	O+M* costs cents/kWh	Well or fuel costs cents/kWh	Total generating costs cents/kWh
Nuclear	1600	3.9	0.2	0.6	4.7
Oil	780	1.9	0.2	1.7–6.0 ($10–36/bbl)	3.8–8.1
Coal	1070	2.6	0.3	1.2 ($30/ton)	4.1
Liquid-dominated hydrothermal	1000	2.4	0.3	2.0–4.0	4.7–6.7
Vapor-dominated hydrothermal‡	800	1.9	0.2	2.2	4.3
HDR¶	1000	2.4	0.2	2.0–6.0	4.6–8.6

	Base load heat supply costs (1984 US $)			
	Annualized capital costs $/10⁶ BTU	Fuel costs $/10⁶ BTU	Heat distribution costs§ $/10⁶ BTU	Total cost $/10⁶ BTU
Nuclear	4.0	1.0	2.0	7.0
Oil	0.50†	2.2–8.0□ ($10–36/bbl)	1.4	4.1–9.9
Coal	0.70†	1.6–5.3● ($30–100/ton)	1.4	3.7–7.4
Liquid-dominated hydrothermal	—	1.0	2.0	3.0
Vapor-dominated hydrothermal	—	0.80	2.2	3.0
HDR¶	—	1.0–2.0	2.0	3.0–4.0

† Furnace capital costs.
* O+M=Operating and maintenance.
**Based on a 17% annual fixed charge rate; 80% load factor (7000 hr/year at capacity).
‡ Current unit cost at the Geysers (DiPippo (1984)).
§ Similar heat distribution system costs assumed to be $2/10⁶ BTU for nuclear and hydrothermal systems and $1.4/10⁶ BTU for oil and coal systems.
¶ Estimated range for a 40 to 60°C/km resource.
□ Approximate furnace efficiency of 80% assumed with 5.5×10^6 BTU/bbl heating value.
● Approximate furnace efficiency of 75% assumed with 25×10^6 BTU/ton heating value.

The question now arises, can heat mining compete with these alternatives? Let us first consider the cost of heat in bulk. Fuel oil in 1984 figures is something of the order of $30 US per barrel which, at 5.5×10^6 BTU/bbl, was equivalent to about $5.45 US/$10^6$ BTU before making any allowance whatsoever for the conversion of the fuel heat into the form of hot water ready for distribution. Heat from coal would undoubtedly be cheaper than from oil, while heat from hydrothermal fluids would probably be cheaper still but only available in comparatively few places. Our estimated range of about $1 to 2/$10^6$ BTU at the wellhead for HDR-supplied heat suggests that heat mining costs for bulk heat supply are likely to be within the range of commercial interest.

As to electricity, busbar costs in 1984 for base load capacity ranged from about 4.1 cents/kWh (average) when generated from coal at $30/ton to a range of 3.8 to 8.1 cents/kWh when generated from oil as compared with our range of estimates of 4.6 to 8.6 cents/kWh for HDR. Here again, these very tentative figures suggest that heat mining favorably compares with oil-fired systems at costs greater than $25/bbl and is within range of nuclear and coal-fired systems.

At this point, one can only surmise that the predicted busbar costs for HDR-produced electricity and heat are reasonably competitive as shown in Fig. 14.17. These results are, of course, subject to several key assumptions with regard to reservoir performance and drilling and stimulation costs. Again, we must emphasize that the costs for the alternatives to HDR must not be regarded as 'cast in stone' either. Large variations in fuel and capital costs could easily alter the picture drastically. Since 1984, oil prices have declined dramatically, and by 1986, HDR could not have competed with oil-fired electricity production. But, as pointed out in Chapter 2, we regard this as a short-term phenomenon. Within the next decade or so, HDR should become increasingly competitive as a supplier both of electricity and of heat.

In addition to these fuel cost fluctuation effects, capital costs for plants have escalated significantly in recent years. For example, DiPippo (1984) reported that new units of 110 MWe capacity to be installed at the Geysers field in California from 1986 to 1991 will cost from $1300 to $2176/kW as compared to a 1984 level of approximately $750 to $850/kW, an increase of about 70%. Similar escalations of capital costs created by introduction of more sophisticated and costly pollution abatement or safety equipment have been observed for fossil-fired and nuclear-fired plants.

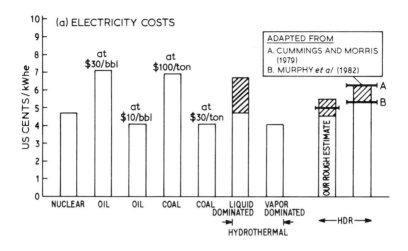

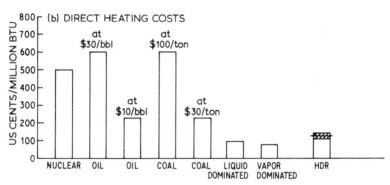

Fig. 14.17 Comparative costs of baseload electricity and heat in bulk from different energy sources in 1984 dollars. (a) base load electricity (b) heat in bulk.

Chapter 15

Prognosis

We have tried to present the case for heat mining objectively, and to show that it could offer a partial solution to the challenging problem of satisfying the world's long-term and medium-term energy needs. Furthermore, given its current state of development, heat mining could become a commercial reality with only a modest extrapolation of demonstrated reservoir performance. We would not yet go so far as to say positively that heat mining is of proven worth as a major source of energy, but we certainly feel justified in believing that its future looks distinctly promising.

At this point, the future course for heat mining needs to be assessed in light of our analysis of relevant technical and economic factors. Certain characteristics of HDR are unassailable and have led to optimism on the part of its advocates and to the investment of a significant amount of money, in excess of 150 million dollars, in supporting an international scientific effort to determine its feasibility over the past fifteen years.

15.1 Size of the resource

The enormous magnitude of the resource and the general accessibility of hot rock throughout the world are unquestionable features of HDR. Even with near-normal gradients, rock temperatures of 150 to 300°C are technically reachable with currently available rotary drilling technology. Where gradients are hyper-normal, the situation is even better. Above average gradients (30–60°C/km) are rather well distributed over the earth's land masses. This decentralizes the resource, making it potentially much more useful in practice than other resources such as wind, wave, tidal, or solar power. Futhermore, HDR is well-suited for base-load applications since it is not subject to seasonal or diurnal fluctuations. On a world scale, the magnitude of the usable energy contained in HDR is vast, literally some orders of magnitude larger than the sum total of all present and possible fossil resources. Given its size and ubiquitous nature, HDR is a perfect candidate for fulfilling electrical and non-electrical needs for a very long time.

440

15.2 Technical feasibility

What milestones have to be reached in order for heat mining to work? Unfortunately, there is not a clear cut definition of what constitutes 'work,' so we cannot say unequivocally at this time that *all* technical feasibility requirements have already been achieved in the field testing programs at Fenton Hill, Rosemanowes, and elsewhere. Nonetheless, many essential questions have been answered regarding our ability to reach zones of hot rock, to create underground reservoirs by fracturing, and actually to mine heat for extended periods without dire practical or environmental consequences. We have not seen the failure of any hypotheses involving basic physical principles as to how effective the heat mining process is. The main technical issue remaining is directed more toward enlargement of two bore-HDR systems to commercially-required sizes. Although this is by no means a trivial task, as both Phase II efforts at Fenton Hill and Rosemanowes have demonstrated, it does not require the enormous number of technical 'quantum jumps' that are frequently necessary before other new technologies, such a photovoltaics or thermonuclear fusion, can become commercially viable energy sources.

Another useful feature of the heat mining concept is that it is amenable to physical extension. If success can be achieved with a single pair of wells and an interconnecting labyrinth of fractures, then this limited system can be treated as a modular unit for scale-up to extract heat from much larger volumes of hot rock. The Phase I reservoir at Fenton Hill has, in fact, served as a module for the design of the Phase II multiple-fracture system. If reservoir stimulation from an inclined set of wellbores can be achieved in as systematic a manner as contemplated, this last technical hurdle is well within reach during the next 5 to 10 years.

15.3 Environmental impact

Heat mining is virtually pollution free. Only when HDR is used to generate electricity are substantial quantities of waste heat involved. However, the heat would be diffusely rejected from widely distributed power plants. Another concern is temporary noise during construction. Apart from these issues, when viewed as a replacement to existing fossil-fired baseload capacity, heat mining offers a positive environmental gain. It would partially displace the ill effects of combustion as well as help to preserve liquid and gaseous fuels as chemical feedstocks and for transportation purposes.

Water loss to the formation presents environmental concerns, directly as a practical issue of consumption in arid regions and indirectly as it may lead to induced seismicity. The magnitude of the seismic risk issue is strongly dependent on the geologic setting of the HDR system as well as on how heat is mined from the reservoir. Field evaluations so far lead to an optimistic outlook. First, water loss rates have been controlled using proper reservoir pressure management strategies such as downhole pumping. And second, the absence of induced seismic hazards under long-term pressurized circulation conditions in several fractured reservoirs has been very encouraging.

15.4 Projected costs

With reasonable assumptions regarding resource quality and reservoir performance, the economic picture looks good: that is to say, if artificial HDR reservoirs are created in areas with above average thermal gradient and if they have productivities comparable to natural hydrothermal systems already in operation in the world. Whether heat mining can compete more universally has yet to be demonstrated. An independent study of HDR economics was started in late 1986 by an industrial group headed by Bechtel, Inc. in California. Interested readers should watch for their report as well as for updated forecasts from the Electric Power Research Institute (EPRI) in the USA and from similar authorities in the UK. Preliminary predictions from EPRI and from a 1986 UK study show HDR-produced heat and electricity to be competitive with oil priced at about $14/bbl for heat and $24/bbl for electricity. This corresponds to an estimated price (in 1986 US dollars) of $2.50 per million BTU for delivered heat or about 6.1 cents per kWh for electricity. These estimates agree quite well with our projections for HDR given in Table 14.4.

15.5 Update on field developments

Progress in the field up to 1985 at Fenton Hill and Rosemanowes has been discussed in considerable detail in Chapters 9 and 10. At both sites, redrilling was required to create a reservoir suitable for prolonged heat extraction and hydraulic testing. Redrilling was successful at both locations and has led to extended phases of testing which were still underway as of January of 1987. This section is intended to update events from 1985 to 1987 in each Phase II development program.

15.5.1 Fenton Hill update

As described in Section 9.4.1(d), early attempts to connect EE-3 and EE-2 using hydraulic fracturing were unsuccessful despite the substantial volume of water injected during massive hydraulic fracturing operations in 1982 to 1983. The last major attempt occurred in December 1983 where over 21 000 m³ of water was injected at a depth of approximately 3500 m in EE-2 at downhole pressures exceeding 80 MPa (11 600 psi) and at an average injection rate of 100 liters/s. Figure 15.1 shows the locations of microseismic events induced by this injection as viewed from South to North. One can see that the stimulated zone is inclined from the vertical and that the zone of seismicity does not intersect

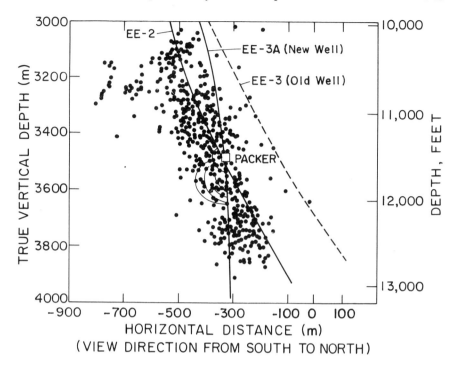

Fig. 15.1 Elevation view of the active region of the Phase II Fenton Hill reservoir with projected microseismic events plotted. Flow paths illustrated below the packer are inferred from joint locations determined from a temperature survey in EE-3A. (Data provided by Whetten, *et al.*, Los Alamos National Laboratory, 1986)

the old EE-3 well. A complete analysis of the microseismic events induced shows that the stimulated zones were three-dimensional rather than planar, having the following approximate dimensions: 800 m high by 900 m wide in the N–S direction by 150 m thick in the E–W direction. This is equivalent to 100 million m^3 of stimulated rock volume, about 5000 times greater than the volume of injected water (see Table 9.4, experiment 2032). Subsequent analysis by House *et al.* (1986) and Fehler *et al.* (1987) concluded that the microseismically active region represents the stimulation of multiple joints rather than a single fracture and that a shear-slippage mechanism was dominant. This agrees with results at Rosemanowes, but is inconsistent with conventional tensile failure theories described in Chapter 8. With no hydraulic connection, the Los Alamos team decided to sidetrack from EE-3 to create a workable reservoir. The microseismic event map was used to design the EE-3A trajectory for maximum penetration into the seismic event cloud as shown in Fig. 15.1.

An early indication of a hydraulic connection was observed during the redrilling of EE-3 when EE-2 was pressurized to about 12 MPa (1750 psi). This connection was improved by fracturing from EE-3A below a tubing string connected to a high temperature open hole packer set at 3520 m as shown in

Fig. 15.1. The streamlines drawn on the figure near the packer location depict potential flow paths to EE-2 from EE-3A which are separated horizontally by about 150 m at these depths. Additional repeated stimulations were attempted lower in EE-3A using openhole packers, with some success. The final result was a reservoir that could be tested. Another point worth noting was the substantial improvement in drilling and completion time associated with the EE-3A campaign. On a comparable basis, the costs in $/m of depth drilled for EE-3A would be less than one-half of those for EE-2 or EE-3.

In March and April of 1986, the new Phase II reservoir was prepared for hydraulic circulation and heat extraction testing by installing a cemented steel liner in EE-3A, finishing the surface piping system and connecting the air-cooled heat exchanger used throughout the Phase I tests. A 31-day field test, referred to as the *I*nitial *C*losed-loop *F*low *T*est (or ICFT) started on 19 May 1986 and was completed on 18 June 1986. Table 15.1 compares major reservoir parameters for this system to those for earlier Fenton Hill and Rosemanowes reservoirs.

The tracer residence time distribution curves of Fig. 15.2 highlight the differences in flow patterns between the Phase I and II systems at Fenton Hill. Note the larger Phase II modal volume and the generally more dispersed nature of the flow with a larger portion distributed in longer residence time flow paths. During the 31 day test, the modal volume increased from 270 to 350 m^3 while the impedance decreased from about 4.5 to 2.1 GPa s/m^3 comparable to that observed in the enlarged Phase I system. With a constant reinjection well temperature of 20°C, the production well temperature steadily increased to about 190°C corresponding to a thermal power level of about 10 MWt. No drawdown was observed. Furthermore, silica and Na-K-Ca geothermeters indicated *in situ* reservoir temperatures of about 242°C and 222°C, respectively. This is consistent with an initial rock temperature of 232°C at an average reservoir depth of 3550 m. Temperature surveys of the production/injection region indicate a multiply-jointed network of fractures connecting EE-2 and EE-3A. Water loss rates of about 28% of the injected flow were observed, but the test was too short to extrapolate to steady state values which would undoubtedly be much lower.

Current plans are to continue testing the redrilled Phase II system at Fenton Hill with a long-term (about 1 year duration) circulation test scheduled for 1987–88 to characterize steady state water losses and potential thermal drawdown by monitoring outlet temperatures and by using reactive tracers.

15.5.2 Rosemanowes update

The interlinking of the RH11 and RH12 wells in the original Phase II system at Rosemanowes was successfully accomplished in 1982. As described earlier in Section 9.4.2, the microseismic events that were mapped during hydraulic stimulation indicated a strong downward growth with a shear-motion domi-

Table 15.1. Comparison of HDR Reservoir Characteristics

	Fenton Hill			Rosemanowes		
	Original Phase I	*Enlarged Phase I*	*Redrilled Phase II*	*Phase I*	*Original Phase II*	*Redrilled Phase II*
Average reservoir depth (m)	2700	2840	3550	100	2100	2100
Average initial rock temperature (°C)	185	190	232	15	80	80
Average wellbore separation distance (m)	100	300	150–300	~40	300	200–300
Modal volume (m³)	11–27	136–187	270–350	50	2500–3600	153–780
Nominal production flow rates (kg/s)	7–16	6–8	12–14	1–10	6–10	12–30
Nominal thermal power level (MWt)	3	3	10	<1	1	1→4
Cumulative operating period (hr)	2500	7320	744	1000	5000	12 500[b]
Surface injection well pressure (MPa) .	10	10	30	6–9	11–14	9–12
Terminal flow impedance (GPa s/m³ (buoyancy corrected))[c]	0.33	1.56	2.1	0.03	1.6	0.60[d] (0.46)[e]
Nominal range of water loss rate as % of injection rate	7→1%	15→10%	35→28%	1–5%	40–60%	25→15%[a]

[a] As of 1 December 1986 results from test in progress, steady decline observed.

[b] As of 1 January 1987, test in progress.

[c] Measured impedances at end of low back pressure portions of test.

[d] As of 1 December 1986.

[e] Minimum value at 50 kg/s injection rate and low back pressure.

nant mechanism. Tracer-determined modal volumes were extremely large ranging from 2500–3500 m³; but, water loss rates and flow impedances were too high to warrant extensive testing. The high water loss rates of 40 to 60% were attributed largely to a stimulation rather than permeation mechanism, characterized by enormous growth in reservoir volume. Furthermore, the potentially large portion of the reservoir in the deeper region below RH11 and RH12 was apparently not readily accessible to recirculated fluids aimed at heat extraction. Thus, a third well RH15 was drilled to explore the lower region of the reservoir and hopefully to provide a lower impedance connection that could be tested. RH15 was completed in January 1985 on schedule and, like RH11

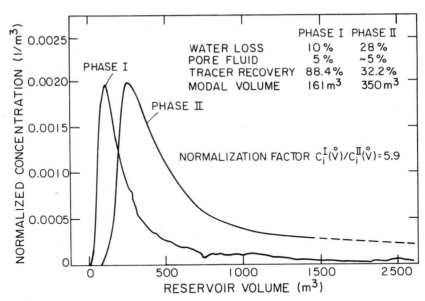

Fig. 15.2 Comparison of residence time distribution (RTD) curves for the Phase I and II systems at Fenton Hill. Phase I test conducted on 9 May 1980, Phase II test on 13 June 1986.

and RH12, without any serious problems. As shown in Fig. 15.3, RH15 penetrated the seismic region to a total vertical depth (TVD) of 2652 m. What is not visible on the figure is the complex spiral trajectory of the well.

Measurements of the *in situ* stresses discussed earlier in Section 9.4.2 and subsequent analysis in Section 9.5 indicated that tensile jacking and fracture extension would be possible if the pressures within the joint could be induced to exceed the effective confining stress. This would avoid the tendency for the joint surfaces to shear as was observed throughout the Phase II stimulations using water injection. A fluid with a viscosity higher than water could be used to generate the necessary *in situ* pressures for tensile jacking. After considerable planning and discussion, a suitable gel formulation was determined and 5500 tons of a commercially available gel were injected into RH15 during an 8-hr period on 4 July 1985. Direct connections were made between RH15 and the RH12 wellbore. Furthermore, subsequent mapping of the RH15 wellbore indicated that a vertical fracture extended upwards toward RH12 in much the same manner as would be predicted from the idealized hydraulic fracturing theories of Chapter 8.

The Rosemanowes Phase II reservoir was then readied for extensive flow testing which began on 7 August 1985 and was still in progress on 1 January 1987 with plans to continue for 2 to 5 years. Primary reservoir parameters are given in Table 15.1 for comparison to other Rosemanowes and Fenton Hill reservoirs. As can be seen, the modal volume of the redrilled system increased from 153 to 780 m³ after 11 000 hr of circulation. The redrilling did result in

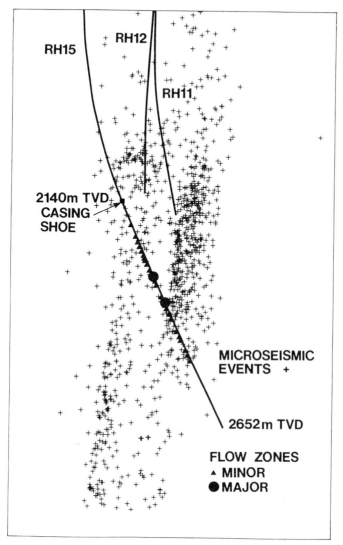

Fig. 15.3 Elevation view of the wellbores with projected microseismic events mapped for the Phase II reservoir at Rosemanowes. Major and minor flow zones indicated along the RH15 wellbore. (Provided by A.S. Batchelor, 1987.)

more than a 3-fold decrease in impedance from about 1.6 to about 0.5 GPa s/m³, while the water loss rates had dropped off to about 15 to 20% after 11 000 hr.

By the end of May 1986, the system had been in continuous circulation for over 7000 hr with no thermal drawdown observed in the production well. By July 1986, it became clear that substantial changes in the reservoir production flow distribution had occurred. An apparent temperature decline in fluid produced in some joints was measured in RH15 but as the flow distribution

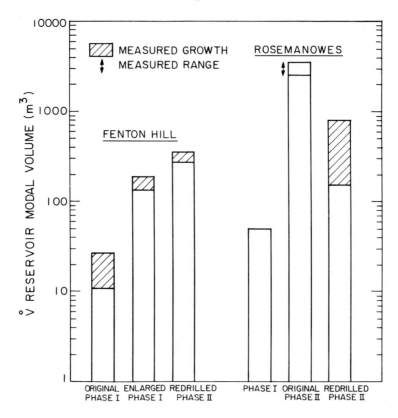

Fig. 15.4 Comparison of tracer-determined reservoir modal volumes for the Phase I and II systems at Fenton Hill and Rosemanowes. Cross-hatched regions indicate growth during heat extraction.

shifted reducing the flow fraction from these joints, the overall thermal drawdown rate was lessened. In fact, given the measured increase in modal volume, suggesting reservoir growth and larger swept areas and volumes for heat transfer, the overall performance picture is indeed bright. In addition, testing with reactive tracers has indicated that the major portion of reservoir temperature distribution is stable. Further testing is necessary before a complete description of the reservoir can be developed and is in progress (Batchelor, Ledingham, Kwakwa and Curtis, 1987).

Figures 15.4 and 15.5 summarize the reservoir modal volume and thermal performance histories for the Phase I and II reservoirs at Rosemanowes and Fenton Hill. Increases in modal volume occurred as a result of stimulation by fracturing over larger wellbore separation distances (original Phase I to enlarged Phase I and to Phase II) as well as by a dynamic growth mechanism induced by pressurization and/or thermal stresses associated with long-term

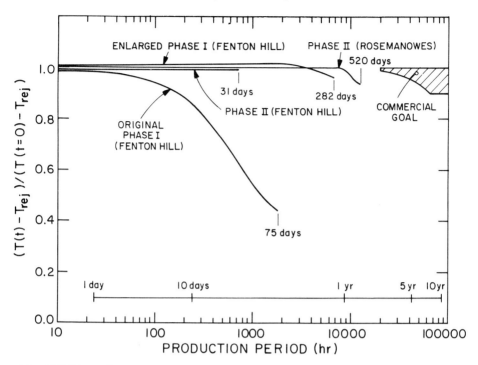

Fig. 15.5 Dimensionless temperature drawdown as a function of operating time for the Fenton Hill and Rosemanowes reservoirs. Note the logarithmic time scale used.

fluid injection. In both cases, the result is positive suggesting that enhanced reservoir thermal performance may result. The dimensionless drawdown curves of Fig. 15.5 exemplify this point by showing the greatly improved performance as the wellbore separation distances increased from about 100 m to 200–300 m implying a 4- to 9-fold increase in reservoir volume or effective area. This correlates well with results obtained for the modal volume which has been shown empirically to be proportional to effective heat transfer area estimated from thermal model fits to actual Phase I drawdown data (see Fig. 10.37).

15.6 Recapitulation

In general, heat mining has certain potential advantages not common to other alternative sources of energy. The vastness of the resource; the wide distribution of the more favorable exploitable sites and the universality of the lower grade sites; the adaptability of the resource to the requirements of bulk heat supply or to the generation of electricity; its continuous availability throughout the assigned life of an exploited zone regardless of season or time of day: and its environmental acceptability; all these qualities provide powerful incentives to persevere with the very active research that is already being done. The fact that every additional meter of attainable drilling depth without extra cost brings such disproportionately large gains in recoverable heat both in quantity and grade gives stimulus to the associated research efforts that are being pursued in drilling technology.

At present, most of the effort in heat mining research and development is concentrated at two sites, in the USA and UK, though other useful work on a smaller scale is being done elsewhere. Interest is increasingly being aroused in several other countries, and it seems likely that research efforts will become intensified throughout the world. However, realistically, the future of HDR, at least in the decade ahead, seems to lie mainly with the fate of developments at Fenton Hill and Rosemanowes. Perhaps too much emphasis has been placed on achieving absolute proof of concept on a fixed time scale of research funding. After all, we are rapidly learning from the tests at Fenton Hill and Rosemanowes how to develop the necessary stimulation and extraction technology. Thus a certain amount of pain is inevitable as we give birth to this new technology. All we can hope for is that the technical achievements and problems associated with the current work at both sites will be evaluated with objective vision rather than politically-motivated shortsightedness.

By the spring of 1985, redrilling at Fenton Hill and Rosemanowes was completed with successful connections made between both sets of boreholes. Extensive reservoir testing is underway as this book undergoes final preparation for publication. This involves long-term heat extraction testing to demonstrate that *commercial-sized* HDR reservoir systems can be created and produced.

Preliminary results at both sites have been very encouraging. For example, tracer-determined modal and mean volume measurements have indicated an enlargement of active reservoir size as the production period increases. Furthermore, this apparent reservoir growth did not result in an increase of water loss or of flow impedance, which both show a general decline as production proceeds. What is strongly suggested is that increased reservoir heat extraction capacity may be induced by sustained injection and circulation of pressurized cold fluid through the reservoir. This would be very good news, as it may lead to reduced thermal drawdown rates as a result of continuous reservoir growth into undepleted regions of hot rock. If this becomes a general phenomenon in other HDR reservoirs at different sites, our earlier conservative estimates of reservoir production capacity would have to be upgraded and the overall economic picture would become considerably brighter.

There are also promising signs that serious attention is already being given to actual commercial projects, though naturally of modest size at first. A Phase III project at Fenton Hill is being considered that could lead to a 5 or 10 MWe electric power plant within ten years. This plant would probably be built and operated by a local rural electrification authority or private company in a joint enterprise with the Los Alamos National Laboratory and the US Department of Energy, so that a foothold may be gained for private enterprise to become involved in a future heat mining industry. If the plant is both technically and commercially successful, it would probably be expanded in capacity to meet demands. In the United Kingdom, consideration is being given to a deeper, hotter commercial-scale heat extraction test at some suitable location within the Carmenellis granite of Cornwall; this system is intended for power generation.

New potential heat mining sites are also being considered in the USA; for example at The Geysers/Clear Lake area in California and in the Roosevelt Hot Springs field in Utah for electricity generation and at Wallops Island, Virginia, for direct heat application.

Like The Geysers and Roosevelt developments, many new sites for HDR will be located near or within unproductive regions of large hydrothermal resources and thus will be of high grade with respect to geothermal gradient. In the USA alone, electricity production from hydrothermal sources is predicted to increase from over 2000 MW(e) in 1987 to possibly more than 6000 MW(e) by the year 2000. This projected growth in hydrothermal resource utilization will help create the large industrial infrastructure involving field developers, producers and end-users needed to enhance the commercialization of HDR.

It would be unwise to expect any commercial heat mining developments to be in full-scale operation during this century, but it is likely that many would be started or planned. If any successful heat mining projects should come into being in the first decade or two of the twenty-first century, they will doubtless catalyze future development at an accelerating pace. Under these assumptions, heat mining as a new form of energy extraction and application should steadily assume an ever-increasing role in the world energy market.

It is not necessary to be a starry-eyed romantic to believe that heat mining may soon become not only a practical reality but also an economically competitive new source of energy. The road ahead will have its rough spots but is definitely negotiable.

Glossary

1 Terms and abbreviations

BHA=Bottom hole assembly of hardware for drilling.

Bit=Device used to drill, located at the end of the drill string.

BOP=Blow out preventer.

BPDM=Baker positive displacement downhole drill motor.

Br^{82}=Gamma emitting isotope of bromine, used as a tracer in reservoir analysis.

CEC=Commission of the European Communities=civil service of the European Communities (The Economic Community (EEC)), The Coal and Steel Community, Euratom.

Christmas tree=Wellhead piping and valving equipment.

Core=Cylindrical section of rock removed from the borehole.

CRF=Capital recovery factor.

CSM=Camborne School of Mines.

DDPDM=Dyna-Drill positive displacement downhole drill motor.

Drillstring=Collected assembly of pipe used to drive the BHA.

DXPDM=Drilex positive displacement downhole drill motor.

EE-1=Energy extraction hole 1 at Fenton Hill, NM (Phase I system).

EE-2=Energy extraction hole 2 at Fenton Hill, NM (injection hole of Phase II system).

EE-3=Energy extraction hole 3 at Fenton Hill, NM (recovery hole of Phase II system).

EEC=European Economic Community or Common Market, more properly called CEC.

ENERGEROC=French HDR research programme (see Section 9.4.3).

EPDM=High temperature elastomer (ethylene-propylene-diene terpolymer).

erf=Error function (see symbol nomenclature).

EW DOT=Eastman Whipstock directional drilling orientation tool.

EW Gyro=Eastman Whipstock gyroscopic survey tool.

EW Singleshot=Eastman Whipstock magnetic survey tool.

EYE=Downhole directional drilling steering device with surface readout available from Scientific Drilling Controls.

FCR=Fixed charge rate.

452

Fish=Any piece of debris that should not be in the hole.

Fishing tool=Device used to remove items from a borehole.

Fishing=Process of using a fish to remove items from a borehole.

FRIP=Fluid Rock Interaction Program=code to model thermal-hydraulic and stress behavior of HDR reservoirs.

GDKZ=Geertsma–de Klerk–Khristianovich–Zheltov model for hydraulic fracturing.

GT-1=Geothermal test hole 1 at Fenton Hill, NM (shallow exploratory hole).

GT-2=Geothermal test hole 2 at Fenton Hill, NM (Phase I system).

Gyro=Gyroscopic survey of a borehole.

Hardbanding=Use of hardened materials such as tungsten carbide on drillpipe to reduce wear.

HDR=Hot dry rock.

HNS=Explosive compound used in high temperature detonations, hexanitrostilbene.

HTC=Hughes Tool Company, manufacturer of drill bits.

IEA=International Energy Agency.

Kelly=Hexagonal or square cross section member for rotating drill string in the rotary table.

KINEFRAC=Commercial process from Kinetech for fracture initiation by rapid gas generation.

KOP=Kick-off point for directional drilling.

LASL=Los Alamos Scientific Laboratory (now the Los Alamos National Laboratory (LANL)).

Locked in=BHA used when hole trajectory is on course during directional drilling.

Log/logging=Method of surveying the interior of a borehole using non-evasive geophysical or geochemical measurements.

MAGES=Man Made Geothermal Energy Systems=international project to study HDR feasibility.

ME15=Maurer Engineering $5\frac{3}{8}$ in. diameter high temperature turbodrill.

ME17=Maurer Engineering $7\frac{3}{4}$ in. diameter high temperature turbodrill.

MHD=Magnetohydrodynamics.

MHF=Massive hydraulic fracture.

NPDM=Navi-drill positive displacement downhole drill motor.

PBR=Polished bore receptacle for downhole emplacement of tubular goods.

PKN=Perkins–Kern–Nordgren model for hydraulic fracturing.

PV=Present value.

P3DH=Pseudo three-dimensional hydrafrac model for hydraulic fracturing.

R-22=Chlorodifluoromethane ($CHClF_2$).
R-32=Difluoromethane (CH_2F_2)
R-114=Dichlorotetrafluoroethane ($C_2Cl_2F_4$).
R-115=Pentafluoromonochloroethane (C_2ClF_5).
R-600a=Isobutane (i-C_4H_{10}).
R-717=Ammonia (NH_3).
RC-318=Octafluorocyclobutane (C_4F_8).
Reaming=Enlarging the diameter of an existing hole.
RGI=Republic Geothermal, Inc.
RH11=Rosemanowes hole 11 at Rosemanowes Quarry in Cornwall. (Phase II system).
RH12=Rosemanowes hole 12 at Rosemanowes Quarry in Cornwall. (Phase II system).
RH15=Rosemanowes hole 15 at Rosemanowes Quarry in Cornwall. (Phase II system).
RTD=Residence time distribution for reservoir flow paths.
Run Segment=A LANL term for an extended field test involving the circulation of fluid through an artificial reservoir.

SEC=Security Division of Dresser Industries, manufacturer of drill bits.
Side tracking=Process of deviating direction from an existing bore.
Single/multishot survey=Magnetic survey of a borehole.
SP/IP=Self- and induced-potential measurements in a borehole.
Spud-in=Start time for drilling operation.
SQUID=Superconducting quantum interference magnetic gradiometer for measuring magnetic fields.
SS Gyro=Sperry Sun gyroscopic survey tool.
SS Hadies=Sperry Sun directional drilling orientation tool.
Stand=Group of pipe joints threaded together and held in place on drilling platform.
STC=Smith Tool Company, manufacturer of drill bits.
Stratapax=Diamond impregnated compacts for drill bits, manufactured by General Electric.
String=Length of pipe joints placed in the hole during drilling.
Sub=Threaded connector to allow assembly of items with threads into the drillstring.
Subterrene=Device developed by Los Alamos for rock drilling by melting.
SUE=Stress Unlocking Experiment=conducted at Fenton Hill after run segment 5.

TATB=Triaminotrinitrobenzene explosive used for fracture initiation.
TD=Total drilled depth.

Trip=Process of emplacing or removing drill pipe.

Turbodrill=Downhole turbine-powered drill operated with drilling fluid.

TV=Television video display for downhole logging.

Twist-off=Mechanical pipe failure resulting in a separation of the drillstring.

UKDOE=United Kingdom Department of Energy.

USDOE=United States Department of Energy.

VAX 11/780=Digital Equipment Corporation mid size computer.

Washover=Process of drilling over cemented-in sections of pipe.

2DL=Two-dimensional Lagrangian code for simulating explosive effects.

2 English symbol nomenclature (SI units, engineering units)

A=fracture or effective heat transfer area (m^2[ft^2]).

a^*=solid surface area per unit fluid volume (m^2/m^3, cm^2/cm^3).

A^*=constant to account for radiogenic heat contribution (equation (6.2)) (W/m^3).

A_c=cross-sectional area of fracture (m^2 [ft^2]).

A_s=seismically defined reservoir area (m^2 [ft^2]).

A_z=exponential constant for depth dependence of well costs (km^{-1}) (equation (11.1)).

b=empirical parameter relating seismic area (A_s) to injected volume (V_{inj}) (mm [in.]).

B=availability function=$H-T_oS$ (J/kg [BTU/lb]).

ΔB=change in availability=$\Delta H-T_o\Delta S$ (J/kg [BTU/lb]).

c=mean reservoir compressibility (Pa^{-1} [psi^{-1}]).

C=species concentration in circulating fluid (kg/m^3 [ppm]).

C^*=empirical constant in equation (10.14).

C_d=well cost parameter for drilling and casing portion of costs (equation (11.1)).

C_i=tracer concentration at particular t_i or V_i (kg/m^3 [ppm]).

C_L=fluid loss coefficient ($cms^{-\frac{1}{2}}$, $ms^{-\frac{1}{2}}$).

C_m=well cost parameter for mobilization charges (equation (11.1)).

$C_{out}(t)$=measured outlet tracer response at time t (kg/m^3 [ppm]).

C_p=fluid heat capacity (J/kg K [BTU/lb°F]).

C_{pr}=rock heat capacity (J/kg K [BTU/lb°F]).

C_{well}=well cost ($ US/well).

C_θ=normalized tracer concentration (equation (10.27)) (kg/m^3 [ppm]).

C^∞=equilibrium saturation concentration in aqueous solution (kg/m^3 [ppm]).

d=depth (m [ft] or km).
D=dispersion coefficient describes degree of mixing (equation (10.22)) (m²/s, cm²/s [ft²/s]).

E=Young's modulus (Pa [psi]).
E_{act}=activation energy for kinetic dissolution processes (equation (10.46)) (J/mol).

$$Ei(z)=\text{exponential integral of } z = -\int_{-z}^{\infty} \frac{e^{-u}}{u} du$$

$$\text{erf}(z)=\text{error function of } z = \frac{2}{\sqrt{\pi}} \int_{0}^{z} e^{-u^2} du$$

$E(t)$=normalized residence time distribution function (equation (10.28)).
ΔE=net thermal energy extracted from HDR reservoir (J [BTU]).

f=fraction of injected flow recovered (equation (10.26)).
F=objective function for fitting model to tracer data (equation (10.25)).
$f*$=enhancement factor for dissolution (equation (10.47)).

g=acceleration of gravity=9.8 m/s² [32 ft/s²].
G=shear modulus (equation (8.3)) (Pa [psi]).
Gr=Grashof number=ratio of buoyancy to viscous forces
  $=(w^{*3}\rho^2 g\beta_T(T-T_i^*)/\mu^2)$ (Section 10.3.2).

H=fracture height (m [ft]).
ΔH=change in enthalpy (J/kg [BTU/lb]).

i=interest rate for borrowed capital.
I=specific flow impedance=$\Delta P/\dot{q}$ (GPa s/m³) [psi/gpm]).

k=permeability (m² or darcy).
K=Bulk modulus=ratio of hydrostatic stress to volumetric strain (Pa [psi]).
$k*$=dissolution reaction rate constant (equation (10.46)) (m/s [ft/s]).
k_0^*=pre-exponential factor (equation (10.46)) (m/s ft/s]).
k_f=fracture permeability=$w^2/12$ for laminar flow in flat crack (equation (10.10)) (m² [ft²]).
k_0=reference state permeability (equation (10.14)) (m² or darcy).
K_b=buoyancy parameter=Gr/Re (Section 10.3.2) (no units).

L=fracture length (m [ft]).
L_j=length of flow path j (m [ft]).

m^*=power-law exponent in P3DH Hydrafrac model (equation (8.36)).
$\dot{m}$=mass flow rate of fluid (kg/s [lb/s]).
$\dot{m}_T$=total mass flow rate (kg/s [lb/s]).
M^*=material constant (equation (9.1)).
m_T=total mass of tracer injected (kg [lb]).

n=number of fractures.
n^*=power-law exponent in P3DH Hydrafrac model (equation (8.36)).
$n°$=depth parameter exponent for well costs (equation (14.3)).

P=pressure (Pa, bar [psi]).
P_{fv}=hydraulic pressure needed to initiate a vertical crack (Pa, bar [psi]).
P_{fh}=hydraulic pressure needed to initiate a horizontal crack (Pa, bar [psi]).
P_{gf}=geothermal fluid wellhead pressure (Pa, bar [psi]).
P_o=reference pressure or pore fluid pressure (Pa, bar [psi]).
Pe=dispersional Peclet number=$\overline{U}L/D$ (equation (10.23)).
Pe_j=dispersional Peclet number in zone j=$\overline{U}_jL/D_j$ (equation (10.24)).
P=power thermal or electric (W, MW [BTU/hr]).
$P(t)$=power as function of time (W, MW [BTU/hr]).
P_{pump}=pumping power (equation (10.8)) (W, MW [hp]).
P_{max}=maximum power possible from specified heat source (W or MWe [BTU/hr]).
P_{net}=net power produced by power cycle (W or MWe [BTU/hr]).
ΔP=pressure drop in fracture (Pa, bar [psi]).
$\overline{\Delta P}$=average pressure difference between pressure in fracture and pore pressure (Pa [psi]).
ΔP_i=P (injection wellhead)$-P_o$ (equation (10.26)) (Pa, bar [psi]).
ΔP_r=P (production wellhead)$-P_o$ (equation (10.26)) (Pa, bar [psi]).

q=total heat flow from mantle (mW/m² [cal/cm²s].
q_{HE}=net heat entering crust from Moho (J/m² [BTU/ft²]).
q_{HL}=net heat leaving crust at surface (J/m² [BTU/ft²]).
$\dot{q}$=volumetric flow rate (m³/s, liters/s [gpm]).
$\dot{q}_L$=fluid loss rate (m³/s, liters/s [gpm]).
q_m=heat flow from mantle (mW/m², cal/cm²s).
q_M=maximum heat flow from mantle (mW/m², cal/cm²s).
$\dot{q}_x$=volumetric flow rate at position x along length of fracture (Figs 8.7 and 8.8) (m³/s, liters/s [gpm]).
$\dot{q}_2$=pore fluid displacement rate.
q_r^*=heat flow due to radioactivity (mW/m² [cal/cm²s]).
q_R^*=maximum heat flow due to radioactivity (mW/m² [cal/cm²s]).

r=radial distance (m [ft]).
R=fracture radius (m [ft]).
Re=Reynolds number=$U\rho L/\mu$ or $\overline{U}\rho w^*/\mu$ (dimensionless).

$R(t)$=fracture radius at time t (m [ft]).
r_d=effective discount rate.
r_w=borehole radius (m [ft]).
R_{eff}=effective fracture radius (m [ft]).
R_{max}=maximum fracture radius (m [ft]).

S=fracture spacing (m [ft]).
S^*=material constant (equation (9.1)).
$S_h=S_Y=\sigma_h-P_o$=least principal horizontal earth stress (Pa, bar [psi]).
$S_H=S_X=\sigma_H-P_o$=larger principal horizontal earth stress (Pa, bar [psi]).
S_i=stress component in i-direction, $S_1>S_2>S_3$ (Pa, bar [psi]).
S_L=lithostatic earth stress (Pa, bar [psi]).
S_r=maximum radial stress on borehole wall (Pa, bar [psi]).
S_θ=maximum tangential stress on borehole wall (Pa, bar [psi]).
S_t=effective tensile breaking strength of rock (Pa, bar [psi]).
$S_v=S_Z=\sigma_v-P_o$=vertical earth stress (Pa, bar [psi]).
ΔS=entropy change (J/kg [BTU/lb]).

t=time (s, hr, min).
T=fluid temperature (K, °C [°F]).
t^*=number of time periods.
t_d=dimensionless time=$(C_p^2/\lambda_r\rho_r\,C_{pr})\,(\dot{m}^*/H)^2t$.
t_r=thermal recovery time (yr).
T_{gf}=geothermal fluid wellhead temperature (K, °C [°F]).
T_i=fluid temperature as it leaves the injection well and enters the formation (K, °C [°F]).
T_i^*=reference fluid temperature for buoyancy effect (equation (10.37)) (K, °C [°F]).
T_o=ambient or dead state temperature (K, °C [°F]).
T_{out}=fluid temperature as it leaves the formation and enters the production well (K, °C [°F]).
T_r=rock temperature (K, °C [°F]).
T_{rej}=fluid surface reinjection temperature (K, °C [°F]).
T_{ri}=initial rock temperature at a particular position x, y, z (equations (10.41) and (10.43)) (K, °C [°F]).
$T_{ri}(Z)$=initial rock temperature at a particular depth Z (K, °C [°F]).
T_{ro}=minimum acceptable rock temperature for heat mining (K, °C [°F]).
T_∞=initial rock temperature (equation (10.30)) (K, °C [°F]).
ΔT=temperature difference in fluid (K, °C [°F]).
ΔT_r=rock temperature difference (K, °C [°F]).
∇T=geothermal temperature gradient (°C/km or K/km [°F/ft]).

u=fluid velocity in x-direction (m/s, cm/s [ft/s]).

U=general fluid velocity vector (m/s, cm/s [ft/s]).
$\bar{U}$=characteristic or average fluid velocity (m/s, cm/s [ft/s]).
$\bar{U}_j$=characteristic or average fluid velocity in zone j (m/s, cm/s [ft/s]).

v=fluid velocity in z-direction (m/s, cm/s [ft/s]).
V=fluid or reservoir volume (m³, liters [ft³, gal]).
V_f=fracture volume (m³, liters [gal]).
V_h=heat capture volume (m³ [ft]).
V_{inj}=injected fluid volume (m³ [ft³]).
V_p=compressive wave velocity (m/s [ft/s]).
V_s=shear wave velocity (m/s [ft/s]).
V_s^*=rock volume defined by seismic activity (m³ [ft³]).
V^{min}=minimum rock volume required for reservoir to last n years (equation (10.3)) (m³).
$\mathring{V}$=modal fracture volume defined at the maximum in the RTD (equation (10.18)) (m³, liters [gal]).
$\Delta\mathring{V}$=change in modal fracture volume from fixed reference point (m³, liters [gal]).
$<V>$=integral mean fracture volume (equation (10.17)) (m³, liters [gal]).
$[V]$=median fracture volume (equation (10.19)) (m³, liters [gal]).

w=crack or fracture aperture (mm, cm [in.]).
w^*=maximum or characteristic fracture aperture for scaling purposes (m, mm, cm [in.]).
$w_{1/2}$=width of tracer RTD at half height (equation (10.21)), measure of degree of dispersion (m³ or s [ft or s]).
w=mean fracture aperture (mm, cm [in.]).
$w(r)$=fracture aperture at a particular fracture radius (m, cm, mm [in.]).
$w(0,t)$=fracture aperture at wellbore as function of time (mm, cm [in.]).
$\dot{W}_{max}$=maximum power possible from specified heat source (W [BTU/hr]).
$\dot{W}_{net}$=net power produced by power cycle (W [BTU/hr]).

x=horizontal distance in the plane of a vertical fracture (Fig. 10.24) (m [ft]).
X=horizontal distance in the earth (m, km [ft]).
x_{ed}=dimensionless distance=C_p/λ_r ($\dot{m}^*/H$) (L/2n).
x_f=distance along fracture (equation (10.35)) (m [ft]).
X_j=dimensionless distance in fracture j (equation (10.24)).

y=horizontal distance perpendicular to vertical fracture plane (Fig. 10.24) (m [ft]).
Y=horizontal distance in the earth (m, km [ft]).
ΔY=distance away from pressurized region (equation (9.6)) (m [ft]).

z=vertical distance in the plane of a vertical fracture (Fig. 10.24)) (m [ft]).
Z=vertical distance or depth in the earth (m, km [ft]).
z^*=distance defining zone of mined heat (m [ft]).

3 Greek symbol nomenclature

$\alpha = A\sqrt{k\beta}$ = hydraulic sizing parameter (m³Pa⁻¹ [ft³ psi⁻¹]).

α^* = empirical constant relating $k = f(\phi)$ (equation (8.37)) (m² [mD]).

α_H = hydraulic diffusivity = $k/\mu c\phi$ (m²/s [ft²/s]).

α_r = rock thermal diffusivity = $\lambda_r/\rho_r C_{pr}$ (m²/s [ft²]).

α_s = specific surface energy (J/m²).

β = bulk rock compressibility = $-\dfrac{1}{V}(\partial V/\partial P)_T$ (Pa⁻¹ [psi⁻¹]).

β_0 = reference state compressibility (equation (10.14)) (Pa⁻¹ [psi⁻¹]).

β_T = fluid volumetric coefficient of thermal expansion

$= \dfrac{1}{V}(\partial V/\partial T)_p$ (K⁻¹ [°F⁻¹]).

β_{Tr} = rock volumetric coefficient of thermal expansion

$= -\dfrac{1}{\rho_r}(\partial \rho_r/\partial T)_p$ (K⁻¹ [°F]).

γ = inclination angle of compressive stress exerted (Fig. 8.1).

γ_1, $\bar{\gamma}_2$, $\bar{\Gamma}_2$ = empirical parameters in P3DH hydraulic fracture model (equation (8.36)).

∇T = geothermal temperature gradient (K/m or °C/km [°F/ft]).

δ_t = thermal penetration depth (equation (10.4)) (m [ft]).

ε_i = strain component in the ith direction.

ζ_f = error function argument in zeroth order model (equation (10.35)).

η_c = cycle efficiency.

η_h = hydraulic efficiency (equation (10.26)).

η_p = pump efficiency.

η_{sweep} = sweep efficiency fraction of total fracture area within circulating fluid paths (equation (10.1)).

η_t = turbine efficiency.

η_u = utilization efficiency.

$\bar{\eta}$ = empirical parameter for P3DH hydraulic fracture model (equation (8.36)).

θ = angle in horizontal plane measured from x-axis along $S_x = S_H$ stress direction (equation (8.3)).

θ_j = dimensionless residence time in zone j (equation (10.24))

λ_r = rock thermal conductivity (W/m K [BTU/ft hr °F] or cal/cm s °C).

μ=fluid viscosity (Pa s [cp]).

v=Poisson's ratio (equation (8.2)).

ξ_j=flow fraction through zone j (equation (10.24)).

ρ=fluid density (kg/m^3 [lb/ft^3 or g/cm^3]).
ρ_r=rock density (kg/m^3 [lb/ft^3 or g/cm^3]).

σ^2=variance of RTD (equation (10.20)).
σ_c=dynamic uniaxial compressive strength (Pa, bar [psi]).
σ_h=minimum horizontal stress (Pa, bar [psi]).
σ_H=maximum horizontal stress (Pa, bar [psi]).
σ_i=stress component in ith direction (Pa, bar [psi]).
σ_n=normal stress (Pa, bar [psi]).
σ_v=vertical stress (Pa, bar [psi]).
σ_t=Brazilian tensile strength (Pa, bar [psi]).
$\Delta\sigma$='deviator stress'=$-(S_r-S_\theta)$ (equation (9.1)) (Pa, bar [psi]).
$\Delta\sigma_{th}$=maximum effective thermal stress (equation (9.2)) (Pa, bar [psi]).

τ, τ_j=shear stress (Pa, bar [psi]).
τ^*=thermal breakthrough time corresponding to heat stored in the fracture aperture (equation (10.34)).
τ_c=characteristic time constant for system (equation (8.36)).
τ_f=total thermal breakthrough time (equation (10.34)).
τ_h=thermal breakthrough time corresponding to heat mined from rock surrounding a fracture (equation (10.34)).
τ_{max}=shear strength, (Pa, bar [psi]).

ϕ=angle of friction.
ϕ=porosity=$\Delta V/V$.
ϕ_f=porosity within a fracture.
ϕ_o=reference state porosity (equation (10.14)).
Φ=total capital investment.
Φ_{dwp}=downhole pump equipment cost.
Φ_e=combined plant equipment cost.
Φ_f=fluid transmission cost.
Φ_s=site development and exploration costs.
Φ_{stim}=stimulation costs (fracturing, downhole valving, etc.).
Φ_w=total drilling and completion cost.

Ψ=thermal physical property group=$\lambda_r\rho_r C_{pr}/C_p$ (equation (10.35)).

Ω=ramp angle of surface irregularities or asperity angle.

References

Aki, K., Fehler, M. *et al.* (1982) Interpretation of seismic data from hydraulic fracturing experiments at Fenton Hill, NM, USA, Hot dry rock site. *J.Geoph. Res.*, **87**, 936–44.

Albright, J.N. and Hanold, R.J. (1976) Seismic mapping of hydraulic fractures made in basement rocks. *Proc ERDA Symp. on enhanced oil and gas recovery*, Vol. 2 gas, Tulsa OK, USA.

Anno, G.H. *et al.* (1977) *Site specific analysis of hybrid geothermal/fossil power plants.* Pacific Sierra Res. Corporation Report 705, Santa Monica, CA, USA, (April 1977).

Anon (1983) Innovative tools could revolutionize drilling. *Pet. Eng. Intl* (Feb. 1983) 92–93.

Armstead, H.C.H. (ed.) (1973) *Geothermal energy: a review of research and development.* UNESCO doc. ISBN 92 3 101063 8.

Armstead, H.C.H. (1978) *Geothermal energy*, 1st edn, E. & F.N. Spon, London.

Armstead, H.C.H. (1980) Effects of growth on natural resources' life *Energy International*, **17**,(4) 25–27.

Armstead, H.C.H. (1983) *Geothermal energy*, 2nd edn, E. & F.N. Spon, London.

Arpaci, V.S. (1966) *Conduction Heat Transfer*, Addison-Wesley, Reading, MA, USA, p. 351.

Barenblatt G.J. (1962) The mathematical theory of equilibrium cracks in brittle fracture. *Adv. in appl. mechs*, **7**, 56.

Basile, P.S. (ed.) (1977) *Energy supply-demand integrations to the year 2000: global and national studies.* Wkp. on alternative energy strategies (WAES), MIT Press, Cambridge, MA, USA.

Batchelor, A.S. (1978) Geothermal energy and its exploitation in South-west England. *Ground Eng.* (March, 1978), p. 39.

Batchelor, A.S. (1982) The creation of hot dry rock systems by combined explosive and hydraulic fracturing. *Proc. Int. Conf. on geothermal energy*, organised and sponsored by BHRA Fluid Eng., England. Florence, Italy (May, 1982), pp. 321–342.

Batchelor, A.S. (1983) *An overview of hot dry rock technology.* Paper presented at the 9th Stanford Wkp. on geothermal reservoir eng., Stanford Univ., Stanford, CA, USA (Dec. 1983).

Batchelor, A.S. (1984) Hot dry rock geothermal exploitation in the United Kingdom. *Modern Geology*, **9**, 1–41.

Batchelor, A.S. *et al.* (1981–84) *Geothermal exploitation res. bi-monthly inf. circs*, Camborne School of Mines Geothermal Energy Project, Redruth, Cornwall, England: No. 6, Aug./Sept. 1981; No. 7, Oct./Nov. 1981; No. 8, Dec. 1981/Jan. 1982; No. 13, Oct./Nov. 1982; No. 14, Dec. 1982/Jan. 1983; No. 15, Feb. 1983/April 1984; No. 16, May 1984/Oct. 1984.

Batchelor, A.S. (1985) Personal communication. Camborne School of Mines, Geothermal Energy Project. Cornwall, England.

Batchelor, A.S., Baria, R. and Hearn, K. (1983) Monitoring the effects of hydraulic stimulation by microseismic event location: a case study. *Soc. Pet. Eng., SPE paper 12109*, 58th annual tech. conf. and exbn., San Francisco, CA, USA (5–8 Oct. 1983).

Batchelor, A.S. and Beswick, A.J. *et al.* (1982) *Well drilling and casing. Vol. 1 Summary Rep.* Camborne School of Mines Geothermal Project, Redruth, Cornwall, England. pp. 2–15.

Batchelor, A.S., Ledingham, P., Kwakwa, K.A. and Curtis, R.H. (1987) Circulation results 1983–1986, Phase IIB report, Camborne School of Mines, Geothermal Energy Project, Cornwall, England, January 1987.

Batchelor, A.S. and Pearson, C.M. (1981) Permeability enhancement studies in Southwest England. *Phase I Report on geothermal exploitation resources*, Camborne School of Mines, Vol. 2. (Dec. 1981). Redruth, Cornwall, England.

Batchelor, A.S., Pearson, C.M. and Halladay, N.P. (1980) The enhancement of the permeability of granite by explosive and hydraulic fracturing. Int. Seminar on the results of the E.C. Geothermal Energy Res., Strasbourg, France (March 1980). *Proc. D. Reidel* (EUR 6872). Dordrecht, Holland, 1009–1031.

Birch, F. and Clark, H. (1940) The thermal conductivity of rocks and its dependence upon temperature and composition. Parts I & II. *Amer. J. Science*, 529–558 and 613–635.

Blair, A., Tester, J.W. and Mortensen, J.J. (eds.) (1976) *Hot dry rock geothermal energy dev. project.* LANL Progress Rep. LA-6625-PR, Los Alamos, NM, USA.

Bodvarsson, G.S. (1974) Geothermal resource energetics. *Geothermics*, **3**,(3), 83.

Bodvarsson, G.S. and Reistad, G.M. (1975) Econometric analysis of forced geo-heat recovery for low temperature uses in the Pacific North-west. *Proc. UN Geothermal Symp.*, San Francisco, CA, USA. (May 1975), **3**, 1559–1564.

Boldiszár, T. (1970) Geothermal energy from porous sediments in Hungary. *Proc. UN Geothermal Symp.*, Pisa, Italy, **2**, Pt. I, p. 101.

Boldiszár, T. (1980) *Assessment of geothermal resources in Hungary.* Draft article enclosed with personal letter to one of the authors.

Brittenham, T.L., Neudecker, J.W., Rowley, J.C. and Williams R.E. (1982) Directional drilling equipment and techniques for deep, hot granite wells. *Int. Pet. Tech.*, **34**, 1421–1430.

Brown, M.C. *et al.* (eds.) (1979) Hot dry rock geothermal development program annual rep. Fiscal Year 1978. LASL Rep. LA-7807-HDR, Los Alamos, NM, USA.

Browning, J.A. (1963) *Flame jet cutting.* US pat. No. 3,103,251 (10 Sept. 1963).

Browning, J.A. (1969) *Flame jet drilling.* US pat. No. 3,76,1984 (4 Nov. 1969).

Browning, J.A. (1982) *Flame jet drilling and chambering to great depths in crystalline rock*, Rep. N00014-82-C-0249.

Browning Engineering Corporation (1981) *Final Report, Flame jet drilling in Conway*, NH, USA, granite.

Bullard, Sir E. (1973) Basic theories in *Geothermal energy: a review of research and development.* (ed. H.C.H. Armstead) UNESCO doc. ISBN 92-3-101063-8, pp. 19–29.

Burnham, J.B. and Stewart, D.H. (1973) Recovery of geothermal energy from hot dry rock with nuclear explosives. In *Geothermal energy*, (eds. P. Kruger and C. Otte) Stanford Univ. Press, Stanford, CA. USA. pp. 223–230.

Calaman, J.J., Bonsall, A.L. and Hoelzlhammer, J. (1966) New Developments in Linde Jet Piercing, *Proceedings of 27th Annual Univ. of Minnesota Mining Symposium*, Minneapolis, Minn. (Jan. 11–12, 1966).

Carson, C.C. and Lin, Y.T. (1981) Geothermal well costs and their sensitivity to changes in drilling and completion operations. *Proc. Int. Conf. on geothermal drilling and completive technology*, Albuquerque, NM, USA, pp. 21–23.

Cepattelli, L. and Ferrara, G.C. (1980) Design and testing of downhole probes for operation in deep and hot environments. *Proc. 2nd Intern. Seminar on the results of the EEC geothermal energy research*. Strasbourg, (eds. A.S. Strub and P. Ungemach) (4–6 March 1980), Reidel Pub. Co., Boston.

Cermak, V. and Ryback, I. (eds.) (1979) *Terrestrial Heat Flow in Europe*, Springer Verlag, Berlin.

Cleary, M.P. (1979) Rate and structure sensitivity in hydraulic fracturing of fluid-saturated porous formations, *Proc. 20th US Rock Mech. Symp.*, Austin, TX, USA, pp. 127–142.

Cleary, M.P. (1980) Analysis of mechanisms and procedures for producing favourable shape of hydraulic fractures. *Soc. Pet. Eng. SPE Paper 920Q*, presented at 55th Annual Fall Tech. Conf. and Exbn., Dallas, TX, USA (Sept. 21–24, 1980).

Cleary, M.P. (1982) *Theory and application of hydraulic fracturing technology*. Invited paper for 1st Japan/US Joint Sem. on hydraulic fracturing. Tokyo, Japan.

Colp, J. (1982) Sandia Laboratory Report. SAND 82-2377 (Oct. 1982).

Cornet, F. (1980a) Analysis of hydraulic fracture propagation: a field experiment. *Proc. 2nd Int. Sem. on the results of EC geothermal energy res.*, Strasburg (March 1980). D. Reidel Pub., Dordrecht, Holland. (EUR. 6862), pp. 1032–43.

Cornet, F. (1980b) Microseismic and acoustic activity associated with hydraulic fracture propagation. *Proc. 2nd Int. Sem. on the results of EC geothermal enregy res.*, Strasburg (March 1980). D. Reidel Pub. Dordrecht, Holland. (EUR. 6862), pp. 967–76.

Cornet, F. (ed.) (1983) *Le projet Énergeroc, Phase II, déscription technique*. ENERGEROC, Compagnie Française des Pétroles, Boulogne Billancourt, France. 2 vols: 233 pp. and tech. annexe.

Crabbe, D. and McBride, R. (1978) *The world energy book*, Kogan Page, London.

Craig, S.B. (1961) Geothermal drilling practice at Wairakei, New Zealand. *Proc. UN Conf. on new sources of energy*. Rome, Aug. 1961, **3**, p. 123.

Cremer, G.M. *et al.* (eds.) (1980) *Hot dry rock geothermal energy development program: Annual Rep., Fiscal Year 1979*. LASL Rep. LA-8280-HDR, Los Alamos, NM, USA.

Cummings, R.G. and Morris, G.E. (1979) Economic modeling of electricity production from hot dry geothermal reservoirs: methodology and analysis. *EPRI Rep. EPRI-EA-630*, Palo Alto, CA, USA.

Cundall, P.A. *et al.* (1978) *Computer modeling of jointed rock masses*. US Army Corps of Eng., Waterways Experiment Station, Vicksburg, MS, USA. Tech. Rep. N-78-4 (1978).

Daneshy, A.A. (1973) On the design of vertical hydraulic fractures. *J. Pet. Tech.* (Jan.), pp. 83–97; and *Trans. AIME*, **255**.

Dash, ZV., Murphy, H.D. *et al.* (1983) Hot dry rock geothermal reservoir testing: 1978–1980. *J. Volcanology and Geothermal Res.*, **15**, 59–99.

Dash, Z.V., Murphy, H.D. and Cremer, G.M. (eds.) (1981) *Hot dry rock geothermal reservoir testing: 1978 to 1980*. LANL Rep. LA-9080-SR, Los Alamos, NM, USA. (Nov. 1981).

Di Pippo, R. (1984) Geothermal power development 1984: overview and update. *GRC Bulletin* (Oct. 1984), pp. 3–12.

Dreesen, D.S. and Miller, J.R. (1985) *Open-hole Packer and Running Procedure for Hot Dry Rock Reservoir Testing*. USDOE Geothermal Review meeting, Washington, DC. (September 11–12, 1985).

Edwards, L.M. *et al.* (eds.) (1982) *Handbook of geothermal energy*. Gulf Pub. Co., Houston, TX, USA.

Edmunds, W.M., Andrews, J.N. *et al.* (1984) The evolution of saline and thermal groundwaters in Carmenellis granite. *Mineralogical Magazine*, **48**, pp. 407–424.

Einarsson, S.S. (1975) Geothermal space heating and cooling. *Proc. UN Geothermal Symp.*, San Francisco, CA, USA, 2117–2126.

Electricité de France (1961) Brochure. *Usine marémotrice de la Rance.*

England, A.H. and Green, A.E. (1963) Some 2-dimensional punch and crack problems in classical elasticity. *Proc. Cambridge Phil. Soc.*, **59**, 489.

ERDA (1977) Hot dry rock geothermal energy: status of exploration and assessment. *Energy Res. and Dev. Admin.* Doc. ERDA-77-74. Washington, DC, USA.

Exxon (1980) *World Energy Outlook*, Exxon Corpn., New York.

Fehler, M.C. (1981) Changes in p wave velocity during operation of a hot dry rock geothermal system. *J. Geoph. Res.*, **86**(B4), 2925–8.

Fehler, M.C., House, L. and Kaieda, H. (1986) *Seismic Monitoring of Hydraulic Fracturing: Techniques for Determining Fluid Flow Paths and State of Stress Away from a Wellbore.* Proc. of 27th US Symp. on Rock Mechanics, University of Alabama, Tuscaloosa, AL, p. 606–14.

Fehler, M.C., House, L. and Kaieda, H. (1987) Determining Planes Along Which Earthquakes Occur: Method and Application to Earthquakes Accompanying Hydraulic Fracturing. *J. Geological Research.*

Fisher, H.N. and Tester, J.W. (1980) *The pressure transient testing of a man-made fractured geothermal reservoir: an examination of fracture versus matrix-dominated flow effects.* LANL Rep. LA-8535-MS, Los Alamos, NM, USA.

Garde–Hansen, H. (1984) Oil industry inflation has become excessive. *World Oil*, pp. 41–45 (March 1981). Joint Ass. Surveys of the US oil and gas producing industry, 1973–1983, Sec. I Drilling costs (1974–1984).

Garnish, J. (1976) Geothermal energy: the case for research in the United Kingdom. *UK Dept. of Energy. Energy Paper No. 9.* HM Stationery Office, London.

Geertsma, J. and de Klerk, F. (1969) A rapid method of predicting width and extent of hydraulically induced fractures, *J. Pet. Tech.*, p. 1571.

Geertsma, J. and Haafkens, R. (1979) A comparison of the theories for predicting width and extent of vertical hydraulically induced fractures. *J. Energy Resources Tech.*, Trans. AIME, **101**, 8.

Gerlach, T.M. (1982) *Analysis of magma-thermal conversion of biomass to gaseous fuel.* SNL Rep. SAND-82-0031 (Feb. 1982).

Glaser, P. (1983) Evolution of the solar power satellite concept: the utilization of energy from space. *Spec. issue on Energy from Space. Space Solar Power Review*, **4**,(1–2), 11.

Goetz, G.J. (1973) *Energy use patterns in US manufacturing for the period 1950–1970.* ORNL Rep. ORNL-MIT-169, Oak Ridge, TN, USA.

Greene, K. and Goodman, L. (1982) Geothermal well drilling and completion in *Handbook of geothermal energy*, (eds. L.M. Edwards *et al.*), Gulf Pubs., Houston, TX, USA.

Griffith, A.A. (1921) The phenomena of rupture and flow in solids. *Phil. Trans. Royal Society* (London), **221A**, p. 163.

Grigsby, C.O. (1983) *Rock–water interactions in hot dry rock geothermal systems: reservoir simulation and modeling.* Master's thesis in Chem. Eng., MIT, Cambridge, MA, USA.

Grigsby, C.O. and Tester, J.W. (1978) Evaluation of the Fenton Hill hot dry rock geothermal reservoir, Pt. II: flow characteristics and geochemistry. *Proc. 4th Wkp. on geothermal reservoir eng.*, Stanford Univ., Stanford, CA, USA.

Grigsby, C.O. and Tester, J.W. (1983) Rock–water interactions in hot dry rock geothermal systems: field investigations of *in situ* geochemical behavior. *J. Volcanology and Geothermal Research*, **15**, 101–136.

Grindgarten, A.C., Witherspoon, P.A. and Ohnishi, Y. (1975) Theory of heat extraction from fractured hot dry rock. *J. Geoph. Res.*, **80**(8), H20.

Haenel, R. (1980) *Atlas of sub-surface temperatures in the European Community.* Schäfer, Hannover, FRG.

Haenel, R. (ed.) (1982) The Urach geothermal project (Swabian Ald. Germany), *E. Schweizerbart's Verlagsbuchhandlung* (Nägele u. Obermiller), Stuttgart, 419 pp.

Haimson, B. (1968) *Hydraulic fracturing in porous and non-porous rock, and its potential for determining in situ at great depth.* Ph.D. thesis, Univ. of Minnesota, Minneapolis, MN, USA.

Haimson, B. (1978) The hydrofracturing stress measuring method and recent field results. *Int. J. Rock Mechanics, Mining Science and Geomechanics*, **15**, 167–178.

Hanold, R.J., Altseimer, J.H. *et al.* (1977) *Rapid excavation by rock melting: LASL Subterrene Program.* LANL Rep. LA-5979-SR, Los Alamos, NM, USA.

Hanold, R.J. and Morris, C.W. (1982) Induced fractured well stimulation through fracturing. *Proc. Fractured geothermal reservoir workshop*, sponsored by the Geothermal Resources Council, Honolulu, HA.

Harlow, F.H. and Demuth, R.B. (1979) *Thermal fracture effects in geothermal energy extraction*, LANL Rep. LA-7963. Los Alamos, NM, USA.

Harlow, F.H. and Pracht, W.E. (1972) A theoretical study of geothermal energy extraction. *J. Geoph. Res.*, **77**, 7038–7048.

Heiken, G., Goff, F. and Cremer, G. (eds.) (1982) *Hot dry rock geothermal resource 1980.* LANL Rep. LA-9295-HDR. (April 1982).

Helmick, C., Koczan, S. and Pettitt, R. (1982) *Planning and drilling of geothermal energy extraction hole EE-2.* LANL Rep. LA-9302-HDR. Los Alamos, NM, USA.

Hoek, E. and Brown, E.T. (1980) *Underground excavation in rock.* Inst. Mineral and Met., 527 pp., London (1980).

House, J. (1986) Locating Microearthquakes Induced by Hydraulic Fracturing in Crystalline Rock. *Geophysical Res. Letters*, (to be submitted).

Howard, G.C. and Fast, C.R. (1970) *Hydraulic fracturing.* Monograph Series, Vol. 2, Soc. Pet. Eng., Richardson, TX, USA.

Hunt, T.M. (1970) Net mass loss from the Wairakei geothermal field, New Zealand. *Proc. UN Geothermal Symp.*, Pisa, Italy. Pt. I, Vol. 2, p. 487.

Jacoby, C.H. and Paul, D.K. (1975) Gulf coast salt domes as possible sources of geothermal energy. *Proc. UN Geothermal Symp.*, San Francisco, CA, USA. (May 1975), **3**, 1673–1679.

Jaeger, J.C. and Cook, N.G.W. (1979) *Fundamentals of rock mechanics*, 3rd edn., Chapman and Hall, London.

Japan's Sunshine Project (1982) *Summary of geothermal Research and Development Report*, pp. 72–78.

Jung, R. (1980) Hydraulische Experimente 1979 in Experimentierfeld bei Falkenberg/Obertalz. Archiv. No. 87151. *Bundesanstalt für Geowissenchaften und Rohstoffe.* Hannover, Fed. Rep. Germany.

Kappelmeyer, O. (1984) *Ermittlung von Basibdatar zur Entwicklung von Technologien für die Erzeugung artifizieller Warmeaustauschflächen in Kristallinen Untergrund durch in situ. Experimente im HDR Experimentierfeld Falkenberg.* Rep. 96/822, 11669/84. Fed. Inst. for Geosciences and Natural Resources. Hannover, Fed. Rep. Germany.

Kappelmeyer, O. and Rummel, F. (1980) Investigations on an aritificially created frac in a shallow and low permeable environment. *Proc. 2nd Int. Sem. on the results of EC Geothermal Energy Res.*, Strasbourg, France, (March 1980). Reidel Pub., Dordrecht, Holland. (EUR 68672), pp. 1033–1053.

Kappelmeyer, O. and Rummel, F. (1984) In situ *experiments in an aritificially created fracture at shallow depth, considering technologies for extraction of terrestial heat from dry rock*, Rep. 03E4150A, Federal Inst. for Geosciences and Natural Resources, Hannover, Fed. Rep. Germany.

Kennedy, G.C. (1964) A proposal for a nuclear power program. 3rd Plowshare Symposium on Experimenting with nuclear explosives. University of California. April 21–23, Davis, 1964.

Keppler, H. (1983) *Seismic observations in the Fenton Hill Phase II hot dry rock reservoir in 1982.* Paper presented at the CEC HDR Wkp., Brussels, Belgium (June 30, 1983).

Kestin, S., Di Pippo, R., Khalija, H.E. and Ryley, D.J. (eds.) (1980) *A sourcebook on the production of electricity from geothermal energy.* US Govt. Printing Office, Washington, DC, USA. (March, 1980).

Kingston, Reynolds, Thom and Allardice (1978) *City of Burbank hybrid geothermal/coal power plant.* Rep. on the geothermal resource requirements and utilization, US DOE Grant EG-77-G-03-1572, Auckland, New Zealand.

Kristianovitch, S.A. and Zheltov, Y.P. (1955) Formation of vertical fractures by means of highly viscous liquid. *Proc. 4th World Pet. Congress*, Sec. II/TOP, Paper 3, pp. 579–586.

Kuriyagawa, M. and Matsunaga, I. (1983) Personal communication. Nat. Res. Inst. for Pollution and Resources, Ministry for Intl. Trade and Industry, Tskuba, Japan. (May 1983).

LANL (1982) *Hot dry rock geothermal resource.* Doc. LA-9295-HDR. Pub. by the LANL.

Laughlin, A.W. and Eddy, A.C. (1977) *Petrography and geochemistry of pre-Cambrian core samples from GT-2 and EE-1.* LASL Rep. LA-6930-MS.

Laughlin, A.W., Eddy, A.C., Laney, R. and Adrich, J.J. (1983) Geology of the Fenton Hill, New Mexico, hot dry rock site. *J. Volcanology and Geothermal Res.*, **15**, 21–41.

Leydecker, G. (1981) Seismische Ortung Hydraulisch Erzeugter Bruche im geothermisch Frac Projekt Falkenberg. Archiv. No. 86549, *Bundesanstalt für Geowissenschaften und Rohstoffe.* Hannover, FRG. (March, 1981).

Leydecker, G. (1982) Seismiche Ortung Hydraulisch Erzeugter Bruche im Französischen Geotermik Projekt le Mayet. Archiv. No. 92297. *Bundesanstalt für Geowissenschaften und Rohstoffe.* Hanover, FRG. (March, 1982).

Linden, H.R. (1980) *1980 assessment of the US and World energy situation and outlook.* Gas Res. Inst., Chicago, IL, USA.

McFarland, R.D. and Murphy, H.D. (1976) Extracting energy from hydraulically fractured geothermal reservoirs. *Proc. 11th Intersociety Energy Convention Eng. Conf.*, p. 828. Reno, NV, USA.

McLennan, J.D. and Roegiers, J.C. (1979) *A synthesis of hydraulic fracturing literature.* Univ. of Toronto, Canada. Pub. 79-13 ISBN 0-7727-7004-2.

MAGES (1979) *Man-made geothermal energy systems.* Int. Energy Agency. OECD, Paris, 2 vols, appendix (unpublished).

Matsunaga, I. and Kuriyagawa, M. (1983) Personal communication. Nat. Res. Inst. for pollution and resources. Ministry for International Trade and Industry. Tskuda, Japan.

Maurer, W.C. (1975) Geothermal drilling technology. *UN Geothermal Symp.* (20–29 May, 1975). San Francisco, CA, USA. Proc. Vol. 2, p. 1509.

Maurer, W.C. (1979) *LASL turbodrill economics.* Maurier Eng. Rep. TR79-1, Houston, TX, USA. (April 1979).

Maurer, W.C. (1980) Advanced drilling techniques. Pet. Pub. Co., Tulsa, OK, USA.

Milora, S.L. and Tester, J.W. (1976) *Geothermal energy as a source of electric power*, MIT Press, Cambridge, MA, USA.

468   *References*

Muffler, L.J.P. and Guffanti, M. (eds.) (1978) *Assessment of geothermal resources in the United States.* USGS Circ. 790.

Murphy, H.D. *et al.* (1977) Preliminary assessment of a geothermal energy reservoir formed by hydraulic fracturing. *Soc. Pet. Eng. J.*, **17**, 317–326.

Murphy, H.D. (ed.) (1980) *Evaluation of the 2nd hot dry rock geothermal energy reservoir: results of Phase I, Run Segment 4.* LANL Rep. LA-8354-MS, Los Alamos, NM, USA.

Murphy, H.D. *et al.* (eds.) (1981) *Relaxation of geothermal stresses induced by heat production.* LANL Rep. LA-8954-MS. Los Alamos, NM, USA.

Murphy, H.D. (1982) Hot dry rock reservoir development and testing in the USA. *Int. Sem. on Rock Mechanics*, Tokyo, Japan.

Murphy, H.D. (1983) *Modeling hydraulic fracturing in jointed rock: preliminary FRIP results.* Tech. Mem. Camborne School of Mines Geothermal Energy Project, HDM/ALB/13/04/01. Redruth, Cornwall, England.

Murphy, H.D. and Aamodt, R.L. (1979) *Hydraulic fracture design calculations.* G-3 Tech. Mem. G-3/79/No. 2, LANL, Los Alamos, NM, USA.

Murphy, H.D., Drake, R., Tester, J.W. and Zyvoloski, G. (1982) *Economics of a 75 MWe hot dry rock geothermal power station based on the design of Phase II reservoir at Fenton Hill.* LANL Rep. LA-9241-MS, Los Alamos, NM, USA.

Murphy, H.D., Keppler, H. and Dash, Z. (1983) Does hydraulic fracturing theory work in jointed rock masses? *GRC Trans.*, **7**, p. 46.

Murphy, H.D., Tester, J.W., Grigsby, C.O. and Potter, R.M. (1981) Energy exraction from fractured geothermal reservoirs in low-permeability crystalline rock. *J. Geoph. Res.*, **86**, 7145–7158.

National Geographic Magazine (1981) Special report in the public interest. Energy: facing up to the problem and getting down to solutions. Washington, DC, USA.

Neudecker, J.W. (1973) *Design description of melting-consolidating prototype subterrene penetrators.* LANL Rep. LA-5212-MS, Los Alamos, NM, USA.

Newson, M.M., Barnett, J.H., Baker, L.E., Varnado, S.G. and Polito, J. (1977) *Geothermal well technology, drilling and completion programme plan.* SNL Rep. SAND 77-1630, Albuquerque, NM, USA.

Nicholson, R., Verity, T. *et al.* (1979) *Industrial assessment of the drilling, completion and workover costs of well and fracture sub-systems of hot dry rock reservoirs.* Republic Geothermal Inc., Santa Fe Springs, CA, USA. (March 1979).

Nierode, D.E. (1983) Comparison of hydraulic fracture design methods to observed field results. *Soc. Pet. Eng., Paper SPE 12059.* 58th Annual Fall Tech. Conf. and Exbn., San Francisco, CA, USA.

Nordgren, R.P. (1972) Propagation of a vertical hydraulic fracture. *Soc. Pet. Eng. J.*, p. 306.

Northrup, C.J.M., Gerlach, T.M., Modreski, P.J. and Galt, J.K. (1978) Magma: a potential source of fuels. *J. Hydrogen Energy*, **3**, pp. 1–10.

Oil and Gas Journal (1981) Energy resource estimate.

Pearson, C.M. (1980) *Permeability enhancement by explosive initiation in South-west granites, with particular reference to hot dry rock energy systems.* Ph.D. thesis, Camborne School of Mines, Redruth, Cornwall, England.

Pearson, C.F. (1981) The relationship between microseismicity and high pore pressures during hydraulic stimulation experiments in low permeability granitic rocks. *J. Geoph. Res.*, **86**(B9), 7855–64.

Pearson, C.F. and Albright, J.N. (1984) Acoustic emissions during hydraulic fracturing in granite. In *Acoustic emission/microseismic activity in geologic structures and materials series on rock and soil mechanics*, Vol. 8, pp. 559–575, (5–7 Oct. 1981) at Penn State Univ. Park, PA, USA.

Pearson, C.F., Fehler, M.C. and Albright, J.N. (1983) Changes in compressional and shear wave velocities and dynamic moduli: during operation of a hot dry rock geothermal system. *J. Geoph. Res.*, **88**(B4), 3468–75.

Perkins, T.K. and Kern, L.R. (1961) Widths of hydraulic fractures. *J. Pet. Tech.*, p. 937. Sept 1961.

Pettitt, R. (1975) *Planning, drilling, testing and logging of geothermal test hole GT-2, Phase I (Jan. 1975), Phase II (March 1975) and Phase III (June 1975).* LANL Reps. LA-5819-PR, LA-5897-PR and LA-5965-PR, Los Alamos, NM, USA.

Pettitt, R. (1977) *Planning, drilling and logging of energy extraction hole EE-1, Phases I and II, Los Alamos.* LANL Rep. LA-6906-MS. Los Alamos, NM, USA.

Pine, R.J. and Batchelor, A.S. (1984) Downward migration of shearing in jointed rock during hydraulic injections. *Int. J. Rock Mechanics and Mineral Sciences and geomechanical abstracts*, 249–65.

Pine, R.J. and Ledingham, P. (1983) *In situ* hydraulic parameter for the Carnmenellis granite hot dry rock geothermal energy research reservoir. *Proc. 58th Annual Tech. Conf. and Exbn, Soc. Pet. Eng.*, San Francisco, CA, USA. SPE paper 12020. (5–8 Oct. 1983).

Pine, R.J., Ledingham, P. and Merrifield, C.M. (1983) *In situ* stress measurements in the Cornmenellis granite, 2-hydrofracture tests at Rosemanowes quarry to depths of 2000 m. *Int. J. Rock Mechanics and Mining Sciences.*

Potter, R.M. (1977) Hydraulic fracture initiation sites in open boreholes identified by geophysical logs. *Proc. Stanford Univ. 3rd Annual Wkp on geothermal reservoir engineering.* Stanford, CA, USA. (14–16 Dec. 1977).

Pruess, K. and Bodvarsson, G.S. (1984) Thermal effects of reinjection in geothermal reservoirs with major vertical fractures. *J. Pet. Tech.*, **36**(10), 1567–78.

Ramey, H.J., Kumar, A. and Gulati, M.S. (1975) *Gas well test analysis under water-drive conditions.* Amer. Gas. Ass., Arlington, VA, USA.

Rauenzahn, R.M. (1986) *Analysis of fluid mechanics and heat transfer effects in the thermal spallation of granite rocks.* Ph.D. thesis, MIT, Cambridge, MA, USA.

Rauenzahn, R.M. and Tester, J.W. (1985) Flame-jet induced thermal Spallation as a Method of rapid drilling and cavity formation. *Proc. of 60th Assn. Tech. Conf. and Exhibition*, Soc. of Petro Engineering paper 14331, Las Vegas, NV.

Rieke, H.H. III and Chilingar, G.V. (1982) Casing and tubular design concepts, in *Handbook of geothermal energy*, (eds. L. Edwards *et al.*). Gulf Publishing, Houston, TX, USA.

Robinson, B.A. (1982) *Quartz dissolution and silica deposition in hot dry rock geothermal systems.* Master's thesis in Chem. Eng., MIT, Cambridge, MA, USA.

Robinson, B.A. and Tester, J.W. (1984) Dispersed fluid flow in fractured reservoirs: an analysis of tracer-determined residence time distributions. *J. Geoph. Res.*, **89** (B12), 10374–84.

Robinson, B.A., Tester, J.W. and Brown, L.F. (1984) *Reservoir sizing, using inert and chemically reacting tracers.* Paper presented at 59th Annual Conf. and Exbn., Soc. Pet. Eng., Houston, TX, USA. (18–19 Sept. 1984).

Rowley, J.C. (1982) Worldwide geothermal resources, in *Handbook of geothermal energy*; (eds. L.H. Edwards *et al.*). Gulf Pubs. Co., Houston, TX, USA.

Rowley, J.C. and Carden, R.S. (1982) *Drilling of hot dry rock geothermal energy extraction well EE-3.* LANL Rep. LA-9512-HDR, Los Alamos, NM, USA.

Rubow, I.H. (1956) Jet-piercing in taconite. *Mines Magazine.*

Rummel, F. (1978) *Hydraulic fracturing stress measurements and geophysical rock properties of the Falkenberg frac project.* Ruhr Univ., Bochum, Institut für Geophysik. Rep. EEC project E 8D, BMFT, ET 4150-A.

Sack, R.A. (1946) Extension of Griffith's theory of rupture to 3 dimensions. *Proc. Phys. Soc. London*, **58,** 729–36.

Schmidt, R.A., Boads, R.R. and Bass, R.C. (1981) A new perspective on well-shooting: behavior of deeply buried explosions and deflagrations. *J. Pet. Tech.*, **33**(7), 1305–11.

Settari, A. and Price, H.S. (1984) Simulations of hydraulic fracturing in low permeability reservoirs. *Soc. Pet. Eng. J.*, **24**(2), 141–152.

Sharman, H. (1975) Wind and water sources of energy in the United Kingdom, in *Energy Options*, (ed. S. Caradoc Evans), Latimer Press, London, p. 55.

Shaw, T.L. (1977) *An environmental approach to the Severn Barrage*, 2nd edn., Univ. of Bristol, England.

Slider, H.C. (1976) *Practical petroleum reservoir engineering methods*. Pet. Pubs. Co., Tulsa, OK, USA, ch. 8.

Smith, M.C. (1979) *The future of hot dry rock geothermal energy systems*. Paper presented at the Pressure Vessels and Piping Conf. Amer. Soc. Mech. Engrs; Paper 79-PVP-35, San Francisco, CA, USA.

Smith, M.C., Aamodt, R.L., Potter, R.M. and Brown, D.W. (1975) Man-made geothermal reservoirs. *Proc. UN Geothermal Symp.*, San Francisco, CA, USA, **3**, 1781–1787.

Smith, M.C., Nunz, G.J. and Ponder, G.M. (eds.) (1983) *Hot dry rock geothermal development program*. Annual Rep. Fiscal Year 1982. LANL Rep. LA-9780-HDR (Sep. 1983).

Smith, R.L. and Shaw, H.R. (1975 and 1979) *Igneous-related geothermal systems*. USGS Circs. 726 [(1975) pp. 58–83] and 790 [(1978) pp. 12–17].

Sneddon, I.N. (1946) The distribution of stress in the neighbourhood of a crack in an elastic solid. *Proc. Royal Soc. London*, A187, 229–260.

Soles, J.A. and Geller, L.B. (1964) *Experimental studies relating mineralogical and petrographic features to the thermal piercing of rocks*. Canada Dept. Mines and Tech. Survey, Ottawa. Rep. TB53.

Sperry Research (1977) *Feasibility demonstration of the Sperry down-well pumping system*. Final Rep. Contract No. EY-76-C-02-2838, USDOE, SCRC-CR-77-48, Sudbury, MA, USA.

Sunshine Project, see Japan.

Swanberg, C.A. (1975) Physical aspects of pollution related to geothermal energy development. *Proc. UN Geothermal Symp.*, San Francisco, CA, USA, **2**, 1435–43.

Tenneco (1980) *Energy: 1980–2000*. Pub. by the Economic and Long Range Planning Dept. of the Tennessee Gas Transmission Co., P.O. Box 2511, Houston, Texas 77001, May 1980.

Tester, J.W. (1982) Energy conversion and economic issues for geothermal energy, in *Handbook of Geothermal Energy*, (eds. L.M. Edwards *et al.*). Gulf Pub. Co., Houston, TX, USA, ch. 10.

Tester, J.W. and Albright, J.N. (eds.) (1979) *Hot dry rock energy extraction field test: 75 days of operation of a prototype reservoir at Fenton Hill*. LANL Rep. LA-7771-MS, Los Alamos, NM, USA.

Tester, J.W., Bivins, R.L. and Potter, R.M. (1982) Interwell tracer analysis of a hydraulically fractured granite geothermal reservoir. *J. Soc. Pet. Eng.*, **22**, pp. 537–54.

Tester, J.W., Morris, G.E., Cummings, R.G. and Bivins, R.L. (1979) Electricity from hot dry rock geothermal energy: technical and economic issues. LANL Rep. LA-7603-MS. (Jan. 1979).

Tester, J.W. and Smith, M.C. (1977) Energy extraction characteristics of hot dry rock geothermal systems. *Proc. 12th Intersociety Energy Conversion Eng. Conf.*, Washington, DC, USA, **1**, p. 818.

Thain, I.A. (1980) *Wairakei: the first 20 years*. New Zealand Electricity Dept. Rep.

UN (1961) *United Nations conference on new sources of energy*, Rome, Italy (21–31 August, 1961). Proc. in 7 vols. Geothermal energy in Vols. 2 and 3.

UN (1970) *United Nations symposium on the development and utilization of geothermal resources*. Proc. in Vol. 1, Vol. 2, Pt. I and Vol. 2, Pt. 2. Pub. in Special Issue 2 of *Geothermics* by the Istituto Internazionale per le Ricerche Geotermiche, Pisa, Italy.

UN (1975) *Second United Nations symposium on the development and use of geothermal resources*. San Francisco, CA, USA. Proc. in 3 vols, prepared by Lawrence Berkeley Lab., Univ. of California for US ERDA, US Nat. Science Foundation and USGS.

UN Yearbooks of world energy statistics (1980) and (1982). Pub. by the UN.

Walsh, J.B. and Brace, W.F. (1964) A fracture criterion for brittle anisotropic rock. *J. Geoph. Res.*, **69**, 3449–56.

Warpinski, N.R. (1985) Measurement of width and pressure in a propagating hydraulic fracture. *Soc. Pet. Engrs J.*, **25**(2), 46–54.

WEC (1974) *World energy conf. Survey of Energy Resources*. Pub. by WEC, London, England.

WEC (1978) *World energy resources: 1985–2020*. Pub. for the WEC by IPC Science and Technology Press, Guildford, England and New York, USA.

WEC (1980) *World Energy Conf. Survey of Energy Resources*. Prepared by BGR Federal Institute for Geosciences of Natural Resources, Hannover, FRG.

Whetten, J. *et al.* (1986) The US Hot Dry Rock Project, LANL Rep. LA-UR-86-1109. Presented at EEC/US Workshop on Geothermal Hot Dry Rock, Brussels, Belgium. May 28–30, 1986.

White, D.F. and Williams, D.L. (1975) *Assessment of geothermal resources of the United States: 1975*. USGS Circ. 726 (1975).

William, B.B., Gidley, J.L. and Schechter, R.S. (1979) *Acidizing fundamentals*. SPE Monograph Series, Dallas, TX, USA.

Wilson, C.R. and Witherspoon, P.A. (1970) *An investigation of laminar flow in fractured porous rocks*. Univ. of California, Dept. of Civil Eng., Rep. No. 70-6, Berkeley, CA, USA.

Witherspoon, P.A. and Wang, J.S.Y. (1974) *Validity of the cubic law for laminar flow in a deformable rock fracture*. Tech. Inf. Rep. 23, Lawrence Berkeley Lab., Univ. of California, Berkeley, CA, USA.

Wunder, R. and Murphy, H.D. (1978) *Thermal drawdown and recovery of singly and multiply fractured hot dry rock reservoirs*. LANL Rep. LA-7219-MS, Los Alamos, NM, USA.

Yagupov, A.B. (1963) On the mechanism of rock destruction in fire well drilling. *Gornyi Zhurnal 2*, pp. 34–38.

Zyvoloski, G. *et al.* (eds.) (1981) *Evaluation of the 2nd hot dry rock geothermal energy reservoir: results of Phase I, Run Segment 5*. LANL Rep. LA-8940-HDR, Los Alamos, NM, USA.

Zyvoloski, G. (1983) Finite element methods for geothermal reservoir simulation. *Int. J. for numerical and analytical methods in geomechanics*, **7**, 75–86.

Index

Page numbers in *italic* refer to figures.

472

478 *Index*

Thermal – *cont.*
cracking 74, 153–6, *154*, 254, 305
gradient 34, 35, *37*, 45, *47*, 83, *87*, *91*, 407
performance of reservoir 280–306, *293*
pollution 76
stresses 163, 169
Thermonuclear fusion 28, 30–1, 32, 55
Thermo-syphon 101, *102*, 105–11, 118, 242
Tidal power 24–7, 32
Time scales for fracturing and propagation 163
Total crustal heat 35, *37*, 38, 46
Total energy complex 62
Tracers, chemical/radioactive 185, *188*, *189*, 258, *259*, *261*, 275, 276, *277*, *278*, *279*, 311, *312*
Tradeoffs 403
Transport of energy 28–9
Transportation 29–30
Tripping 324
Tritium 31
Turbo drill 333–7, *336*

Unavoidable temperature drops 117–20
UN energy statistics 5
Union Carbide jet piercing drill 371, *372*
United Kingdom, *see* Rosemanowes
Urach 162, 228
Uranium 14, 17
USA, *see* Fenton Hill
USSR 9, 16, 230, 381

Valles Caldera, NM, USA 191, *191*
Vapor pressure, water 104, *106*, 107, *111*
Visco-elasticity 127, 133
Viscosity 104
Viscous gel 222
Voids
survival of 155–60
volume of 69, 100
Volcano formation 381–3, *382*
Volume of active rock 69, 100, 264
Volumetric fracturing model 233, *236*

WAES (workshop on alternative energy strategies) 8
Wairakei, New Zealand 73, 77–8
Water drive 100, *252*, 253, *253*, *264*, *274*
Water losses 103, 138, 139, 141, 145, 156, 231, 242, 243, 254, 256, 263, 265, 272–5, 280, 315–17
Water treatment 103
Wave power 23–4, *32*
Wedge action 137
World Energy Conference (WEC) 5, 8, 9, 14, 15
Well casings 325, *325*
Well costs 359–68, *362*, 363, *364*, 412–13, *413*
Wind power 22–3, *32*
Working fluid 116–17

Yakedake, Japan 212, 229
Young's modulus 73, 143, 154